水通道蛋白与中枢神经系统疾病

主 编 孙善全 杨 美 孙晓川 李燕华

科学出版社

北 京

内 容 简 介

水通道蛋白（AQP）家族是新近发现的一类可介导水分子及甘油、乳酸等小分子物质转运的整合膜蛋白，在各系统的组织细胞内有着广泛的分布，在中枢神经系统的分布尤为丰富。AQP与中枢神经系统的发育、水和电解质平衡、能量代谢、认知功能，以及脑水肿、神经胶质瘤等疾病的发生发展密切相关。本书主要介绍两部分内容：AQP的分子生物学特征、家族的进化、转运特征、定位分布与功能及其表达调控的信号通路；AQP在中枢神经系统疾病发生发展过程中的作用及其调控机制等。本书为AQP的深入研究提供了翔实的资料，并为中枢神经系统疾病治疗策略的制定提供了新思路和新治疗靶点。

本书可供神经科、眼科、影像科等临床医生，科研工作者、医学生及其他生命科学相关领域人员使用。

图书在版编目（CIP）数据

水通道蛋白与中枢神经系统疾病 / 孙善全等主编 . -- 北京 : 科学出版社，2024.12. --ISBN 978-7-03-079955-5

Ⅰ. Q51；R741

中国国家版本馆CIP数据核字第20244Z5P81号

责任编辑：丁慧颖　戚东桂 / 责任校对：张小霞
责任印制：肖　兴 / 封面设计：吴朝洪

科学出版社 出版
北京东黄城根北街16号
邮政编码：100717
http://www.sciencep.com

三河市骏杰印刷有限公司印刷
科学出版社发行　各地新华书店经销

*

2024年12月第 一 版　开本：787×1092 1/16
2024年12月第一次印刷　印张：15 1/2
字数：355 000
定价：98.00元
（如有印装质量问题，我社负责调换）

《水通道蛋白与中枢神经系统疾病》编写人员

主　编　孙善全　杨　美　孙晓川　李燕华
副主编　卓　飞
编　者　（按姓氏拼音排序）

陈　鸿　陈　哲　陈玉琴　程崇杰
甘胜伟　何骏驰　黄　娟　黄波月
李燕华　廖玉慧　刘　辉　刘　茜
陆蔚天　骆世芳　彭雪华　邱国平
冉建华　任　丽　孙善全　孙晓川
万珊珊　徐　进　许士叶　杨　美
易耀兴　张兴业　朱淑娟　卓　飞

主编简介

孙善全 医学博士，二级教授，博士研究生导师，日本滋贺医科大学分子神经研究中心客座教授。曾任重庆市重点学科学术带头人，重庆市学位评定委员会委员，中国解剖学会理事，重庆市解剖学会副理事长，重庆市神经科学学会常务理事，美国神经科学学会（SFN）会员，《局解手术学杂志》副主编，《解剖学杂志》《中国临床解剖学杂志》《解剖与临床》等学术期刊编委，国家自然科学基金委学科评审专家，国务院学位委员会学位评审专家。首届重庆市名师。

长期从事本科生、硕士研究生、博士研究生和留学生的解剖学、神经解剖学等学科的教学工作。负责重庆市重点、校级重点教改课题各一项。荣获重庆医科大学教学成果奖，重庆医科大学师资培养优秀奖等奖项。获"重庆医科大学优秀教师""重庆医科大学优秀研究生导师"称号。获省级自然科学奖二等奖1项，省级科技进步奖4项。

首先提出了"喉上神经袢"和"喉返神经神经袢"的新概念，在国际上具有重要影响。目前，主要从事中枢神经系统水和电解质平衡运输机制、脊髓损伤修复和神经元化学可塑性机制的研究。2000年以来，作为负责人先后承担国家自然科学基金项目6项。在国际、国内刊物公开发表论文100余篇，其中SCI论文30余篇。作为主编、副主编参与编写解剖学本科教材4部，作为副主编、编者参与编写国家精品课程教材（双语版）及教育部规划教材各2部。

杨 美 医学博士，教授，博士研究生导师。教育部课程思政教学名师，重庆市首批青年骨干教师，重庆市首批高校"双带头人"教师党支部书记工作室负责人，高等学校基础医学创新创业协同育人工作组常委，重庆市解剖学会理事。

教学方面，注重学生创新精神和实践能力的培养，两次获"重庆医科大学优秀教师"称号。作为项目负责人主持及开展了重庆医科大学留学生解剖试题库的建设和运用，多次获重庆医科大学教学成果一、二等奖，发表教学论文10余篇，作为副主编、编者参与编写国家级规划教材10余部。指导学生积极参与创新创业活动，发表多篇论文，获全国大学生基础医学创新论坛暨实验设计大赛一等奖、中国国际"互联网+"大学生创新创业大赛重庆赛区铜奖等。

科研方面，长期从事水通道蛋白的功能研究和神经系统疾病的调控机制研究，担任国家自然科学基金委学科评审专家、教育部学位中心通讯评议专家，主持国家自然科学基金项目3项、教育部博士点新教师基金项目，重庆市科委、重庆市教委、重庆医科大学等课题，作为通讯作者发表SCI论文10余篇。

孙晓川 医学博士、二级教授、主任医师、博士研究生导师。现任中国医师协会神经外科医师分会副会长、中国医促会神经外科学分会副主任委员、中国神经科学学会理事、国家神经疾病医学中心脑胶质瘤MDT专科联盟副理事长、重庆市医师协会神经外科医师分会会长、重庆市神经科学学会副理事长、重庆市抗癌协会神经肿瘤专业委员会主任委员。曾任中华医学会神经外科学分会委员、重庆市医学会神经外科学分会主任委员。担任《中华神经外科杂志》（中文版和英文版）、《中华创伤杂志》（中文版和英文版）等国内学术期刊编委，以及 Journal of Neuroscience、Cerebrovascular Diseases 等国际学术期刊审稿人。荣获"重庆英才计划"名家名师称号、2021年度王忠诚中国神经外科医师学术成就奖、重庆市五一劳动奖章。多次荣获"重庆医科大学优秀研究生导师"称号。

从事脑血管疾病与颅脑创伤的临床诊疗和基础研究30余年，对复杂脑动脉瘤及重型颅脑损伤的诊治具有丰富的经验。先后主持国家自然科学基金面上项目4项，以第一主研人参与国家自然科学基金项目5项，承担国家"十一五""十二五""十三五"科技支撑项目子课题5项，荣获省部级科技进步一等奖1项、二等奖及三等奖各2项。发表论文220余篇，其中SCI论文60余篇。主编专著2部，参编专著10余部。

李燕华 医学博士，主任医师，博士研究生导师，广西医学科学院广西壮族自治区人民医院神经内科副主任。任广西医学会神经病学分会副主任委员，广西医师协会神经内科分会副主任委员，中国人体健康科技促进会神经急诊重症监护专业委员会常务委员，中国人体健康科技促进会重症脑损伤专业委员会委员，中国研究型医院学会临床神经电生理专业委员会委员，中国卒中学会免疫分会委员，广西医学会神经病学分会重症学组组长，广西科技专家库专家，广西劳动能力鉴定委员会医疗卫生专家库专家,《中国临床新医学》期刊审稿专家。

长期从事神经内科的临床、教学和科研工作,具有扎实的专业基础理论知识和丰富的临床经验。熟练掌握神经内科常见、疑难和危重症疾病的诊断和治疗,擅长神经系统急危重疾病的诊治。

主持国家自然科学基金项目2项,广西自然科学基金面上项目3项,广西卫生健康委员会科研基金项目3项,参与多项省部级科研项目,发表专业学术论文30余篇。作为第一完成人获广西科技进步三等奖1项;多次荣获院级"科技拔尖人才奖""优秀带教老师""优秀医师"等称号。

前　言

　　水既是参与人体新陈代谢的基本物质之一，同时又是新陈代谢的重要产物之一。水占成人体重的70%～75%；而对于正常的婴幼儿，水占其体重则高达80%。在不同的器官中，由于结构和功能间的差异，其含水量也存在差异，脑的含水量仅次于眼球和血液，约占其总量的80%。如此大量的水分，分布于组织细胞之内，充斥在细胞之间。在正常情况下，细胞内外水分子的分布保持着某种动态平衡，这种平衡对于维持细胞的正常功能具有十分重要的意义。因此，水分子的跨膜转运及其机制，一直是基础与临床工作者十分关注的科学问题。

　　长期以来，研究者认为水分子的跨膜转运是水分子逆浓度梯度简单扩散引起的。20世纪80年代末期，Peter Agre及其同事克隆出了第一个水通道蛋白（aquaporin 1，AQP1）。自此，翻开了水分子跨膜转运及其分子机制研究的新篇章。水通道蛋白（AQP）是一组跨越细胞双层脂质膜，对水及甘油、尿素等小分子物质特异性通透的膜蛋白，故又有水通道整合膜蛋白（water channel integral membrane protein）之称。迄今为止在哺乳动物体内共克隆出13个AQP成员（AQP0～AQP12）。通过大量的研究，目前已明确了某些AQP的基因组构、蛋白晶体结构、表达定位及其生理功能。近年来，对AQP敲除鼠的研究结果提示，AQP可能与胚胎的发育、细胞的增殖生长、某些疾病的发生发展密切相关，AQP可作为某些选择性抑制剂的特异性靶点。然而，迄今为止有关哺乳动物AQP对水和其他小分子物质转运的生理功能、调控机制及其与临床联系的研究还远未结束。本书的撰写旨在介绍既往的相关成果，以激发对AQP研究的热情，并为AQP研究提供新的视角，开辟新的研究领域。本书简要介绍了近三十年来哺乳动物AQP在生物物理、遗传学、蛋白结构、分子生物学方面的研究进展，并聚焦于AQP在中枢神经系统（CNS）中的定位分布，以及AQP表达异常与CNS疾病发生发展的关系。

　　CNS是哺乳动物最重要、最复杂的一个系统，而神经组织细胞内外水和电解质的平衡，对维持中枢神经系统的正常形态和功能显得尤为重要。迄今为止，在CNS中已克隆出来的AQP多达7种，仅分布于星形胶质细胞的就有

5种；星形胶质细胞AQP既与水的转运有关，也与脑的能量代谢底物乳酸、甘油等小分子物质的转运有关。众所周知，在神经组织细胞中星形胶质细胞数量多、功能复杂，多个AQP同源基因在星形胶质细胞上均有表达，表明星形胶质细胞内、外水和电解质平衡调节的复杂性与功能的多样性。中枢神经系统AQP的这种特异性的分布模式，给研究人员以启示，在哺乳动物的各器官、系统中，其功能越复杂，水的转运和调控机制也越复杂、越精细。在正常情况下，AQP涉及神经发育阶段的细胞分裂、凋亡和细胞迁移，与神经网络的形成有关。反之，如AQP的表达异常，则与某些神经系统的先天性疾病的发生有关，也与脑水肿的发生发展有关；AQP涉及神经肿瘤的增殖、生长和转移；AQP还与多发性硬化等神经免疫性疾病有关。本书对上述相关领域的研究进展进行了介绍，以期加深读者对相关疾病病理机制的理解。

本书对以AQP为靶点治疗相关疾病的研究现状进行了系统回顾，特别是以AQP的亚细胞定位作为靶点的相关研究进行了总结，以期为治疗相关疾病新药的开发提供新的方案。

本书受国家自然科学基金项目的资助。本书的编写人员从20世纪末至今，长期从事AQP与相关神经疾病的研究。在本书的撰写过程中，使用了编写人员所在课题组公开发表的研究结果和图表，并结合了其他相关的研究结果。本书力图全面反映AQP的研究现状，但由于编者水平有限，在撰写过程中难免挂一漏万，敬请谅解。

孙善全

2024年8月

目 录

第一章 水通道蛋白家族概述 ··· 1
  第一节 水通道蛋白的分子生物学特征 ··· 1
    一、水通道蛋白的发现 ·· 1
    二、水通道蛋白的分类 ·· 1
    三、水通道蛋白的基因结构 ··· 6
    四、水通道蛋白的晶体结构 ··· 7
    五、水通道蛋白的蛋白质修饰 ··· 9
    六、水通道蛋白的表达调控 ··· 15
  第二节 水通道蛋白家族的进化 ·· 31
    一、引言 ·· 31
    二、水通道蛋白家族的亚家族 ··· 33
    三、水通道蛋白的聚类与演化 ··· 38
    四、展望 ·· 42
  第三节 水通道蛋白的转运特征 ·· 48
    一、水通道蛋白介导的水转运 ··· 48
    二、水通道蛋白介导的甘油转运 ··· 49
    三、水通道蛋白介导的尿素转运 ··· 51
    四、水通道蛋白介导的气体转运 ··· 51
    五、水通道蛋白介导的其他分子转运 ······································ 52
  第四节 哺乳动物水通道蛋白的定位分布与功能 ·························· 54
    一、心血管系统中的水通道蛋白 ··· 54
    二、呼吸系统中的水通道蛋白 ··· 58
    三、消化系统中的水通道蛋白 ··· 62
    四、泌尿系统中的水通道蛋白 ··· 65
  第五节 水通道蛋白的表达调控机制 ··· 86
    一、AQP1的表达调控机制 ·· 86
    二、AQP2的表达调控机制 ·· 87
    三、AQP3的表达调控机制 ·· 87
    四、AQP4的表达调控机制 ·· 87

五、AQP5 的表达调控机制 ············· 93
六、AQP6 的表达调控机制 ············· 93
七、AQP7 的表达调控机制 ············· 93
八、AQP8 的表达调控机制 ············· 94
九、AQP9 的表达调控机制 ············· 94
十、AQP10 的表达调控机制 ············ 94
十一、AQP11 的表达调控机制 ·········· 95
十二、AQP12 的表达调控机制 ·········· 95

第二章 水通道蛋白与中枢神经系统 ············· 103
第一节 水通道蛋白在中枢神经系统的定位分布 ········· 103
一、水通道蛋白在脑组织中的表达定位 ········· 103
二、水通道蛋白在脊髓中的表达定位 ··········· 106
第二节 水通道蛋白在中枢神经系统的锚定 ··········· 110
一、AQP4 的极性表达模式 ··········· 110
二、AQP4 极性分布的锚定机制 ······· 111
第三节 水通道蛋白与脑损伤 ········· 115
一、水通道蛋白在缺血性脑水肿发生发展中的表达变化及其调控机制 ··· 115
二、水通道蛋白与缺血再灌注损伤及其调控机制 ······· 121
三、水通道蛋白在出血性脑水肿发生发展中的作用及其调控机制 ······ 125
四、AQP4 与烧伤后继发性脑水肿及其调控机制 ······· 128
五、水通道蛋白与创伤性脑损伤 ······· 132
六、水通道蛋白在颅内感染疾病中的表达变化及其调控机制 ········· 137
七、高眼压状态下水通道蛋白的表达变化及其调控机制 ··········· 141
八、水通道蛋白与脑积水 ············· 144
第四节 水通道蛋白与脑胶质瘤 ······· 163
第五节 水通道蛋白与神经退行性疾病 ··········· 169
一、水通道蛋白与阿尔茨海默病 ······· 169
二、水通道蛋白与帕金森病 ··········· 171
三、水通道蛋白与肌萎缩侧索硬化 ····· 173
第六节 水通道蛋白与精神疾病 ······· 182
一、AQP4 与自闭症 ················· 182
二、AQP4 与抑郁症 ················· 183
三、AQP3、AQP4 与精神分裂症 ········ 183
第七节 水通道蛋白与脊髓疾病 ······· 186
一、AQP4 与视神经脊髓炎 ············ 186

二、水通道蛋白与脊髓损伤 ……………………………………………………… 187
　第八节　水通道蛋白与脑影像 …………………………………………………………… 194
　　一、脑水肿成像技术研究进展 …………………………………………………… 194
　　二、AQP4与弥散加权成像 ……………………………………………………… 195
　　三、AQP4与缺血半暗带 ………………………………………………………… 198

第三章　以水通道蛋白为治疗靶点的研究现状 …………………………………………… 203
　第一节　以水通道蛋白家族为靶点的小分子和生物制剂的研究现状 ………………… 203
　　一、水通道蛋白抑制剂的概况 …………………………………………………… 203
　　二、水通道蛋白的功能测定 ……………………………………………………… 204
　　三、水通道蛋白抑制剂的研究现状 ……………………………………………… 207
　　四、展望 …………………………………………………………………………… 212
　第二节　以AQP4的亚细胞定位为靶点的研究现状 …………………………………… 215
　　一、AQP4的极性分布 …………………………………………………………… 215
　　二、神经损伤后AQP4极性分布被破坏 ………………………………………… 216
　　三、靶向AQP4的亚细胞定位 …………………………………………………… 217
　　四、小结 …………………………………………………………………………… 221

第四章　其他膜蛋白介导的水转运 ………………………………………………………… 224
　第一节　概述 ……………………………………………………………………………… 224
　第二节　尿素通道蛋白B ………………………………………………………………… 225
　第三节　囊性纤维跨膜转导调节因子 …………………………………………………… 225
　第四节　协同转运蛋白 …………………………………………………………………… 226
　　一、K^+-Cl^-协同转运蛋白 ………………………………………………………… 226
　　二、Na^+-K^+-Cl^-协同转运蛋白 …………………………………………………… 227
　　三、单羧酸转运体 ………………………………………………………………… 227
　　四、GABA转运蛋白 ……………………………………………………………… 228
　　五、Na^+偶联谷氨酸转运蛋白 ………………………………………………… 228
　　六、Na^+-葡萄糖共转运蛋白 …………………………………………………… 228
　　七、硼酸钠协同转运蛋白 ………………………………………………………… 229
　　八、Na^+双羧酸协同转运蛋白1 ………………………………………………… 229
　　九、葡萄糖转运体 ………………………………………………………………… 229

第一章　水通道蛋白家族概述

第一节　水通道蛋白的分子生物学特征

一、水通道蛋白的发现

长期以来科学家们一直推测，细胞膜上可能存在着一类与水转运直接相关的通道蛋白。20世纪80年代初，有人认为红细胞膜电泳图谱上第3条带是水和电解质的共同通道蛋白[1]。直到20世纪80年代末至90年代初，Peter Agre开创性地鉴定出水通道蛋白1（AQP1），细胞膜上的水通道才被真正认识。Agre和同事们偶然从红细胞膜中纯化出一种新型蛋白[2]，它含有28kDa的非糖基化成分和35～60kDa的糖基化组分，28kDa多肽以四聚体寡聚蛋白形式存在并具有一定的生化特征。随后，该蛋白的氨基酸序列被确定[3]，也被克隆[4]。最初，这种新蛋白被称为CHIP28（28kDa类通道整合膜蛋白），随后被重新命名为aquaporin 1或AQP1[5]。

研究者将AQP1的染色体RNA（cRNA）注入非洲爪蟾卵母细胞后发现，该细胞表现出显著的水通透性，细胞在低渗缓冲液中迅速膨胀至破裂[4]。为了验证AQP1作为水分子通道的作用，研究者将磷脂和人红细胞AQP1蛋白重组为蛋白质脂质体，与不含AQP1的脂质体进行比较[5-7]。与对照组相比，AQP1脂质体的单位透水能力极高，但不能通透多种小分子溶质，表明AQP1具有水选择性（后续研究发现AQP1还可通透气体）。上述结果证实了AQP1具有水分子通道的功能，进而提示AQP1在其所表达组织的跨膜或跨细胞水转运中具有重要意义。同时，AQP1的发现也为通过同源克隆和其他方法鉴定水通道家族成员提供了思路和方法，从而将人们对水通道蛋白（AQP）在组织水转运中生理功能及机制的认识提升到一个新的高度。

二、水通道蛋白的分类

自AQP1被发现后，大量证据表明原核生物和真核生物中的AQP具有多样性[8,9]。迄今已发现300多种的AQP，在人类中已鉴定出13种水通道蛋白亚型（AQP0～AQP12）。AQP作为一种完整的、疏水跨膜蛋白，主要受细胞膜两侧渗透压差的驱使，促进水的被动转运。随后的研究表明，AQP不仅可以转运水分子，还能顺浓度梯度转运甘油、尿素等其他不带电荷的小分子。

AQP的结构分析表明，它们具有共同的结构特征。功能性水通道蛋白单元是同源四聚体，水通道蛋白的特征性序列有6个α螺旋跨膜结构域[含2个嵌入细胞膜的保守天冬酰胺-脯氨酸-丙氨酸（NPA）基序]，5个连接环（A～E）。人AQP的氨基酸序列30%～50%相同。AQP借助构象变化促进尿素、甘油、过氧化氢（H_2O_2）、氨（NH_3）、二氧化碳（CO_2）等小分子通过细胞膜。

根据结构和功能的相似性，AQP最初被分为两个亚家族，即经典AQP（水选择性）和水甘油通道蛋白（aquaglyceroporin，又称甘油通道，Glps）。然而，这一分类方法近期受到一些研究结果的质疑。有证据表明，上述两个亚家族在功能上可相互重叠。例如，一些所谓的经典AQP除转运水以外，还可转运甘油等其他小分子溶质。此外，最新发现的一类AQP，其结构与以往的AQP差异很大，尤其是在AQP的NPA框周围，其差别则更大[10-12]。该亚家族因与前两个亚家族的同源性较低，被命名为超级水通道蛋白，又称非经典水通道蛋白[11]。这一种分类方法已被公认，因此在后续内容中，将按此种分类方法进行介绍。根据系统发育树（图1-1）或由贝叶斯推断的系统发育拓扑学，AQP可分为四类：经典水通道蛋白、AQP8型水通道蛋白、水甘油通道蛋白和超级水通道蛋白[8, 12]。

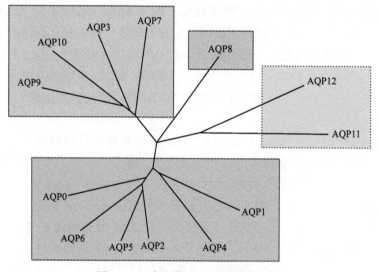

图1-1　13个人类AQP的系统发育树

树状图显示经典水通道蛋白分类（AQP0、AQP1、AQP2、AQP4、AQP5、AQP6）；AQP8又称为AQP8型AQP，因其在系统发育上与其他经典AQP不同而独立出来；水甘油通道蛋白（AQP3、AQP7、AQP9、AQP10）；超级水通道蛋白（AQP11、AQP12）

（一）水通道蛋白亚型分类

第一个亚家族是经典水通道蛋白，最初包括AQP0、AQP1、AQP2、AQP4、AQP5、AQP6和AQP8。对这一亚家族的广泛研究，将有助于加深对AQP的表达调节机制及其在病理生理条件下潜在作用的认识[13, 14]。然而，最新的文献表明，由于AQP8独特的系统发育过程，AQP8被独立出来[15, 16]。

第二个亚家族以通透水、氨、尿素和甘油等不带电荷小分子的水甘油通道蛋白为代表，也称为水甘油通道蛋白。这类成员还促进亚砷酸盐和亚锑酸盐的扩散，并在体内的类金属稳态调节过程中发挥关键作用[17]。该亚家族包括AQP3、AQP7、AQP9和AQP10，可根据氨基酸序列对其进行区分[18]。AQP3是首个被克隆出来的哺乳动物水甘油通道蛋白，对甘油和水具有通透性[19, 20]。AQP7、AQP9和AQP10在非洲爪蟾卵母细胞中表达，可转运水、甘油和尿素[21-23]。AQP9对卵母细胞中的多种溶质具有通透性[23]。然而，人们目前对于多数水甘油通道蛋白的了解尚不充分。

第三个亚家族在NPA框周围的氨基酸序列保守度较低，因而无法归类到前两个亚家族[11]。哺乳动物AQP11和AQP12是这个亚家族中的成员，被称为超级水通道蛋白或非经典水通道蛋白。二者的NPA结构与其他经典AQP的同源性小于20%，表明其属于AQP的超基因家族。目前，人们对AQP11和AQP12的结构和功能还知之甚少。

（二）水通道蛋白亚型成员

迄今为止，在人体内发现有13种AQP亚型（表1-1）。在发现首个AQP后的30年里，学者们对这些蛋白质的生物学功能进行了深入研究。尽管还存在一些尚未明确的问题，但这些哺乳动物AQP的结构、细胞定位、生物学功能和潜在的病理生理学意义已得到系统阐述。

表1-1 哺乳动物AQP亚型成员及其分布

	运输	分布
经典水通道蛋白		
AQP0	水	眼
AQP1	水	脑，眼，肾，心，肺，胃肠道，唾液腺，肝，卵巢，睾丸，肌肉，红细胞，脾
AQP2	水	肾，耳，输精管
AQP4	水	脑，肾，唾液腺，心，胃肠道，肌肉
AQP5	水	唾液腺，肺，胃肠道，卵巢，眼，肾
AQP6	水，尿素（+/-），阴离子	脑，肾
AQP8型水通道蛋白		
AQP8	水，尿素（+/-），氨	睾丸，肝，胰腺，卵巢，肺，肾
水甘油通道蛋白		
AQP3	水，尿素，甘油，氨	肾，心，卵巢，眼，唾液腺，胃肠道，呼吸道，脑，红细胞，脂肪
AQP7	水，尿素，甘油，氨	睾丸，心，肾，卵巢，脂肪
AQP9	水，尿素，甘油	肝，脾，睾丸，卵巢，白细胞
AQP10	水，尿素，甘油	胃肠道
超级水通道蛋白		
AQP11	水?	睾丸，心，肾，卵巢，肌肉，胃肠道，白细胞，肝，脑
AQP12	未知	胰腺

1. 经典AQP的亚型

（1）AQP0：AQP0表达于眼球晶状体纤维细胞，具有维持晶状体内稳态和透明性的作用[24, 25]。AQP0透水性比AQP1低，约为其透水性的1/40[26]。AQP0的水转运功能受C端裂解情况[27]、pH和Ca^{2+}/钙调蛋白（CaM）的调节[28]。降低Ca^{2+}浓度或抑制钙调蛋白，均可增加AQP0的透水性。最新的分子动力学和功能突变研究表明，钙调蛋白与AQP0结合后，通过变构关闭胞膜上AQP0的通道门孔，从而抑制AQP0的透水性[29]。蛋白激酶C（protein kinase C，PKC）依赖的Ser235磷酸化可诱导AQP0向细胞膜转移[30]。

（2）AQP1：AQP1是首个被发现的水通道蛋白[2, 4, 31]，也是首个被发现可通透气体的水通道蛋白[32, 33]。AQP1在体内分布广泛[34]，在相关组织的水转运调节过程中发挥关键作用，如血管生成、细胞迁移和细胞生长过程[35]，AQP1的表达下调可阻断血管生成并延缓肿瘤的进展[36]。除了促进水的转运外，研究发现AQP1可提高CO_2和NH_3的通透性[13, 37]，并可作为细胞内环磷酸鸟苷（cyclic guanosine monophosphate，cGMP）激活的非选择性单价阳离子通道发挥作用[38]。C端酪氨酸Y253的磷酸化可参与AQP1作为cGMP门控阳离子通道的调控[39]。早期证据表明，苏氨酸和丝氨酸激酶活性也参与AQP1离子通道活性的调节[40]。

（3）AQP2：AQP2是迄今研究最透彻的受精氨酸加压素（arginine vasopressin，AVP）调节的水通道蛋白，除水分子外，对其他小分子无通透性。AQP2主要表达于肾脏集合管主细胞顶膜及其下方的囊泡中[41-43]，在尿浓缩过程中发挥重要作用。AQP2从胞内腔室到主细胞顶膜的转位过程依赖于血管升压素与位于主细胞基底外侧膜上V2受体的结合[42, 43]，因此血管升压素可增加AQP2的透水性。对AQP2的表达调控、转录后修饰及转运机制的研究将在后文中详细阐述。有研究报道，心房钠尿肽（ANP）[44-46]、一氧化氮（NO）[44]、前列腺素E_2[47]、血管紧张素Ⅱ[36, 48]、催产素[49]、嘌呤[50]和他汀类[36, 51]等多种局部或全身性化学物质参与AQP2的表达或转运调控。

（4）AQP4：AQP4是中枢神经系统最主要的水通道蛋白，可通透水[52, 53]和CO_2[13]。AQP4的调控和转运机制均已被研究揭示，血管升压素[54]或组胺暴露[55]可诱导AQP4的转位。AQP4在胞质内丝氨酸残基的Ser111和Ser180位点磷酸化可通过门控机制调节其透水性[56]；但近期体内的实验数据无法证实Ser111磷酸化的作用[57]。研究表明，PKC诱导的AQP4 C端磷酸化是高尔基体转位所必需的[57]。

（5）AQP5：AQP5表达于腺上皮、肺泡上皮和一些分泌腺，参与唾液、眼泪和肺分泌物的生成[58, 59]；可通透水和CO_2[13, 60]。蛋白激酶A（PKA）可直接磷酸化AQP5胞质环和C端的Ser156和Thr259位点[61, 62]。然而，诱导AQP5向胞膜的转运依赖于细胞内Ca^{2+}的增加，而非PKA所诱导的磷酸化[63, 64]。

（6）AQP6：AQP6在肾集合管A型间质细胞内与H^+-ATP酶共定位于胞内囊泡[14]，提示AQP6可能与囊泡H^+-ATP酶之间存在相互作用，调节囊泡内的pH。研究发现，当酸碱度变化时，间质细胞内的H^+-ATP酶从胞质内的囊泡迁移到顶端细胞膜[65]，但并未见AQP6发生转位，提示AQP6仅在细胞内位点发挥作用。AQP6的N端对蛋白转运到胞内位点和囊泡的定位至关重要[66]。有趣的是，钙调蛋白可在N端以钙依赖的方式与AQP6结合[67]，提示钙信号可能参与了AQP6的内化。AQP6不通透水[13, 68]，但在$HgCl_2$存在或酸性环境

（pH＜5.5）中，卵母细胞中AQP6对水和阴离子的通透性迅速增加[14]。此外，AQP6还能够转运尿素、甘油和硝酸盐[69, 70]。

2. AQP8型水通道蛋白 AQP8首次被发现位于肾脏近端小管和集合管细胞的胞内结构域[71]。多项研究表明，AQP8可转运水[13, 72]和NH_3[13, 73]。尽管超微结构显示，AQP8位于肝脏线粒体内膜（inner mitochondrial membrane，IMM）[72]；但后期研究发现，野生型和AQP8缺失小鼠肝细胞线粒体内膜的透水能力无差异[74]。近期研究表明，线粒体AQP8在肾近端小管对酸中毒的适应性反应过程中发挥重要作用。AQP8促进谷氨酰胺从IMM释放并分泌至管腔[75]；尤其是在代谢性酸中毒时，NH_3可缓冲上皮细胞分泌至管腔中的酸[76]。AQP8在活性氧生成，如电子传递链大量减少时，也可促进H_2O_2跨线粒体膜的扩散[77, 78]。

3. 水甘油通道蛋白

（1）AQP3：AQP3组织分布广泛，可通透水、甘油和尿素。近期有研究采用人红细胞模型发现，人AQP3的pH门控对水和甘油的通透率均有影响[79]。AQP3是皮肤表达最丰富的水甘油通道蛋白，可促进水和甘油的转运，在保持哺乳动物皮肤表皮层含水量及角质细胞的增殖和分化过程中发挥重要作用[80-82]。在肾脏中，AQP3表达于皮质和外髓集合管主细胞的基底外侧膜[83]，而经由主细胞顶膜的AQP2进入的水分子在基底外侧膜经由AQP3排出。长期血管升压素刺激后，主细胞顶膜的AQP2和基底外侧膜的AQP3均表达增加，从而提高了集合管主细胞跨上皮转运水的能力。然而，与AQP2不同的是，由于主细胞基底外侧膜下方无含AQP3的囊泡存在，提示AQP3缺乏胞内穿梭过程[84]。有研究表明，AQP3可能受到cAMP（环磷酸腺苷）-PKA信号通路的短期调控[85-87]。在肾脏内，胞内cAMP升高可改变AQP3与细胞膜中其他蛋白质或脂质的相互作用，进而引起基底外侧膜上AQP3的扩散增加，这一变化可能是对顶膜AQP2所介导的水分子流动增加的生理性适应[87]。近期实验证实，AQP3通过细胞膜转运H_2O_2[78, 88]，可能在细胞迁移[89]、炎症[90]和癌症进展[91, 92]的胞内信号始动过程中扮演重要角色。

（2）AQP7：AQP7可促进水、甘油、尿素、亚砷酸盐和NH_3的转运[13, 21, 93]，在脂肪组织中大量表达，介导新生成甘油的释放。甘油的调节异常是代谢性疾病发生的重要因素，AQP7缺乏会导致肥胖[94, 95]和胰岛素抵抗[96]。最近有报道显示，白色脂肪细胞中脂肪分解活化可诱导AQP7从外膜到内膜的转位[97]；这一结论与之前研究认为的AQP7转位方向相反[98]，这可能是研究采用的技术和实验条件不同所致。已有数据分析出AQP7蛋白磷酸化的6个潜在位点[99]，但PKA的直接调节作用及机制仍有待阐明。

（3）AQP9：AQP9表达于肝血窦的肝细胞膜，作为吸收NH_3并介导新合成尿素排出的通道[100]；也可作为甘油通道，促进肝脏组织对甘油的摄取。AQP9还可以通透水、甘油、尿素（碳酰胺）、CO_2等物质，并通过转运亚锑酸盐和亚砷酸盐在类金属的稳态调节中发挥至关重要的作用[9, 101]。有趣的是，AQP9还可以转运乳酸、嘌呤、嘧啶等分子量更大的底物[9, 23]，原因在于其3D结构的分析结果显示其具有更大的孔径[102]。最新的报道显示，AQP9能促进人和小鼠细胞中H_2O_2的跨膜转运；AQP9的缺失可减弱人和小鼠细胞中H_2O_2的细胞毒作用，这一结果表明AQP9介导的H_2O_2的跨膜转运可调节氧化还原调控的下游细胞信号[103]。目前的研究已经鉴定出人AQP9在Asn142处有潜在的N-糖基化位点，在

Ser11和Ser222处有潜在的PKC磷酸化位点,在Ser28处有潜在的酪蛋白激酶Ⅱ磷酸化位点[23, 104]。然而,人们对AQP9的短期调节机制还知之甚少。

(4)AQP10:AQP10是一种仅在人胃肠道中表达的水甘油通道蛋白[22, 105],其在鼠小肠中已被证实是假基因,故而无表达。表达AQP10的非洲爪蟾卵母细胞可转运水、甘油和尿素[22]。最近的研究表明,AQP10是人脂肪细胞细胞膜上表达的另一个甘油通道蛋白[106]。沉默人分化脂肪细胞中的AQP10可导致其甘油和水的渗透系数降低50%,提示AQP10与AQP7对将脂肪细胞内甘油含量保持在正常或低水平极为重要,该保护机制可避免肥胖的发生[106]。有研究预测了AQP10的3个潜在性糖基化位点,其中位于AQP10胞外环Asn133的位点已被证实;Asn133的糖基化作用可提高AQP10在低温条件下的热稳定性,表明N-连接聚糖具有稳定蛋白的作用[107]。

4. 超级水通道蛋白

(1)AQP11:AQP11具有保守的N端天冬酰胺-脯氨酸-丙氨酸(NPA)特征性基序以及独特的包含天冬酰胺-脯氨酸-半胱氨酸(NPC)基序的氨基酸序列,后者是AQP11分子功能完全表达所必需的结构基础[10]。最近的证据表明,AQP11的Cys227对其四级结构的形成和蛋白功能至关重要[108]。AQP11重组囊泡的研究结果表明,其转运水分子的功能可媲美AQP1[109, 110]。尽管AQP11亚细胞定位尚不清楚,但在瞬时转染的细胞和采用透明质酸(HA)标记的AQP11转基因小鼠肾脏细胞中[111]发现,AQP11与内质网的标志物共定位[112]。AQP11缺失与肾近端小管的内质网应激及凋亡发生有关[112]。

(2)AQP12:与AQP11相比,AQP12与其他水通道蛋白的关系更加密切。在模式基序方面,AQP12的第1个NPA基序被NPC基序取代,而其C端的NPA基序则保守存在[12, 113]。AQP12特异性表达于胰腺的腺泡细胞胞质内,但其生理作用还有待阐明[12]。研究认为,在受到快速而强烈的刺激时,AQP12发挥控制胰液适当分泌的作用[114]。

三、水通道蛋白的基因结构

*AQP0*基因全长3.6kb,包含4个外显子,位于人单倍体基因组中;其转录从TATA盒下游第26个核苷酸开始[115]。

基因组DNA印迹(Southern印迹)分析表明*AQP1*单个基因的存在,原位杂交分析确定该基因定位于人类7号染色体短臂1区4带[5, 116, 117]。将大鼠肾集合管顶膜水通道蛋白克隆为AQP2互补DNA(cDNA)[41],其氨基酸序列与AQP1有42%的同源性。人*AQP2*编码的蛋白与大鼠AQP2蛋白的氨基酸同源性为89.7%~91%[116, 118-120]。原位杂交结果显示,*AQP2*定位于人类12号染色体长臂1区3带[118, 120],与主要同源性蛋白(MIP)的位点非常接近[25]。

Ishibashi[121]利用大鼠*AQP3*探针筛选人肾脏的cDNA文库,证实*AQP3*基因定位于人类9号染色体短臂1区3带上,为单拷贝基因,有6个外显子,转录起始位点位于第一个ATG密码子上游64bp处。5′端侧翼区包含一个TATA盒,2个Sp1序列,以及一些包括AP-2位点的共有序列[122]。

从胎儿脑cDNA文库中克隆出人AQP4（最初称为汞敏感水通道，MIWC）cDNA，其最长的开放阅读框编码301个氨基酸，与大鼠*AQP4*有94%的同源性。MIWC基因组分析结果表明，其含有2个不同但重叠的转录单位，其中多个MIWC的mRNA被转录。后续研究表明，*AQP4*基因由4个外显子组成，分别编码第127、55、27和92位氨基酸，由0.8kb、0.3kb和5.2kb的内含子分隔开。基因组Southern分析表明，单个MIWC基因位于18号染色体长臂上[123, 52]。

从人下颌下腺文库中分离和鉴定出人*AQP5*的cDNA和基因，其中包含一个795bp的开放阅读框，编码含265个氨基酸的多肽，转录起始位点为起始甲硫氨酸上游518bp；*AQP5*基因定位于12号染色体长臂1区3带[124]。

研究人员使用简并聚合酶链式反应（PCR）技术从人肾脏cDNA文库中分离出*AQP6*的cDNA，其具有4个外显子，与*AQP0*和*AQP2*相似，随后被称为*AQP6*，定位于12号染色体长臂1区3带[125, 126]。

人*AQP7*基因包含10个外显子，其启动子区域内有几种不同转录因子的Alu重复序列和结合位点，包括1个过氧化物酶体增殖物反应元件（PPRE）和1个可能参与能量代谢的胰岛素反应元件[127]。

与非水选择性水通道蛋白的基因一样，*AQP8*基因包含6个外显子，但其外显子和内含子边界与其他水通道蛋白基因不同。*AQP8*基因定位于16号染色体短臂1区2带[15, 128]。

研究者采用逆转录聚合酶链式反应（RT-PCR）方法分离出白细胞AQP9部分cDNA[129]。AQP9与AQP3、AQP7的序列同源性高于其他家族成员，表明这3种蛋白属于同一亚家族。

编码AQP10的cDNA来源于空肠cDNA文库。序列预测分析结果显示，该蛋白含有264个氨基酸，与AQP3和AQP9的同源性约为53%；RNA印迹（Northern印迹）分析显示，*AQP10*基因在空肠中表达[130]。

人*AQP11*基因包含3个外显子，全长8kb，定位于11号染色体长臂1区4带。人*AQP12*基因包含4个外显子，仅在胰腺中表达[131, 111]。

AQP的基因变异可能导致AQP分子结构和转运功能受到影响，包括四聚体或阵列的形成受到干扰、蛋白质错误折叠等功能障碍[75]。

四、水通道蛋白的晶体结构

总体而言，尽管AQP氨基酸序列有显著差异，但AQP的结构在不同亚型和种属间保守存在，其结构研究为动物AQP同源四聚体的形成，以及四级结构发挥水分子转运功能的特定要求提供了一个新的视角[3, 132]。AQP的3D结构分析证实了以往通过序列分析提出的沙漏理论[133]。

AQP是由4个相同的30kDa单体组成的蛋白四聚体，每个单体具有独立的水分子通道功能。AQP单体有6个跨膜螺旋（H1～H6，相对于正常膜倾斜约30°），由5个环（A～E）连接，亲水的氨基端和羧基端始终位于细胞质中[134]。AQP有3个胞外环（A、C和E）和两个胞内环（B和D），NPA序列高度保守[134-147]（图1-2和图1-3）。

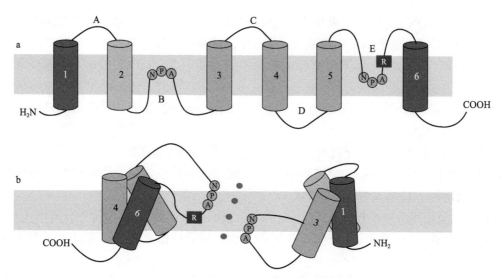

图1-2 AQP分子的二级结构和拓扑结构

a：AQP1单体有6个跨膜区（1~6），5个环（A~E），具有细胞内氨基和羧基末端及内部串联重复序列。b：在单体中，亲水性B环和E环被弯曲回到腔中并在中间相遇，形成包含两个一致的NPA基序的假定水选择通道。芳香/精氨酸（Ar/R）区域显示在靠近孔隙入口的位置

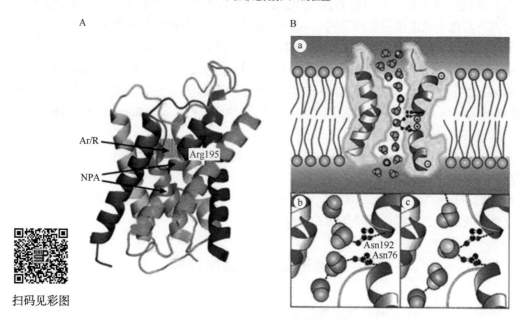

扫码见彩图

图1-3 AQP1的结构示意图

A：AQP1的彩色带状模型，从蓝色（N端）到红色（C端）。AQP1孔中最窄的区域以前称为Ar/R，位于孔的细胞外入口附近。Arg195和NPA基序分别以洋红色和浅蓝色显示[141]。B：AQP1亚单位内通道的结构示意（矢状截面）。a图水分子通过孔隙收缩的示意图。b图和c图所示的4个水分子代表与Asn76和（或）Asn192的瞬时相互作用。两个螺旋部分在通道的中点相交，提供带正电的偶极子，当水分子穿过该点时，偶极子可重新定向水分子。Arg. 精氨酸；Asn. 天冬酰胺

AQP作为狭窄、跨膜的水分子通道，被6个跨膜结构域形成的束包围。6个跨膜α螺旋包围1个狭窄的水分子孔[138]，其中由B环和E环形成的半跨膜螺旋（NPA特征基序）从膜的相对侧折叠进入通道，B环和E环的位置通过离子配对和与相邻跨膜螺旋的氢键结合

保持稳定。该孔具有静电相互作用，在细胞外环境中水分子进行随机的布朗运动，使AQP外壁处于疏水状态，导致排斥作用[139]（图1-2、图1-3A）。

两个短螺旋的B环和E环与紧密压缩的NPA基序共同构成AQP的中心收缩孔，内衬疏水氨基酸残基，孔的直径约为3Å，水分子只能单个通过收缩孔；通过与Asn残基形成氢键以降低水分子进入收缩孔的能量壁垒，进而驱使水分子通过[140, 141]。水通道蛋白的分子动力学模拟实验揭示了水分子是以单次的方式被动通过[138]，并在通道原子形成的局部电场中调整方向。当水分子从细胞外进入通道时，以氧原子朝下的方式沿着通道向下运动；在中心收缩孔的水平上颠倒方向，以氧原子朝上的方式沿着通道向下运动[140, 142]（图1-3B）。

AQP家族的另一个保守结构特征是位于通道外侧的芳香/精氨酸（Ar/R）缩窄位点，该位点含有高度保守的芳香族氨基酸和精氨酸残基[140]，可作为选择性过滤器发挥作用。通道孔径为3Å，仅略大于水分子2.8Å的直径，孔隙收缩可阻止所有直径大于水的分子通透[140, 142]。因此，Ar/R缩窄位点也被称为"选择性过滤器"。与AQP水通道相比，水甘油通道具有更大的选择孔径，可达3.4Å[143]（图1-2、图1-3B）。

AQP的大多数成员在E环中含有半胱氨酸残基，定位于对汞敏感的孔隙附近[134]。在AQP1中，Cys189残基是与汞结合并抑制水转运的部位[133, 31]。这种抑制机制最近通过分子动力学模拟实验得以阐明[144]，是由汞敏感半胱氨酸残基位于Ar/R区域的构象变化导致孔隙塌陷而抑制水的转运。然而，并非所有的AQP都会受到$HgCl_2$的抑制，如AQP4[53]和AQP6，并且汞的存在还增强了AQP6的透水性[14]。

这种蛋白质可自组装成一个四聚体生物单位并借此另外形成一个四聚体孔的机制，被认为是引起溶解气体和离子转位的原因[145]。另一方面，AQP4在细胞膜中形成较大的正交阵列（膜内粒子簇形成的特殊几何组织）[53]，表明AQP4在体内膜连接形成中可能发挥作用[146]。

五、水通道蛋白的蛋白质修饰

蛋白质在翻译后可以通过可逆或不可逆地添加功能基团（如磷酸化、乙酰化和甲基化）、多肽[如泛素化、小分子泛素相关修饰物蛋白（SUMO）化]或其他复杂大分子（如糖基化）的修饰进行调节。这些翻译后修饰（post-translational modification，PTM）通过改变蛋白质构象调节其底物蛋白的定位、稳定性、活性和相互作用，从而在细胞内信号传递、蛋白质成熟和蛋白质折叠中发挥关键作用。PTM的具体作用取决于共价修饰的性质、化学反应对底物和残基的靶向性[147, 148]。这里重点介绍AVP调节AQP2的PTM，并讨论其他AQP的修饰。

（一）磷酸化

细胞膜中蛋白质丰度的调节需要多种分选信号和PTM的参与，进而实现蛋白质从胞内转位到细胞膜的精细调节。磷酸化是目前研究最深入的PTM之一，常参与蛋白质功能和细胞分布的调节，而AQP2磷酸化是其中最具特征的代表。

AQP2表达于肾脏集合管的主细胞[43, 116, 149]，其在细胞内的分布受到AVP的精准调节。AVP刺激导致AQP2从顶膜下方转位到顶膜，使得顶膜对水的高度通透是尿浓缩形成的关键事件，因而可调控机体内的水平衡。这个由AVP介导的AQP2在主细胞内的再分布过程与AQP2磷酸化/去磷酸化和（或）泛素化调节密切相关。

AQP2在C端多个位点的磷酸化调控其从胞内小泡向顶膜的转运过程[150, 151]。生物信息学分析表明，AQP2含有多种蛋白激酶的磷酸化位点，如PKA、PKG、PKC和酪蛋白激酶Ⅱ[152]。AQP2的C端第256位（S256）丝氨酸的磷酸化最早被发现，也是AQP2特征的磷酸化位点[41, 151, 153]。大规模的磷酸化蛋白质组学分析结果证实，除S256磷酸化位点外，AQP2的多磷酸化区域还包含S261、S264和S269位点（图1-4A）。AQP2的C端多肽的体外磷酸化实验证实，S256-AQP2是PKA诱导的磷酸化靶点[154]。PKA以外的激酶也可能参与AQP2的C端磷酸化。PKG可能参与AQP2的转运调节过程；PKG的激动剂cGMP可以介导转染AQP2的LLC-PK1细胞和离体肾脏切片中AQP2的转位[44]。ANP、L-精氨酸、cGMP磷酸二酯酶5型（PDE 5）抑制剂西地那非柠檬酸作为cGMP通路的激活剂，可提高胞内cGMP水平而促进AQP2在细胞膜中聚集[44, 46, 155]，这些数据表明PKG在AQP2转运方面发挥了积极作用。也有研究表明，ANP和NO信号可下调S256-AQP2的磷酸化水平，减少细胞膜中的AQP2，拮抗AVP所介导的肾脏内髓集合管细胞的透水性[156]。此外，PKC通路的激活还介导AQP2的内吞作用，这与AQP2在S256的磷酸化状态无关[151]。最近有研究证明，丝氨酸/苏氨酸激酶（AKT，又称PKB）可以调节AVP刺激的AQP2在细胞膜聚集[157]。然而，这些蛋白激酶对S261、S264和S269磷酸化的调节和作用更为复杂[150, 154, 158, 159]。

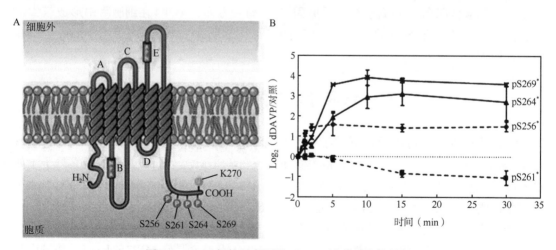

图1-4　AQP2拓扑示意图及AQP2 C端磷酸化的研究

A：AQP2的拓扑结构及AQP2的C端磷酸化位点（S256，S261，S264，S269）和泛素化位点（K270）的示意图[162]。B：1nm dDAVP（1-去氨基-8-D-精氨酸加压素，一种V2R激动剂）对大鼠内髓集合管（IMCD）AQP2磷酸化位点S256、S261、S264和S269的调节作用。注意：在S256处的最大磷酸化发生得很快，而在其他S264和S269处的最大磷酸化发生的时间更长。只要激动剂存在，S256、S264和S269位点的磷酸化水平仍然很高。相比之下，dDAVP刺激导致S261位点磷酸化水平下降

肾脏集合管细胞的胞内囊泡和顶膜中均可检测到S256位点磷酸化的AQP2（pS256），

其含量随AVP处理而相应增加[160]。S264位点磷酸化的AQP2在细胞膜相关区域和早期内源性细胞通路中均有表达；dDAVP急性处理后，pS264-AQP2在主细胞的顶膜和基底外侧膜中大量增加[161]。与AQP2的S256位点磷酸化相似，S269位点磷酸化也与AQP2的膜聚集有关，提示磷酸化在AQP2靶向细胞膜定位中的作用[154, 162-164]。pS261-AQP2主要定位于内质网、高尔基体和溶酶体不同的膜性结构[158]。

在肾脏内髓集合管（IMCD）的小管悬液中，特异性V2R激动剂dDAVP或外源性cAMP可增加AQP2在S256、S264和S269位点的磷酸化并保持在较高水平[154]。S256位点的磷酸化最先发生，并且迅速达到峰值，而其他位点（S264和S269）达到最大磷酸化水平相对缓慢。与之相反，dDAVP的刺激导致S261位点的磷酸化水平降低[154]（图1-4B）。

S256位点的磷酸化对S264和S269的磷酸化有明显的促进作用[154]，是由于在S256突变型的AQP2或S256突变为亮氨酸的小鼠肾脏切片中未检测到AQP2在S264和S269位点的磷酸化[154, 165]。最新的证据表明，单独的S256磷酸化对AVP（或cAMP）所诱导的AQP2调控是必要的，且不依赖于C端S264、S269等任何其他位点的磷酸化状态[166]。这些结果表明，S256位点的磷酸化是S264和S269位点磷酸化的启动事件，在AQP2细胞内转运中发挥关键作用。S256和S269位点的磷酸化参与了AQP2在顶膜中插入的过程[162]，而pS269-AQP2的磷酸化形式在顶膜中有独特的细胞定位[165]。AQP2在S264位点的磷酸化在AQP2亚细胞分布中的作用尚不清楚[161]。早期的研究表明，AVP刺激大鼠IMCD时，AQP2在S256位点的单磷酸化增强与AQP2在S261位点的磷酸化水平降低相一致，并且与细胞内囊泡的分布有关，提示S256和S261磷酸化可能是反向调控AQP2的转运过程[150, 158, 161, 167]。

（二）泛素化

哺乳动物细胞中有两条主要的蛋白质降解途径，即泛素-蛋白酶体途径和溶酶体蛋白水解途径。泛素（Ub）是由76个氨基酸组成的多肽，在蛋白酶体介导的蛋白质降解中发挥关键作用。泛素是由泛素激活酶（E1）激活、泛素结合酶（E2）结合和泛素连接酶（E3）所组成的共轭系统标记蛋白，其与蛋白质结合后可由胞质蛋白酶体复合物进行降解。某些细胞膜蛋白的泛素化可通过内吞途径促进其内化，然后在溶酶体中降解[56]。蛋白质泛素化可被去泛素化酶（deubiquitinase，DUB）逆转，从而维持细胞内稳态。

首个AQP泛素化的研究表明，AQP1能够被蛋白酶体泛素化并降解。暴露于高渗溶液可导致AQP1泛素化减少，AQP1的蛋白稳定性显著增强[168]。

据推测AQP2有3个潜在的泛素化位点，分别在胞质赖氨酸残端的228、238和270位点；但位点突变实验结果表明，K270是泛素化的唯一底物[169]（图1-4A）。细胞膜上AQP2的泛素化将导致AQP2的内化，转运到胞内完成蛋白酶体降解过程[169]。转录组学和液相色谱串联质谱蛋白质组学分析结果表明，5种常见的E3连接酶亚型（UBR4、UHRF1、NEDD 4、BRE1B和Cullin-5）与dDAVP诱导的AQP2调控有关[170]。例如，血管升压素激活的钙动员受体Cullin-5，其为E3复合体支架蛋白的*cullin*基因家族成员[171]，在dDAVP撤药过程中表达上调，与细胞内囊泡碎片中AQP2的泛素化水平增加相关。这一发现提示，Cullin-5可能在泛素与AQP2的连接过程中发挥作用，在dDAVP撤药后可能通过溶酶体和（或）蛋白小体的降解作用，引起AQP2的泛素化、AQP2的内化及AQP2水平降低[170]。

AQP2在S256位点的磷酸化和在S261位点的去磷酸化导致AQP2从胞内小泡转移到顶膜，而AQP2在K270处的泛素化则导致其发生内化和溶酶体降解，或通过胞吐释放外泌体进入尿液。磷酸化和泛素化在细胞中具有高度的动态性，二者之间的相互作用已有研究[172]，磷酸化和泛素化可能通过协同作用提高对蛋白质功能的精细调节。结合细胞膜靶向信号的S256、S264和S269位点磷酸化，胞内S261位点磷酸化，以及K270的泛素化共同参与AQP2亚细胞分布的精细调节过程（图1-5）。

最近2项研究揭示了AQP2多泛素化和多磷酸化之间的潜在相互作用[167, 173]。dDAVP或Forskolin（毛喉素）刺激可诱导AQP2单体S256位点的磷酸化，随后增加S269和S264位点的磷酸化，减少S261位点的磷酸化，促使AQP2从囊泡到顶膜的转运保持稳态；而AQP2的泛素化增加则导致AQP2内吞并向胞内小泡的转运以保持稳态。有趣的是，在泛素介导的内吞作用后，在AQP2的S261位点发生磷酸化，这表明S261位点的磷酸化本身并不能诱导AQP2泛素化，而可能是稳定泛素化的AQP2（图1-5）[170]。

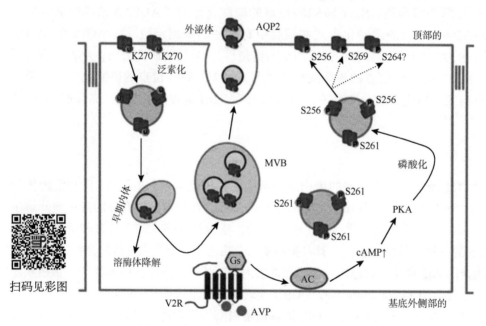

图1-5　磷酸化和泛素化决定了AQP2在细胞内的定位

AVP诱导AQP2单体S256位点磷酸化，随后S269和S264位点磷酸化增加，S261位点磷酸化减少，导致AQP2从胞内囊泡稳定重新分布到顶端细胞膜。AQP2在K270位点被一个或多个泛素蛋白泛素化。在去除AVP刺激后，泛素化发生在细胞膜上，并介导AQP2稳定再分布到细胞内囊泡。AQP2的泛素化可能被分类到多泡体（MVB）中，在溶酶体中降解，或者通过胞吐作用至外泌体中再释放到尿液中。V2R. 精氨酸加压素受体2；cAMP. 环磷酸腺苷；PKA. 蛋白激酶A；AC. 腺苷酸环化酶；Gs. G蛋白

磷酸化是泛素化发生的启动事件，泛素化可以通过调节激酶活性而反馈调节蛋白的磷酸化[172]。AQP2磷酸化被证实可涵盖K63所连接的主要多泛素化内源性信号。在极化的上皮细胞和肾组织中，AQP2在细胞膜上的分布受S256和S269位点磷酸化的调控。在S269位点磷酸化时，AQP2内吞速率随磷酸化时间的延长而降低。AQP2在S269位点的磷酸化

与在K270位点的泛素化可同时发生,当K270多泛素化水平达到最大值时,S269的磷酸化水平增加。AQP2内吞减少,S269位点的磷酸化水平增加,K270出现泛素化[173]。以上研究表明,位点特异性的磷酸化可以抵消多泛素化作用,并最终确定其定位。

(三)类泛素化

除了泛素外,目前研究得最多的泛素类蛋白是小泛素类修饰剂[小分子泛素相关修饰物蛋白(SUMO)]。类泛素化是一种可逆的PTM,其中SUMO与靶蛋白中赖氨酸残基共价结合,类似于泛素化,参与多个细胞核内过程,包括复制、转录和DNA修复等。类泛素化蛋白在调节通道蛋白活性、受体功能、G蛋白信号、细胞骨架、细胞胞吐和自噬等方面也发挥着重要作用[173]。迄今为止,还缺乏类泛素化参与AQP表达调控的直接证据。

(四)谷胱甘肽化

作为一种重要形式的PTM,S-谷胱甘肽化对半胱氨酸残基在氧化还原失衡过程中的不可逆氧化具有保护作用。AQP2与S-谷胱甘肽化的关系具有潜在性意义,原因在于活性氧物质(ROS)可能影响包括AQP在内的不同转运蛋白和通道蛋白的表达和活性。最近的证据表明,在mpkCCD细胞中,血管升压素刺激7种谷胱甘肽S-转移酶(GST)蛋白的翻译,促进三肽谷胱甘肽与半胱氨酸等底物的结合[174]。作为主要的细胞抗氧化分子之一,谷胱甘肽不断转化为还原型谷胱甘肽。AQP2的拓扑分析提示,胞质B环上的Cys75和Cys79可能是S-谷胱甘肽化的作用靶点[175]。随后的研究表明,AQP2在肾组织和稳定表达AQP2的HEK细胞中均发生S-谷胱甘肽化反应,并且ROS含量变化可精确调控AQP2的S-谷胱甘肽化水平,尤其是氧化剂诱导ROS含量增加可导致AQP2的S-谷胱甘肽化程度显著增加,而S-谷胱甘肽化是否影响AQP2的定位和活性尚未见报道[175]。

(五)糖基化

AQP的胞外环中含有N-连接糖基化共识别位点,其中一些位点在寡糖糖基转移酶合成蛋白质的过程中不能被有效识别,形成了糖基化和非糖基化的混合物。N-糖基化在AQP的转运功能中可能无足轻重。AQP1中N-糖基化的位点是Asn42,Asn42突变体的研究结果表明,AQP1的非糖基化对卵母细胞的透水性没有影响[134]。在AQP2中,糖基化形式的半衰期短于非糖基化形式,说明N-连接糖基化对AQP2的稳定性并非必需[174]。糖基化可能对AQP2上膜表达很重要[176],但在cAMP增加时抑制糖基化并不能阻止AQP2向细胞膜转运[177]。此外,糖基化对于内质网中AQP2的四聚体化并不重要,主要在于四聚体的部分与1个或多个非糖基化的AQP2分子交联[176]。然而,早期的数据结果显示,添加N-连接低聚糖可以部分缓解由疾病相关基因突变引起的内质网折叠缺陷[178]。

(六)其他几种PTM

N端乙酰化已被公认为是决定蛋白质稳定性的重要因素[179]。AQP2是受N端乙酰化修饰的蛋白质之一[174]。近年的研究发现,在AQP0的N端氨基酸残基上出现了N-连接乙酰化、碳基化和油酸化等修饰[180]。虽然这些PTM的生物学和生理学意义尚未明晰，但可能在蛋白质与蛋白质的相互作用中发挥潜在作用，从而调节水的通透性。另有研究表明，AQP4的N端半胱氨酸经棕榈酸翻译后修饰可抑制转染AQP4的中国仓鼠卵巢细胞中AQP4方形阵列的形成[181]。

肿瘤坏死因子-α（TNF-α）诱导AQP5组蛋白H4乙酰化水平降低，导致AQP5表达下调，唾液腺泡细胞的分泌减少[182]。胰岛素治疗可显著降低肝细胞内*AQP9*启动子区组蛋白H3的乙酰化程度而显著增加其甲基化程度，并与AQP9蛋白表达下调有关[183]。然而，关于乙酰化和甲基化的研究主要集中在组蛋白及其在调节染色质结构和基因转录中的作用，属于表观遗传调控。

（七）其他几种AQP的PTM

以下是几个关于代表性AQP磷酸化调控的举例，特别是AQP1、AQP4、AQP5和AQP8，均与引发膜的特异性转运有关。

1. AQP1 AQP1的透水性受激素调节。在非洲爪蟾卵母细胞表达系统中，血管升压素可增加AQP1的透水性，ANP可降低AQP1的透水性[184]。早期的体内外研究表明，PKA磷酸化导致AQP1从细胞内转运到顶膜[185-187]。此外，PKC通过磷酸化Thr157和Thr239位点对AQP1通道的透水性和离子电导均有正向调节作用[40]。近期研究表明，信号分子cAMP和cGMP可促进AQP1向近端管状细胞刷状缘转运[188]。另有研究报道，cAMP和cGMP都降低了AQP1的泛素化水平，提高了AQP1蛋白的稳定性，并明确了AQP1氨基酸序列中两个潜在的泛素化位点Lys243和Lys267[56, 168]。

最近的研究表明，AQP1的低渗介导的易位快速发生是由Ca^{2+}/钙调蛋白、PKC和微管依赖性引发[24, 189]。此外，高张力暴露也增加了AQP1在培养的肾近端和内髓细胞中的表达水平[190, 191]。水通道蛋白高渗性的分子生物学效应可能是通过启动子介导的AQP1合成激活[192]和抑制AQP1蛋白降解[168]介导的。

2. AQP4 研究表明，AQP4的透水性可通过可逆性蛋白磷酸化进行调节。AQP4在PKA、PKC、PKG、酪蛋白激酶（CK）和Ca^{2+}/钙调蛋白依赖性蛋白激酶（CaMK）中有潜在的磷酸化位点。

AQP4的Ser111残基是PKA磷酸化和钙依赖性CaMKⅡ磷酸化的潜在位点。PKA磷酸化Ser111可提高AQP4的透水性[56, 193, 194]。有研究发现，Forskolin、AVP和V2受体激动剂通过刺激cAMP的产生，增加转染AQP4肾细胞的透水性[56]。Ca^{2+}/CaMKⅡ抑制剂可逆转由Ser111位点磷酸化介导的转染AQP4 cDNA的星形胶质细胞株细胞膜透水性的增加，提示Ser111位点磷酸化通过Ca^{2+}/CaMKⅡ磷酸化可提高AQP4的透水性[195]。以上结果表明，Ser111在肾脏细胞中被PKA磷酸化，CaMKⅡ在星形胶质细胞中磷酸化是具有共性的，两种磷酸化均可导致AQP4通透性的增加。此外，Ser111也可以被PKG通过CaMKⅡ-NO-cGMP-PKG信号通路磷酸化[193]。相对于Ser111位点的磷酸化，PKC对Ser180位点的磷酸化下调了AQP4在非洲爪蟾卵母细胞表达系统和体外培养肾上皮细胞中的透水性[185, 194]，此前被认为是门控效应所致，原因在于基础条件下和激素刺激的细胞中AQP4的表达都可以忽略[195]。

然而，最近通过晶体结构、功能研究和分子动力学模拟的证据否定了磷酸化依赖的AQP4门控作用发生在Ser111和Ser180位点[54, 139, 196]。最新的质谱研究表明，AQP4细胞膜转运或通道门控作用不受羧基末端丝氨酸残基磷酸化的调控[196]。

3. AQP5 AQP5膜转运已被证明是以PKA依赖的方式受到cAMP调控[197, 198]。胞内cAMP升高具有显著的急性效应和慢性效应，导致AQP5细胞膜的峰度在短期（分钟）内降低，而AQP5总蛋白在长期（小时）内增加[199]。AQP5中2个可被磷酸化的PKA位点已被鉴定，分别位于细胞质D环的Ser156位点[200]和羧基末端的Thr259位点[61, 197]。然而，这些磷酸化位点突变的AQP5细胞膜水平与野生型AQP5细胞膜水平一致，提示在基础条件下可能不会发生磷酸化调控。相比之下，近期发现cAMP-PKA在Thr259处磷酸化的AQP5与扩散有关，可能参与调节腺体分泌物中的水分子流动[201]。最近的数据表明，AQP5质膜表达受S156位点磷酸化水平的影响，这可能是由于靶点增加、内化减少或两者兼而有之[202]。

4. AQP8 AQP8主要位于肝细胞内的小泡[203, 204]和线粒体[205]。在基础条件下，AQP8在细胞表面的表达水平很低[204, 206]，而胰高血糖素或其第二信使cAMP则强烈诱导AQP8从胞内向胞膜进行再分布[206]，从而提高了细胞膜的透水性，促进了渗透性水转运。这些研究表明，PKA和PI3K（磷脂酰肌醇-3-羟激酶）通路都参与了胰高血糖素所诱导的AQP8转运[206, 207]。

六、水通道蛋白的表达调控

AQP在各种组织中广泛表达，它们通常位于细胞的特定空间区域。AQP介导了渗透压梯度驱使的水的双向转运。门控、构象变化或特定细胞膜区域AQP的密度变化都可以调控AQP介导的水转运。AQP介导的转运在转录或翻译水平都受到了调控，此外，AQP在膜内囊泡和目的细胞膜区域之间的穿梭也可以调控AQP介导的转运。翻译后修饰，尤其是磷酸化，也是一种重要的AQP重新分布的调节机制。AQP的调控，无论是门控还是转运，都是以组织依赖性方式进行的特定且迅速的调节。另外一种相对长期的调控则是通过增加或减小AQP蛋白浓度来实现，AQP蛋白浓度的增减可能受到体循环激素的影响（比如血管升压素、胰岛素、心房利钠肽、血管紧张素Ⅱ），或局部分子的影响（比如嘌呤[50]、前列腺素[207]、多巴胺[208]），或其他微环境信号包括pH[209]、二价阳离子浓度[194]、渗透压的影响。AQP的这些调控方式都和特定的生理或病理生理情况有关。以下基于AQP2的诸多研究结果，对AQP2的调控机制进行探讨。

（一）AQP2的门控

在植物和酵母菌中，环境变化可诱发表达于细胞膜上的AQP的门控调节[210]。对于哺乳动物，门控以pH和Ca^{2+}/钙调蛋白依赖的方式调节AQP0对水的通透性[211]。也有研究报道AQP4经由磷酸化的门控调节[192]，而AQP2经由磷酸化的门控调节目前仍存在争议[149, 212]。事实上，哺乳动物AQP的主要调节机制是转运，AQP2的转运尤为经典。

（二）AQP2 的转运

跨上皮的易化运输过程需要跨膜运输蛋白如AQP的参与，AQP在上皮是从上皮游离面到基底面呈现极性分布的。AQP通过与上皮游离面或基底面质膜域相连的囊泡进行的运输，称为转运（trafficking），包括胞吐、胞吞、分选、群集，以及维持细胞膜蛋白的完整性[213]。

AQP2在翻译后折叠形成单体构型，随后在内质网内形成四聚体。这些四聚体则运输至高尔基复合体，在此处两个单体的氨基端糖基化，再通过跨高尔基网络运输到不同的亚细胞结构[214]。跨高尔基网络内的多数AQP2都位于内体囊泡内，在相关刺激（如AVP）作用下运输并表达于细胞膜的游离面[149]。内含AQP2的囊泡转运到细胞膜、在细胞膜定位、囊泡和细胞膜的融合（胞吐）以及AQP2以胞吞方式被运输至胞内，这些过程都影响AQP2在细胞膜上的分布及数量[215]。

1. 加压素对AQP2转运的影响由环磷酸腺苷（cAMP）介导 在AVP作用下，位于肾集合管主细胞上的AQP2表达密度和细胞内定位决定了该部位水的重吸收能力[149]。在缺乏AVP的情况下，AQP2位于细胞游离面下的囊泡内；在AVP的作用下，AQP2则大量表达于细胞膜的游离面。通常，AVP可以和位于细胞基底面的加压素V2受体（V2R）结合；V2R则经由异三聚体G蛋白（Gs）和腺苷酸环化酶（AC）偶联。加压素和受体的结合可以使得G蛋白的α亚单位释放鸟苷二磷酸（GDP），结合鸟苷三磷酸（GTP），并导致β及γ亚单位的解离。Gα-GTP复合体则进一步激活AC并合成cAMP，再由cAMP激活蛋白激酶A（PKA）。PKA可以直接或间接使得AQP2的S256磷酸化，从而增加AQP2从存储的胞内囊泡运输或转运至细胞游离面[214, 215]。因为PKA可以作用于细胞内的多个蛋白，所以目的蛋白在特定时间和空间里有效的磷酸化依赖于PKA定位于目的蛋白的特定位置。这个过程是由PKA锚定蛋白（AKAP）介导的。针对AQP2的磷酸化过程，AKAP将PKA锚定在AQP2附近区域是先决条件[216]。研究发现相关的剪接变异体AKAP有AKAP18、AKAP18δ和AKAP220[217]。

2. Ca^{2+} 在加压素诱导的AQP2转运中的作用 有研究显示在加压素诱导的AQP2转运中，细胞内Ca^{2+}的动员也发挥了作用。加压素和V2R结合后，可以引起内髓集合管（IMCD）细胞内Ca^{2+}浓度和钙离子振荡短暂增高[218, 219]。实验中将大鼠的IMCD分离，去除外环境Ca^{2+}，再灌注后发现加压素导致的细胞内Ca^{2+}升高不受抑制，但持续的钙离子振荡可被抑制[219]。在原代培养的IMCD细胞中，Ryanodine抑制剂（Ryanodine可诱导胞内钙库释放钙）、钙调蛋白抑制剂或细胞内Ca^{2+}螯合剂都显示出可以阻止加压素诱导的AQP2膜转位和水通透性增加[220]。这些研究结果都证实了加压素诱导的细胞内Ca^{2+}增加在AQP2转位至细胞游离面过程中发挥着重要的作用。这既包括细胞内Ca^{2+}从Ryanodine敏感的钙库释放，也包括细胞外Ca^{2+}的内流。相应的，也有研究表明在原代培养的肾内髓质上皮细胞，仅cAMP信号足以激发AQP2胞吐至细胞游离面，并不需要加压素来诱导细胞内Ca^{2+}增加[221]。

3. AQP2经由囊泡运输至细胞膜 AQP的囊泡协同转运是沿着细胞骨架被运输至特定区域的。以AQP2为例，微管和肌动蛋白细胞骨架的再组装在AQP2的转运中极为重要。

肌动蛋白细胞骨架形成的"笼子"可与AQP2锚定，从而避免了无刺激（如无加压素刺激）的情况下出现胞吐转位。当加压素和V2R结合后，集合管细胞内F肌动蛋白解聚，促使含有AQP2的囊泡转运至细胞游离面并与细胞膜融合[222]。实际上，仅AQP2可以直接调节肌动蛋白细胞骨架的解聚及随后的胞吐过程。PKA诱导的AQP2的S256磷酸化可以减少AQP2和G肌动蛋白的直接结合，从而增加了AQP2和肌球蛋白Vb的亲和性，后者是AQP2膜转运中的重要调节因子之一。S256磷酸化后的AQP2与肌球蛋白Vb结合会促使F肌动蛋白网络中的肌球蛋白Vb释放，从而使得F肌动蛋白网络不稳定并解聚，进而促使AQP2囊泡转位至细胞膜[223]。

PKA锚定蛋白220（AKAP220）在囊泡和膜相关的锚定蛋白中广泛表达，并在细胞膜突起中正向调节肌动蛋白的聚合及微管的稳定性[224]。早期研究显示在肾集合管主细胞中AKAP220与AQP2有关联。近期有研究显示缺乏AKAP220会导致AQP2在细胞膜的聚集并减低水中毒情况下的尿稀释能力[224]。这项研究也支持肌动蛋白网格的再组装在肾脏AQP2亚细胞定位中发挥作用。

值得注意的是，在AQP2羧基端或在含有AQP2的囊泡内的一些结合蛋白，可介导AQP2的分选和最终运输。大规模的蛋白质组分析已经明确了180种此类蛋白，包括SNARE蛋白（介导转运小泡与细胞器特异性融合）、跨高尔基网络标志物、马达蛋白等。这些蛋白和AQP2经由直接结合、间接连接形成蛋白复合体或共定位在囊泡内，积极参与AQP2的运输调节[225]。

4. 内含AQP2的囊泡在细胞膜游离面的定位和融合（胞吐） 在加压素调节的水转运过程中，内含AQP2的囊泡和细胞膜的融合是最终的关键步骤。内含AQP2的囊泡在细胞膜的定位和融合是由SNARE（可溶性N-乙基马来酰亚胺敏感因子结合蛋白受体）机制介导的。细胞膜的融合需要一种可溶性的细胞质融合蛋白NSF（N-乙基马来酰亚胺敏感因子）和SNAP（可溶性NSF结合蛋白），NSF和SNAP负责介导不同类型的细胞膜和囊泡的融合，无明显特异性。而每种转运囊泡上均有特异性v-SNARE（SNAP受体），能识别并与靶膜上的t-SNARE相互作用[214]。在集合管主细胞中已经发现了SNARE系统的多种因子，比如在含AQP2的囊泡中发现了VAMP-2、VAMP-3（v-SNARE的囊泡相关膜蛋白）[226]；在主细胞细胞膜的游离面，发现了syntaxin4、syntaxin3、SNAP23和SNAP25等t-SNARE[227]。Snapin是一种中间支架分子，可促使AQP2和t-SNARE复合物连接，从而辅助AQP2从储存囊泡转运到细胞膜游离面[228]。实验显示在破伤风毒素作用下，VAMP-2被剪切，导致AVP介导的AQP2细胞膜转位过程受抑制，证实了v-SNARE在AQP2细胞膜转运中的作用[229]。在敲除SNARE抑制蛋白Munc18后，AQP2在细胞膜的表达增加，然而敲除VAMP-2、VAMP-3、syntaxin3和SNAP23则可以阻止AQP2囊泡在细胞膜游离面的融合[230]。这些研究结果均证实，在AQP2与细胞膜的定位和融合过程中SNARE发挥了重要作用。此外，还有研究表明也有一些其他的蛋白（比如annexin-2、鸟苷三磷酸酶等）同样在AQP2的转运和胞吐过程中发挥作用，但是具体作用机制尚不清楚[214]。

5. AQP2从细胞膜的移除（胞吞）和降解 调节AQP2的胞吞可以极大地影响AQP2在细胞膜的表达。阻止胞吞可以增加AQP2在细胞膜游离面的表达，这也是肾脏增强集合管水通透性的一个重要途径[231]。在胞吞过程中，AQP2聚集在网格蛋白有被小窝内，进而经由网格蛋白介导发动蛋白（dynamin）依赖性的方式内化至细胞内[232]。热休克同源蛋白70（hsc70）被证实在AQP2的内吞过程中是必不可少的，可与未磷酸化的AQP2羧基端结合，参与网格蛋白有被小泡的脱包被过程。研究发现AQP2的S256和S269位点被磷酸化，可导致AQP2和网格蛋白、hsc70及发动蛋白的相互作用减弱，因而减少了AQP2胞吞的发生[233]。AQP2的磷酸化可能改变了有被小窝和网格蛋白有被小泡的形成，从而调节了AQP2经由网格蛋白有被小窝的内化进程[234]。

小窝的主要结构蛋白小窝蛋白-1参与了受体介导的胞吞过程，近期研究表明AQP2和小窝蛋白-1可相互作用。当去除Forskolin后，AQP2可依赖小窝蛋白-1而发生内化[235]。此外，研究也证实了膜脂筏在AQP2胞吞中的调节作用。通过试剂去除膜胆固醇后，在体和离体试验都发现了AQP2在细胞膜上的表达增加，这也有可能是AQP2的内化减少所致[231]。

AQP2从细胞膜胞吞至细胞内后，经由PI3K机制进入了早期内体，随后转运至Rab11阳性小泡储存[236]。还有一些AQP2会转运至多泡体，经尿液排出外体[237]。

（三）AQP2的蛋白合成

除了细胞内转运和转录后修饰外，AQP2蛋白的调节也发生在转录水平。参与转录水平调节的转录因子包括CREB、AP1[238]、NFAT家族蛋白（TonEBP和NFATc）及NF-γB[239]。

一段时间的加压素或失水处理可以促使集合管水通透性增加，这种处理是一种长期调节。该过程是在受到外界刺激后，经加压素-V2R信号级联途径，*AQP2*基因的转录被激活，最终导致AQP2蛋白增加[240]。基因测序显示，*AQP2*的5′侧翼序列含有可能的顺式作用因子结构域：cAMP应答元件（CRE）和SP-1位点[241]，其中CRE与*AQP2*的转录调控有关。在高渗透压时，渗透压应答增强子（TonE）转录因子（TonEBP）敲除后，小鼠的AQP2蛋白表达下调，证实TonE/TonEBP在*AQP2*转录中起作用[242]。针对AQP2在肾集合管的细胞特异性的基因表达分析显示，还有许多转录调节子和转录调节子的结合因子参与了*AQP2*基因的转录调控[243]。

（冉建华　陈　哲　徐　进）

参 考 文 献

[1] Solomon AK, Chasan B, Dix JA, et al. The aqueous pore in the red cell membrane: band 3 as a channel for anions, cations, nonelectrolytes, and water[J]. Ann N Y Acad Sci, 1983, 414(1): 97-124.

[2] Denker BM, Smith BL, Kuhajda FP, et al. Identification, purification, and partial characterization of a novel Mr 28,000 integral membrane protein from erythrocytes and renal tubules[J]. J Biol Chem, 1988, 263(30): 15634-15642.

[3] Smith BL, Agre P. Erythrocyte Mr 28,000 transmembrane protein exists as a multisubunit oligomer similar to channel proteins[J]. J Biol Chem, 1991, 266(10): 6407-6415.

[4] Preston GM, Agre P. Isolation of the cDNA for erythrocyte integral membrane protein of 28 kilodaltons: member of an ancient channel family[J]. Proc Natl Acad Sci U S A, 1991, 88(24): 11110-11114.

[5] Agre P, Preston GM, Smith BL, et al. Aquaporin CHIP: the archetypal molecular water channel[J]. Am J

Physiol, 1993, 265(4 Pt 2): F463-F476.

[6] Zeidel ML, Ambudkar SV, Smith BL. Reconstitution of functional water channels in liposomes containing purified red cell CHIP28 protein[J]. Biochemistry, 31(33): 7436-7440.

[7] Zeidel ML, Nielsen S, Smith BL, et al. Ultrastructure, pharmacologic inhibition, and transport selectivity of aquaporin channel-forming integral protein in proteoliposomes[J]. Biochemistry, 1994, 33(6): 1606-1615.

[8] Abascal F, Irisarri I, Zardoya R. Diversity and evolution of membrane intrinsic proteins[J]. Biochim Biophys Acta, 2014, 1840(5): 1468-1481.

[9] Finn RN, Cerdà J. Evolution and functional diversity of aquaporins[J]. Biol Bull, 2015, 229(1): 6-23.

[10] Ikeda M, Andoo A, Shimono M, et al. The NPC motif of aquaporin-11, unlike the NPA motif of known aquaporins, is essential for full expression of molecular function[J]. J Biol Chem, 2011, 286(5): 3342-3350.

[11] Ishibashi K, Tanaka Y, Morishita Y. The role of mammalian superaquaporins inside the cell[J]. Biochim Biophys Acta, 2014, 1840(5): 1507-1512.

[12] Itoh T, Rai T, Kuwahara M, et al. Identification of a novel aquaporin, AQP12, expressed in pancreatic acinar cells[J]. Biochem Biophys Res Commun, 2005, 330(3): 832-838.

[13] Geyer RR, Musa-Aziz R, Qin X, et al. Relative CO_2/NH_3 selectivities of mammalian aquaporins 0-9[J]. Am J Phys Cell Physiol, 2013, 304(10): C985-C994.

[14] Yasui M, Hazama A, Kwon TH, et al. Rapid gating and anion permeability of an intracellular aquaporin[J]. Nature, 1999, 402(6758): 184-187.

[15] Koyama N, Ishibashi K, Kuwahara M, et al. Cloning and functional expression of human aquaporin8 cDNA and analysis of its gene[J]. Genomics, 1998, 54(1): 169-172.

[16] Michalek K. Aquaglyceroporins in the kidney: present state of knowledge and prospects[J]. J Physiol Pharmacol, 2016, 67(2): 185-193.

[17] Bienert GP, Thorsen M, Schüssler MD, et al. A subgroup of plant aquaporins facilitate the bi-directional diffusion of $As(OH)_3$ and $Sb(OH)_3$ across membranes[J]. BMC Biol, 2008, 6: 26.

[18] Borgnia M, Nielsen S, Engel A, et al. Cellular and molecular biology of the aquaporin water channels[J]. Annu Rev Biochem, 1999, 68: 425-458.

[19] Echevarria M, Windhager EE, Tate SS, et al. Cloning and expression of AQP3, a water channel from the medullary collecting duct of rat kidney[J]. Proc Natl Acad Sci USA, 1994, 91(23): 10997-11001.

[20] Yang B, Verkman AS. Water and glycerol permeabilities of aquaporins 1-5 and MIP determined quantitatively by expression of epitope-tagged constructs in *Xenopus* oocytes[J]. J Biol Chem, 1997, 272(26): 16140-16146.

[21] Ishibashi K, Kuwahara M, Gu Y, et al. Cloning and functional expression of a new water channel abundantly expressed in the testis permeable to water, glycerol, and urea[J]. J Biol Chem, 1997, 272(33): 20782-20786.

[22] Ishibashi K, Morinaga T, Kuwahara M, et al. Cloning and identification of a new member of water channel (AQP10) as an aquaglyceroporin[J]. Biochim Biophys Acta, 2002, 1576(3): 335-340.

[23] Tsukaguchi H, Weremowicz S, Morton CC, et al. Functional and molecular characterization of the human neutral solute channel aquaporin-9[J]. Am J Phys, 1999, 277(5): F685-F696.

[24] Conner MT, Conner AC, Brown JE, et al. Membrane trafficking of aquaporin 1 is mediated by protein kinase C via microtubules and regulated by tonicity[J]. Biochemistry, 2010, 49(5): 821-823.

[25] Gorin MB, Yancey SB, Cline J, et al. The major intrinsic protein (MIP) of the bovine lens fiber membrane: characterization and structure based on cDNA cloning[J]. Cell, 1984, 39(1): 49-59.

[26] Chandy G, Zampighi GA, Kreman M, et al. Comparison of the water transporting properties of MIP and AQP1[J]. J Membr Biol, 1997, 159(1): 29-39.

[27] Gonen T, Cheng Y, Kistler J, et al. Aquaporin-0 membrane junctions form upon proteolytic cleavage[J]. J Mol Biol, 2004, 342(4): 1337-1345.

[28] Németh-Cahalan KL, Kalman K, Hall JE. Molecular basis of pH and Ca^{2+} regulation of aquaporin water permeability[J]. J Gen Physiol, 2004, 123(5): 573-580.

[29] Reichow SL, Clemens DM, Freites JA, et al. Allosteric mechanism of water-channel gating by Ca^{2+}-calmodulin[J]. Nat Struct Mol Biol, 2013, 20(9): 1085-1092.

[30] Golestaneh N, Fan J, Zelenka P, et al. PKC putative phosphorylation site Ser235 is required for MIP/AQP0 translocation to the plasma membrane[J]. Mol Vis, 2008, 14: 1006-1014.

[31] Preston GM, Carroll TP, Guggino WB, et al. Appearance of water channels in *Xenopus* oocytes expressing red cell CHIP28 protein[J]. Science, 1992, 256(5055): 385-387.

[32] Nakhoul NL, Davis BA, Romero MF, et al. Effect of expressing the water channel aquaporin-1 on the CO_2 permeability of *Xenopus* oocytes[J]. Am J Phys, 1998, 274(2): C543-C548.

[33] Prasad GV, Coury LA, Finn F, et al. Reconstituted aquaporin 1 water channels transport CO_2 across membranes[J]. J Biol Chem, 1998, 273(50): 33123-33126.

[34] Day RE, Kitchen P, Owen DS, et al. Human aquaporins: regulators of transcellular water flow[J]. Biochim Biophys Acta, 2014, 1840(5): 1492-1506.

[35] Nico B, Ribatti D. Aquaporins in tumor growth and angiogenesis[J]. Cancer Lett, 2010, 294(2): 135-138.

[36] Kong B, Zhao SP. Acetazolamide inhibits aquaporin-1 expression and colon cancer xenograft tumor growth[J]. Hepatogastroenterology, 2011, 58(110/111): 1502-1506.

[37] Ripoche P, Goossens D, Devuyst O, et al. Role of RhAG and AQP1 in NH_3 and CO_2 gas transport in red cell ghosts: a stopped-flow analysis[J]. Transfus Clin Biol, 2006, 13(1/2): 117-122.

[38] Anthony TL, Brooks HL, Boassa D, et al. Cloned human aquaporin-1 is a cyclic GMP-gated ion channel[J]. Mol Pharmacol, 2000, 57(3): 576-588.

[39] Campbell EM, Birdsell DN, Yool AJ. The activity of human aquaporin 1 as a cGMP-gated cation channel is regulated by tyrosine phosphorylation in the carboxyl-terminal domain[J]. Mol Pharmacol, 2012, 81(1): 97-105.

[40] Zhang W, Zitron E, Hömme M, et al. Aquaporin-1 channel function is positively regulated by protein kinase C[J]. J Biol Chem, 2007, 282(29): 20933-20940.

[41] Fushimi K, Uchida S, Hara Y, et al. Cloning and expression of apical membrane water channel of rat kidney collecting tubule[J]. Nature, 1993, 361(6412): 549-552.

[42] Marples D, Knepper MA, Christensen EI, et al. Redistribution of aquaporin-2 water channels induced by vasopressin in rat kidney inner medullary collecting duct[J]. Am J Phys, 1995, 269(3 Pt 1): C655-C664.

[43] Nielsen S, DiGiovanni SR, Christensen EI, et al. Cellular and subcellular immunolocalization of vasopressin-regulated water channel in rat kidney[J]. Proc Natl Acad Sci USA, 1993, 90(24): 11663-11667.

[44] Bouley R, Breton S, Sun T, et al. Nitric oxide and atrial natriuretic factor stimulate cGMP-dependent membrane insertion of aquaporin 2 in renal epithelial cells[J]. J Clin Invest, 2000, 106(9): 1115-1126.

[45] Bouley R, Pastor-Soler N, Cohen O, et al. Stimulation of AQP2 membrane insertion in renal epithelial cells in vitro and in vivo by the cGMP phosphodiesterase inhibitor sildenafil citrate(Viagra)[J]. Am J Physiol Ren Physiol, 2005, 288(6): F1103-F1112.

[46] Wang W, Li C, Nejsum LN, et al. Biphasic effects of ANP infusion in conscious, euvolumic rats: roles of AQP2 and ENaC trafficking[J]. Am J Physiol Ren Physiol, 2006, 290(2): F530-F541.

[47] Zelenina M, Christensen BM, Palmér J, et al. Prostaglandin E(2)interaction with AVP: effects on AQP2 phosphorylation and distribution[J]. Am J Physiol Ren Physiol, 2000, 278(3): F388-F394.

[48] Li C, Wang W, Rivard CJ, et al. Molecular mechanisms of angiotensin II stimulation on aquaporin-2 expression and trafficking[J]. Am J Physiol Ren Physiol, 2011, 300(5): F1255-F1261.

[49] Li C, Wang W, Summer SN, et al. Molecular mechanisms of antidiuretic effect of oxytocin[J]. J Am Soc

Nephrol, 2008, 19(2): 225-232.

[50] Zhang Y, Peti-Peterdi J, Müller CE, et al. P2Y12 receptor localizes in the renal collecting duct and its blockade augments arginine vasopressin action and alleviates nephrogenic diabetes insipidus[J]. J Am Soc Nephrol, 2015, 26(12): 2978-2987.

[51] Procino G, Barbieri C, Carmosino M, et al. Lovastatin-induced cholesterol depletion affects both apical sorting and endocytosis of aquaporin-2 in renal cells[J]. Am J Physiol Ren Physiol, 2010, 298(2): F266-F278.

[52] Yang B, Ma T, Verkman AS. cDNA cloning, gene organization, and chromosomal localization of a human mercurial insensitive water channel. Evidence for distinct transcriptional units[J]. J Biol Chem, 1995, 270(39): 22907-22913.

[53] Yang B, Brown D, Verkman AS. The mercurial insensitive water channel (AQP-4) forms orthogonal arrays in stably transfected Chinese hamster ovary cells[J]. J Biol Chem, 1996, 271(9): 4577-4580.

[54] Moeller HB, Fenton RA, Zeuthen T, et al. Vasopressin-dependent short-term regulation of aquaporin 4 expressed in *Xenopus* oocytes[J]. Neuroscience, 2009, 164(4): 1674-1684.

[55] Carmosino M, Procino G, Tamma G, et al. Trafficking and phosphorylation dynamics of AQP4 in histamine-treated human gastric cells[J]. Biol Cell, 2007, 99(1): 25-36.

[56] Gunnarson E, Zelenina M, Aperia A. Regulation of brain aquaporins[J]. Neuroscience, 2004, 129(4): 947-955.

[57] Assentoft M, Kaptan S, Fenton RA, et al. Phosphorylation of rat aquaporin-4 at Ser111 is not required for channel gating[J]. Glia, 2013, 61(7): 1101-1112.

[58] Song Y, Verkman AS. Aquaporin-5 dependent fluid secretion in airway submucosal glands[J]. J Biol Chem, 2001, 276(44): 41288-41292.

[59] Song Y, Sonawane N, Verkman AS. Localization of aquaporin-5 in sweat glands and functional analysis using knockout mice[J]. J Physiol, 2002, 541(Pt 2): 561-568.

[60] Musa-Aziz R, Chen LM, Pelletier MF, et al. Relative CO_2/NH_3 selectivities of AQP1, AQP4, AQP5, AmtB, and RhAG[J]. Proc Natl Acad Sci USA, 2009, 106(13): 5406-5411.

[61] Hasegawa T, Azlina A, Javkhlan P, et al. Novel phosphorylation of aquaporin-5 at its threonine 259 through cAMP signaling in salivary gland cells[J]. Am J Phys Cell Physiol, 2011, 301(3): C667-C678.

[62] Woo J, Lee J, Kim MS, et al. The effect of aquaporin 5 overexpression on the Ras signaling pathway[J]. Biochem Biophys Res Commun, 2008, 367(2): 291-298.

[63] Ishikawa Y, Skowronski MT, Inoue N, et al. Alpha(1)-adrenoceptor-induced trafficking of aquaporin-5 to the apical plasma membrane of rat parotid cells[J]. Biochem Biophys Res Commun, 1999, 265(1): 94-100.

[64] Tada J, Sawa T, Yamanaka N, et al. Involvement of vesicle-cytoskeleton interaction in AQP5 trafficking in AQP5-gene-transfected HSG cells[J]. Biochem Biophys Res Commun, 1999, 266(2): 443-447.

[65] Verlander JW, Madsen KM, Tisher CC. Effect of acute respiratory acidosis on two populations of intercalated cells in rat cortical collecting duct[J]. Am J Phys, 1987, 253(6 Pt 2): F1142-F1156.

[66] Beitz E, Liu K, Ikeda M, et al. Determinants of AQP6 trafficking to intracellular sites versus the plasma membrane in transfected mammalian cells[J]. Biol Cell, 2006, 98(2): 101-109.

[67] Rabaud NE, Song L, Wang Y, et al. Aquaporin 6 binds calmodulin in a calcium-dependent manner[J]. Biochem Biophys Res Commun, 2009, 383(1): 54-57.

[68] Liu K, Kozono D, Kato Y, et al. Conversion of aquaporin 6 from an anion channel to a water-selective channel by a single amino acid substitution[J]. Proc Natl Acad Sci USA, 2005, 102(6): 2192-2197.

[69] Holm LM, Klaerke DA, Zeuthen T. Aquaporin 6 is permeable to glycerol and urea[J]. Pflugers Arch,

2004, 448(2): 181-186.

[70] Ikeda M, Beitz E, Kozono D, et al. Characterization of aquaporin-6 as a nitrate channel in mammalian cells. Requirement of pore-lining residue threonine 63[J]. J Biol Chem, 2002, 277(42): 39873-39879.

[71] Elkjaer ML, Nejsum LN, Gresz V, et al. Immunolocalization of aquaporin-8 in rat kidney, gastrointestinal tract, testis, and airways[J]. Am J Physiol Ren Physiol, 2001, 281(6): F1047-F1057.

[72] Calamita G, Ferri D, Gena P, et al. The inner mitochondrial membrane has aquaporin-8 water channels and is highly permeable to water[J]. J Biol Chem, 2005, 280(17): 17149-17153.

[73] Saparov SM, Liu K, Agre P, et al. Fast and selective ammonia transport by aquaporin-8[J]. J Biol Chem, 2007, 282(8): 5296-5301.

[74] Yang B, Zhao D, Verkman AS. Evidence against functionally significant aquaporin expression in mitochondria[J]. J Biol Chem, 2006, 281(24): 16202-16206.

[75] Soria LR, Fanelli E, Altamura N, et al. Aquaporin-8-facilitated mitochondrial ammonia transport[J]. Biochem Biophys Res Commun, 2010, 393(2): 217-221.

[76] Molinas SM, Trumper L, Marinelli RA. Mitochondrial aquaporin-8 in renal proximal tubule cells: evidence for a role in the response to metabolic acidosis[J]. Am J Physiol Ren Physio, 2012, 303(3): F458-F466.

[77] Bienert GP, Møller AL, Kristiansen KA, et al. Specific aquaporins facilitate the diffusion of hydrogen peroxide across membranes[J]. J Biol Chem, 2007, 282(2): 1183-1192.

[78] Bienert GP, Chaumont F. Aquaporin-facilitated transmembrane diffusion of hydrogen peroxide[J]. Biochim Biophys Acta, 2014, 1840(5): 1596-1604.

[79] de Almeida A, Martins AP, Mósca AF, et al. Exploring the gating mechanisms of aquaporin-3: new clues for the design of inhibitors?[J]. Mol BioSyst, 2016, 12(5): 1564-1573.

[80] Bollag WB, Xie D, Zheng X, et al. A potential role for the phospholipase D2-aquaporin-3 signaling module in early keratinocyte differentiation: production of a phosphatidylglycerol signaling lipid[J]. J Invest Dermatol, 2007, 127(12): 2823-2831.

[81] Boury-Jamot M, Sougrat R, Tailhardat M, et al. Expression and function of aquaporins in human skin: is aquaporin-3 just a glycerol transporter?[J]. Biochim Biophys Acta, 2006, 1758(8): 1034-1042.

[82] Nakahigashi K, Kabashima K, Ikoma A, et al. Upregulation of aquaporin-3 is involved in keratinocyte proliferation and epidermal hyperplasia[J]. J Invest Dermatol, 2011, 131(4): 865-873.

[83] Ishibashi K, Sasaki S, Fushimi K, et al. Immunolocalization and effect of dehydration on AQP3, a basolateral water channel of kidney collecting ducts[J]. Am J Phys, 1997, 272(2 Pt 2): F235-F241.

[84] Ecelbarger CA, Terris J, Frindt G, et al. Aquaporin-3 water channel localization and regulation in rat kidney[J]. Am J Phys, 1995, 269(5 Pt 2): F663-F672.

[85] Hua Y, Ding S, Zhang W, et al. Expression of AQP3 protein in hAECs is regulated by Camp-PKA-CREB signalling pathway[J]. Front Biosci (Landmark Ed), 2015, 20(7): 1047-1055.

[86] Jourdain P, Becq F, Lengacher S, et al. The human CFTR protein expressed in CHO cells activates aquaporin-3 in a cAMP-dependent pathway: study by digital holographic microscopy[J]. J Cell Sci, 2014, 127(Pt 3): 546-556.

[87] Marlar S, Arnspang EC, Koffman JS, et al. Elevated cAMP increases aquaporin-3 plasma membrane diffusion[J]. Am J Phys Cell Physiol, 2014, 306(6): C598-C606.

[88] Almasalmeh A, Krenc D, Wu B, et al. Structural determinants of the hydrogen peroxide permeability of aquaporins[J]. FEBS J, 2014, 281(3): 647-656.

[89] Hara-Chikuma M, Chikuma S, Sugiyama Y, et al. Chemokine-dependent T cell migration requires aquaporin-3-mediated hydrogen peroxide uptake[J]. J Exp Med, 2012, 209(10): 1743-1752.

[90] Hara-Chikuma M, Satooka H, Watanabe S, et al. Aquaporin-3-mediated hydrogen peroxide transport is required for NF-κB signalling in keratinocytes and development of psoriasis[J]. Nat Commun, 2015, 6: 7454.

[91] Hara-Chikuma M, Watanabe S, Satooka H. Involvement of aquaporin-3 in epidermal growth factor receptor signaling via hydrogen peroxide transport in cancer cells[J]. Biochem Biophys Res Commun, 2016, 471(4): 603-609.

[92] Satooka H, Hara-Chikuma M. Aquaporin-3 controls breast cancer cell migration by regulating hydrogen peroxide transport and its downstream cell signaling[J]. Mol Cell Biol, 2016, 36(7): 1206-1218.

[93] Liu Z, Shen J, Carbrey JM, et al. Arsenite transport by mammalian aquaglyceroporins AQP7 and AQP9[J]. Proc Natl Acad Sci USA, 2002, 99(9): 6053-6058.

[94] Hara-Chikuma M, Sohara E, Rai T, et al. Progressive adipocyte hypertrophy in aquaporin-7-deficient mice: adipocyte glycerol permeability as a novel regulator of fat accumulation[J]. J Biol Chem, 2005, 280(16): 15493-15496.

[95] Hibuse T, Maeda N, Funahashi T, et al. Aquaporin 7 deficiency is associated with development of obesity through activation of adipose glycerol kinase[J]. Proc Natl Acad Sci USA, 2005, 102(31): 10993-10998.

[96] Rodríguez A, Catalán V, Gómez-Ambrosi J, et al. Aquaglyceroporins serve as metabolic gateways in adiposity and insulin resistance control[J]. Cell Cycle, 2011, 10(10): 1548-1556.

[97] Miyauchi T, Yamamoto H, Abe Y, et al. Dynamic subcellular localization of aquaporin-7 in white adipocytes[J]. FEBS Letter, 2015, 589(5): 608-614.

[98] Kishida K, Kuriyama H, Funahashi T, et al. Aquaporin adipose, a putative glycerol channel in adipocytes[J]. J Biol Chem, 2000, 275(27): 20896-20902.

[99] Maeda N. Implications of aquaglyceroporins 7 and 9 in glycerol metabolism and metabolic syndrome[J]. Mol Asp Med, 2012, 33(5/6): 665-675.

[100] Elkjaer M, Vajda Z, Nejsum LN, et al. Immunolocalization of AQP9 in liver, epididymis, testis, spleen, and brain[J]. Biochem Biophys Res Commun, 2000, 276(3): 1118-1128.

[101] Rojek A, Praetorius J, Frøkiaer J, et al. A current view of the mammalian aquaglyceroporins[J]. Annu Rev Physiol, 2008, 70: 301-327.

[102] Viadiu H, Gonen T, Walz T. Projection map of aquaporin-9 at 7 Å resolution[J]. J Mol Biol, 2007, 367(1): 80-88.

[103] Watanabe S, Moniaga CS, Nielsen S, et al. Aquaporin-9 facilitates membrane transport of hydrogen peroxide in mammalian cells[J]. Biochem Biophys Res Commun, 2016, 471(1): 191-197.

[104] Loitto VM, Huang C, Sigal YJ, et al. Filopodia are induced by aquaporin-9 expression[J]. Exp Cell Res, 2007, 313(7): 1295-1306.

[105] Morinaga T, Nakakoshi M, Hirao A, et al. Mouse aquaporin 10 gene(AQP10)is a pseudogene[J]. Biochem Biophys Res Commun, 2002, 294(3): 630-634.

[106] Laforenza U, Scaffino MF. Gastaldi G. Aquaporin-10 represents an alternative pathway for glycerol efflux from human adipocytes[J]. PLoS One, 2013, 8(1): e54474.

[107] Öberg F, Sjöhamn J, Fischer G, et al. Glycosylation increases the thermostability of human aquaporin 10 protein[J]. J Biol Chem, 2011, 286(36): 31915-31923.

[108] Takahashi S, Muta K, Sonoda H, et al. The role of Cysteine 227 in subcellular localization, water permeability, and multimerization of aquaporin-11[J]. FEBS Open Bio, 2014, 4: 315-320.

[109] Yakata K, Hiroaki Y, Ishibashi K, et al. Aquaporin-11 containing a divergent NPA motif has normal water channel activity[J]. Biochim Biophys Acta, 2007, 1768(3): 688-693.

[110] Yakata K, Tani K, Fujiyoshi Y. Water permeability and characterization of aquaporin-11[J]. J Struct Biol,

2011, 174（2）：315-320.

[111] Morishita Y, Matsuzaki T, Hara-chikuma M, et al. Disruption of aquaporin-11 produces polycystic kidneys following vacuolization of the proximal tubule[J]. Mol Cell Biol, 2005, 25（17）：7770-7779.

[112] Inoue Y, Sohara E, Kobayashi K, et al. Aberrant glycosylation and localization of polycystin-1 cause polycystic kidney in an AQP11 knockout model[J]. J Am Soc Nephrol, 2014, 25（12）：2789-2799.

[113] Calvanese L, Pellegrini-Calace M, Oliva R. In silico study of human aquaporin AQP11 and AQP12 channels[J]. Protein Sci, 2013, 22（4）：455-466.

[114] Ohta E, Itoh T, Nemoto T, et al. Pancreas-specific aquaporin 12 null mice showed increased susceptibility to caerulein-induced acute pancreatitis[J]. Am J Phys Cell Physiol, 2009, 297（6）：C1368-C1378.

[115] Pisano MM, Chepelinsky AB. Genomic cloning, complete nucleotide sequence, and structure of the human gene encoding the major intrinsic protein（MIP）of the lens[J]. Genomics, 1991, 11（4）：981-990.

[116] Deen PM, Verdijk MA, Knoers NV, et al. Requirement of human renal water channel aquaporin-2 for vasopressin-dependent concentration of urine[J]. Science, 1994, 264（5155）：92-95.

[117] van Lieburg AF, Verdijk MA, Knoers VV, et al. Patients with autosomal nephrogenic diabetes insipidus homozygous for mutations in the aquaporin 2 water-channel gene[J]. Am J Hum Genet, 1994, 55（4）：648-652.

[118] Deen PM, Weghuis DO, Sinke RJ, et al. Assignment of the human gene for the water channel of renal collecting duct aquaporin 2（AQP2）to chromosome 12 region Q12：>q13[J]. Cytogenet Cell Genet, 1994, 66（4）：260-262.

[119] Matsumura Y, Uchida S, Rai T, et al. Transcriptional regulation of aquaporin-2 water channel gene by cAMP[J]. J Am Soc Nephrol, 1997, 8（6）：861-867.

[120] Sasaki S, Fushimi K, Saito H, et al. Cloning, characterization, and chromosomal mapping of human aquaporin of collecting duct[J]. J Clin Invest, 1994, 93（3）：1250-1256.

[121] Ishibashi K, Sasaki S, Saito F, et al. Structure and chromosomal localization of a human water channel（AQP3）gene[J]. Genomics, 1995, 27（2）：352-354.

[122] Inase N, Fushimi K, Ishibashi K, et al. Isolation of human aquaporin 3 gene[J]. J Biol Chem, 1995, 270（30）：17913-17916.

[123] Lu M, Lee MD, Smith BL, et al. The human AQP4 gene: definition of the locus encoding two water channel polypeptides in brain[J]. Proc Natl Acad Sci USA, 1996, 93（20）：10908-10912.

[124] Lee MD, Bhakta KY, Raina S, et al. The human Aquaporin-5 gene. Molecular characterization and chromosomal localization[J]. J Biol Chem, 1996, 271（15）：8599-8604.

[125] Ma T, Yang B, Kuo WL, et al. cDNA cloning and gene structure of a novel water channel expressed exclusively in human kidney: evidence for a gene cluster of aquaporins at chromosome locus 12q13[J]. Genomics, 1996, 35（3）：543-550.

[126] Ma T, Yang B, Umenishi F, et al. Closely spaced tandem arrangement of AQP2, AQP5, and AQP6 genes in a 27-kilobase segment at chromosome locus 12q13[J]. Genomics, 1997, 43（3）：387-389.

[127] Kondo H, Shimomura I, Kishida K, et al. Human aquaporin adipose（AQPap）gene[J]. Eur J Biochem, 2002, 269（7）：1814-1826.

[128] Viggiano L, Rocchi M, Svelto M, et al. Assignment of the aquaporin-8 water channel gene（AQP8）to human chromosome 16p12[J]. Cytogenet Cell Genet, 1999, 84（3/4）：208-210.

[129] Ishibashi K, Kuwahara M, Gu Y, et al. Cloning and functional expression of a new aquaporin（AQP9）abundantly expressed in the peripheral leukocytes permeable to water and urea, but not to glycerol[J]. Biochem Biophys Res Commun, 1998, 244（1）：268-274.

[130] Hatakeyama S, Yoshida Y, Tani T, et al. Cloning of a new aquaporin(AQP10)abundantly expressed in duodenum and jejunum[J]. Biochem Biophys Res Commun, 2001, 287(4): 814-819.

[131] Ishibashi K, Kuwahara M, Kageyama Y, et al. Molecular cloning of a new aquaporin superfamily in mammals: AQPX1 and AQPX2 [M]//Hohmann S, Nielsen S. Molecular Biology and Physiology of Water and Solute Transport. New York: Kluwer Academic/Plenum Publishers, 2000: 123-126.

[132] Mathai JC, Agre P. Hourglass pore-forming domains restrict aquaporin-1 tetramer assembly[J]. Biochemistry, 1999, 38(3): 923-928.

[133] Jung JS, Preston GM, Smith BL, et al. Molecular structure of the water channel through aquaporin CHIP. The hourglass model[J]. J Biol Chem, 1994, 269(20): 14648-14654.

[134] Preston GM, Jung JS, Guggino WB, et al. The mercury-sensitive residue at cysteine 189 in the CHIP28 water channel[J]. J Biol Chem, 1993, 268(1): 17-20.

[135] Agre P, King LS, Yasui M, et al. Aquaporin water channels: from atomic structure to clinical medicine[J]. J Physiol, 2002, 542(Pt 1): 3-16.

[136] Nielsen S, Frøkiaer J, Marples D, et al. Aquaporins in the kidney: from molecules to medicine[J]. Physiol Rev, 2002, 82(1): 205-244.

[137] Verkman AS, Mitra AK. Structure and function of aquaporin water channels[J]. Am J Physiol Ren Physiol, 2000, 278(1): F13-F28.

[138] Sui H, Han BG, Lee JK, et al. Structural basis of water-specific transport through the AQP1 water channel[J]. Nature, 2001, 414(6866): 872-878.

[139] Sachdeva R, Singh B. Phosphorylation of Ser-180 of rat aquaporin-4 shows marginal affect on regulation of water permeability: molecular dynamics study[J]. J Biomol Struct Dyn, 2014, 32(4): 555-566.

[140] Benga G. The first discovered water channel protein, later called aquaporin 1: molecular characteristics, functions and medical implications[J]. Mol Asp Med, 2012, 33(5/6): 518-534.

[141] Tani K, Mitsuma T, Hiroaki Y, et al. Mechanism of aquaporin-4's fast and highly selective water conduction and proton exclusion[J]. J Mol Bio, 2009, 389(4): 694-706.

[142] Murata K, Mitsuoka K, Hirai T, et al. Structural determinants of water permeation through aquaporin-1[J]. Nature, 2000, 407(6804): 599-605.

[143] Sales AD, Lobo CH, Carvalho AA, et al. Review Structure, function, and localization of aquaporins: their possible implications on gamete cryopreservation[J]. Genet Mol Res, 2013, 12(4): 6718-6732.

[144] Hirano Y, Okimoto N, Kadohira I, et al. Molecular mechanisms of how mercury inhibits water permeation through aquaporin-1: understanding by molecular dynamics simulation[J]. Biophys J, 2010, 98(8): 1512-1519.

[145] Huber VJ, Tsujita M, Nakada T. Aquaporins in drug discovery and pharmacotherapy[J]. Mol Asp Med, 2012, 33(5/6): 691-703.

[146] Wspalz T, Fujiyoshi Y, Engel A. The AQP structure and functional implications[M]//Beitz E. Aquaporins. Berlin, Heidelberg: Springer, 2009: 31-56.

[147] Zeuthen T. How water molecules pass through aquaporins[J]. Trends Biochem Sci, 2001, 26(2): 77-79.

[148] Mateo Sánchez S, Freeman S D, Delacroix L, et al. The role of post-translational modifications in hearing and deafness[J]. Cell Mol Life Sci, 2016, 73(18): 3521-3533.

[149] Nielsen S, Chou CL, Marples D, et al. Vasopressin increases water permeability of kidney collecting duct by inducing translocation of aquaporin-CD water channels to plasma membrane[J]. Proc Natl Acad Sci USA, 1995, 92(4): 1013-1017.

[150] Hoffert JD, Pisitkun T, Wang G, et al. Quantitative phosphoproteomics of vasopressin-sensitive renal cells: regulation of aquaporin-2 phosphorylation at two sites[J]. Proc Natl Acad Sci USA, 2006, 103(18):

7159-7164.

[151] van Balkom BW, Savelkoul PJ, Markovich D, et al. The role of putative phosphorylation sites in the targeting and shuttling of the aquaporin-2 water channel[J]. J Biol Chem, 2002, 277(44): 41473-41479.

[152] Brown D. The ins and outs of aquaporin-2 trafficking[J]. Am J Physiol Ren Physiol, 2003, 284(5): F893-F901.

[153] Katsura T, Gustafson CE, Ausiello DA, et al. Protein kinase A phosphorylation is involved in regulated exocytosis of aquaporin-2 in transfected LLC-PK1 cells[J]. Am J Phys, 1997, 272(6 Pt 2): F817-F822.

[154] Hoffert JD, Fenton RA, Moeller HB, et al. Vasopressin-stimulated increase in phosphorylation at Ser269 potentiates plasma membrane retention of aquaporin-2[J]. J Biol Chem, 2008, 283(36): 24617-24627.

[155] Boone M, Kortenoeven M, Robben JH, et al. Effect of the cGMP pathway on AQP2 expression and translocation: potential implications for nephrogenic diabetes insipidus[J]. Nephrol Dial Transplant, 2010, 25(1): 48-54.

[156] Klokkers J, Langehanenberg P, Kemper B, et al. Atrial natriuretic peptide and nitric oxide signaling antagonizes vasopressin-mediated water permeability in inner medullary collecting duct cells[J]. Am J Physiol Ren Physiol, 2009, 297(3): F693-F703.

[157] Kim HY, Choi HJ, Lim JS, et al. Emerging role of Akt substrate protein AS160 in the regulation of AQP2 translocation[J]. Am J Physiol Ren Physiol, 2011, 301(1): F151-F161.

[158] Hoffert JD, Nielsen J, Yu MJ, et al. Dynamics of aquaporin-2 serine-261 phosphorylation in response to short-term vasopressin treatment in collecting duct[J]. Am J Physiol Ren Physiol, 2007, 292(2): F691-F700.

[159] Rinschen MM, Yu MJ, Wang G, et al. Quantitative phosphoproteomic analysis reveals vasopressin V2-receptor-dependent signaling pathways in renal collecting duct cells[J]. Proc Natl Acad Sci USA, 2010, 107(8): 3882-3887.

[160] Christensen BM, Zelenina M, Aperia A, et al. Localization and regulation of PKA-phosphorylated AQP2 in response to V(2)-receptor agonist/antagonist treatment[J]. Am J Physiol Ren Physiol, 2000, 278(1): F29-F42.

[161] Fenton RA, Moeller HB, Hoffert JD, et al. Acute regulation of aquaporin-2 phosphorylation at Ser-264 by vasopressin[J]. Proc Natl Acad Sci USA, 2008, 105(8): 3134-3139.

[162] Moeller HB, Olesen ET, Fenton RA. Regulation of the water channel aquaporin-2 by posttranslational modification[J]. Am J Physiol Ren Physiol, 2011, 300(5): F1062-F1073.

[163] Lu HJ, Matsuzaki T, Bouley R, et al. The phosphorylation state of serine 256 is dominant over that of serine 261 in the regulation of AQP2 trafficking in renal epithelial cells[J]. Am J Physiol Ren Physiol, 2008, 295(1): F290-F294.

[164] Moeller HB, Praetorius J, Rützler MR, et al. Phosphorylation of aquaporin-2 regulates its endocytosis and protein-protein interactions[J]. Proc Natl Acad Sci USA, 2010, 107(1): 424-429.

[165] Moeller HB, Knepper MA, Fenton RA. Serine 269 phosphorylated aquaporin-2 is targeted to the apical membrane of collecting duct principal cells[J]. Kidney Int, 2009, 75(3): 295-303.

[166] Arthur J, Huang J, Nomura N, et al. Characterization of the putative phosphorylation sites of the AQP2 C terminus and their role in AQP2 trafficking in LLC-PK1 cells[J]. Am J Physiol Ren Physiol, 2015, 309(8): F673-F679.

[167] Tamma G, Robben JH, Trimpert C, et al. Regulation of AQP2 localization by S256 and S261 phosphorylation and ubiquitination[J]. Am J Phys Cell Physiol, 2011, 300(3): C636-C646.

[168] Leitch V, Agre P, King LS. Altered ubiquitination and stability of aquaporin-1 in hypertonic stress[J]. Proc Natl Acad Sci USA, 2001, 98(5): 2894-2898.

[169] Kamsteeg EJ, Hendriks G, Boone M, et al. Short-chain ubiquitination mediates the regulated endocytosis of the aquaporin-2 water channel[J]. Proc Natl Acad Sci USA, 2006, 103(48): 18344-18349.

[170] Lee YJ, Lee JE, Choi HJ, et al. E3 ubiquitin-protein ligases in rat kidney collecting duct: response to vasopressin stimulation and withdrawal[J]. Am J Physiol Ren Physiol, 2011, 301(4): F883-F896.

[171] Petroski MD, Deshaies RJ. Function and regulation of cullin-RING ubiquitin ligases[J]. Nat Rev Mol Cell Biol, 2005, 6(1): 9-20.

[172] Hunter T. The age of crosstalk: phosphorylation, ubiquitination, and beyond[J]. Mol Cell, 2007, 28(5): 730-738.

[173] Moeller HB, Aroankins TS, Slengerik-Hansen J, et al. Phosphorylation and ubiquitylation are opposing processes that regulate endocytosis of the water channel aquaporin-2[J]. J Cell Sci, 2014, 127(Pt 14): 3174-3183.

[174] Sandoval PC, Slentz DH, Pisitkun T, et al. Proteome-wide measurement of protein half-lives and translation rates in vasopressin-sensitive collecting duct cells[J]. J Am Soc Nephrol, 2013, 24(11): 1793-1805.

[175] Tamma G, Ranieri M, Di Mise A, et al. Glutathionylation of the aquaporin-2 water channel: a novel post-translational modification modulated by the oxidative stress[J]. J Biol Chem, 2014, 289(40): 27807-27813.

[176] Hendriks G, Koudijs M, van Balkom BW, et al. Glycosylation is important for cell surface expression of the water channel aquaporin-2 but is not essential for tetramerization in the endoplasmic reticulum[J]. J Biol Chem, 2004, 279(4): 2975-2983.

[177] Baumgarten R, Van De Pol MH, Wetzels JF, et al. Glycosylation is not essential for vasopressin-dependent routing of aquaporin-2 in transfected Madin-Darby canine kidney cells[J]. J Am Soc Nephrol, 1998, 9(9): 1553-1559.

[178] Buck TM, Eledge J, Skach WR. Evidence for stabilization of aquaporin-2 folding mutants by N-linked glycosylation in endoplasmic reticulum[J]. Am J Phys Cell Physiol, 2004, 287(5): C1292-C1299.

[179] Hwang CS, Shemorry A, Varshavsky A. N-terminal acetylation of cellular proteins creates specific degradation signals[J]. Science, 2010, 327(5968): 973-977.

[180] Gutierrez DB, Garland D, Schey KL. Spatial analysis of human lens aquaporin-0 post-translational modifications by MALDI mass spectrometry tissue profiling[J]. Exp Eye Res, 2011, 93(6): 912-920.

[181] Suzuki H, Nishikawa K, Hiroaki Y, et al. Formation of aquaporin-4 arrays is inhibited by palmitoylation of N-terminal cysteine residues[J]. Biochim Biophys Acta, 2008, 1778(4): 1181-1189.

[182] Yamamura Y, Motegi K, Kani K, et al. TNF-α inhibits aquaporin 5 expression in human salivary gland acinar cells via suppression of histone H4 acetylation[J]. J Cell Mol Med, 2012, 16(8): 1766-1775.

[183] Qiu LW, Gu LY, Lü L, et al. FOXO1-mediated epigenetic modifications are involved in the insulin-mediated repression of hepatocyte aquaporin 9 expression[J]. Mol Med Rep, 2015, 11(4): 3064-3068.

[184] Patil RV, Han Z, Wax MB. Regulation of water channel activity of aquaporin 1 by arginine vasopressin and atrial natriuretic peptide[J]. Biochem Biophys Res Commun, 1997, 238(2): 392-396.

[185] Han Z, Patil RV. Protein kinase A-dependent phosphorylation of aquaporin-1[J]. Biochem Biophys Res Commun, 2000, 273(1): 328-332.

[186] Marinelli RA, Pham L, Agre P, et al. Secretin promotes osmotic water transport in rat cholangiocytes by increasing aquaporin-1 water channels in plasma membrane. Evidence for a secretin-induced vesicular translocation of aquaporin-1[J]. J Biol Chem, 1997, 272(20): 12984-12988.

[187] Marinelli RA, Tietz PS, Pham LD, et al. Secretin induces the apical insertion of aquaporin-1 water channels in rat cholangiocytes[J]. Am J Phys, 1999, 276(1): G280-G286.

[188] Pohl M, Shan Q, Petsch T, et al. Short-term functional adaptation of aquaporin-1 surface expression in

the proximal tubule, a component of glomerulotubular balance[J]. J Am Soc Nephrol, 2015, 26(6): 1269-1278.

[189] Conner MT, Conner AC, Bland CE, et al. Rapid aquaporin translocation regulates cellular water flow: mechanism of hypotonicity-induced subcellular localization of aquaporin 1 water channel[J]. J Biol Chem, 2012, 287(14): 11516-11525.

[190] Jenq W, Cooper DR, Bittle P, et al. Aquaporin-1 expression in proximal tubule epithelial cells of human kidney is regulated by hyperosmolarity and contrast agents[J]. Biochem Biophys Res Commun, 1999, 256(1): 240-248.

[191] Umenishi F, Schrier RW. Hypertonicity-induced aquaporin-1(AQP1)expression is mediated by the activation of MAPK pathways and hypertonicity-responsive element in the AQP1 gene[J]. J Biol Chem, 2003, 278(18): 15765-15770.

[192] Umenishi F, Schrier RW. Identification and characterization of a novel hypertonicity-responsive element in the human aquaporin-1 gene[J]. Biochem Biophys Res Commun, 2002, 292(3): 771-775.

[193] Gunnarson E, Zelenina M, Axehult G, et al. Identification of a molecular target for glutamate regulation of astrocyte water permeability[J]. Glia, 2008, 56(6): 587-596.

[194] Zelenina M, Zelenin S, Bondar AA, et al. Water permeability of aquaporin-4 is decreased by protein kinase C and dopamine[J]. Am J Physiol Ren Physiol, 2002, 283(2): F309-F318.

[195] Gunnarson E, Axehult G, Baturina G, et al. Lead induces increased water permeability in astrocytes expressing aquaporin 4[J]. Neuroscience, 2005, 136(1): 105-114.

[196] Assentoft M, Larsen BR, Olesen ET, et al. AQP4 plasma membrane trafficking or channel gating is not significantly modulated by phosphorylation at COOH- terminal serine residues[J]. Am J Phys Cell Physiol, 2014, 307(10): C957-C965.

[197] Kosugi-Tanaka C, Li X, Yao C, et al. Protein kinase A-regulated membrane trafficking of a green fluorescent protein-aquaporin 5 chimera in MDCK cells[J]. Biochim Biophys Acta, 2006, 1763(4): 337-344.

[198] Yang F, Kawedia JD, Menon AG. Cyclic AMP regulates aquaporin 5 expression at both transcriptional and post-transcriptional levels through a protein kinase A pathway[J]. J Biol Chem, 2003, 278(34): 32173-32180.

[199] Sidhaye V, Hoffert JD, King LS. cAMP has distinct acute and chronic effects on aquaporin-5 in lung epithelial cells[J]. J Biol Chem, 2005, 280(5): 3590-3596.

[200] Chae YK, Kang SK, Kim MS, et al. Human AQP5 plays a role in the progression of chronic myelogenous leukemia(CML)[J]. PLoS One, 2008, 3(7): e2594.

[201] Koffman JS, Arnspang EC, Marlar S, et al. Opposing effects of cAMP and T259 phosphorylation on plasma membrane diffusion of the water channel aquaporin-5 in Madin-Darby canine kidney cells[J]. PLoS One, 2015, 10(7): e0133324.

[202] Kitchen P, Öberg F, Sjöhamn J, et al. Plasma membrane abundance of human aquaporin 5 is dynamically regulated by multiple pathways[J]. PLoS One, 2015, 10(11): e0143027.

[203] Calamita G, Mazzone A, Bizzoca A, et al. Expression and immunolocalization of the aquaporin-8 water channel in rat gastrointestinal tract[J]. Eur J Cell Biol, 2001, 80(11): 711-719.

[204] García F, Kierbel A, Larocca MC, et al. The water channel aquaporin-8 is mainly intracellular in rat hepatocytes, and its plasma membrane insertion is stimulated by cyclic AMP[J]. J Biol Chem, 2001, 276(15): 12147-12152.

[205] Ferri D, Mazzone A, Liquori GE, et al. Ontogeny, distribution, and possible functional implications of an unusual aquaporin, AQP8, in mouse liver[J]. Hepatology, 2003, 38(4): 947-957.

[206] Gradilone SA, García F, Huebert RC, et al. Glucagon induces the plasma membrane insertion of functional aquaporin-8 water channels in isolated rat hepatocytes[J]. Hepatology, 2003, 37(6): 1435-1441.

[207] Gradilone SA, Carreras FI, Lehmann GL, et al. Phosphoinositide 3-kinase is involved in the glucagon-induced translocation of aquaporin-8 to hepatocyte plasma membrane[J]. Biol Cell, 2005, 97(11): 831-836.

[208] Boone M, Kortenoeven ML, Robben JH, et al. Counteracting vasopressin-mediated water reabsorption by ATP, dopamine, and phorbol esters: mechanisms of action[J]. Am J Physiol Renal Physiol, 2011, 300(3): F761-F771.

[209] Choi HJ, Jung HJ, Kwon TH. Extracellular pH affects phosphorylation and intracellular trafficking of AQP2 in inner medullary collecting duct cells[J]. Am J Physiol Renal Physiol, 2015, 308(7): F737-F748.

[210] Kreida S, Törnroth-Horsefield S. Structural insights into aquaporin selectivity and regulation[J]. Curr Opin Struct Biol, 2015, 33: 126-134.

[211] Németh-Cahalan KL, Clemens DM, Hall JE. Regulation of AQP0 water permeability is enhanced by cooperativity[J]. J Gen Physiol, 2013, 141(3): 287-295.

[212] Eto K, Noda Y, Horikawa S, et al. Phosphorylation of aquaporin-2 regulates its water permeability[J]. J Biol Chem, 2010, 285(52): 40777-40784.

[213] Edemir B, Pavenstädt H, Schlatter E, et al. Mechanisms of cell polarity and aquaporin sorting in the nephron[J]. Pflugers Arch, 2011, 461(6): 607-621.

[214] Moeller HB, Fenton RA. Cell biology of vasopressin-regulated aquaporin-2 trafficking[J]. Pflugers Arch, 2012, 464(2): 133-144.

[215] Nishimoto G, Zelenina M, Li D, et al. Arginine vasopressin stimulates phosphorylation of aquaporin-2 in rat renal tissue[J]. Am J Physiol, 1999, 276(2): F254-F259.

[216] Noda Y, Sasaki S. Regulation of aquaporin-2 trafficking and its binding protein complex[J]. Biochim Biophys Acta, 2006, 1758(8): 1117-1125.

[217] Klussmann E, Maric K, Wiesner B, et al. Protein kinase A anchoring proteins are required for vasopressin-mediated translocation of aquaporin-2 into cell membranes of renal principal cells[J]. J Biol Chem, 1999, 274(8): 4934-4938.

[218] Yip KP. Coupling of vasopressin-induced intracellular Ca^{2+} mobilization and apical exocytosis in perfused rat kidney collecting duct[J]. J Physiol, 2002, 538(3): 891-899.

[219] Champigneulle A, Siga E, Vassent G, et al. V2-like vasopressin receptor mobilizes intracellular Ca^{2+} in rat medullary collecting tubules[J]. Am J Physiol, 1993, 265(1 Pt 2): F35-F45.

[220] Chou CL, Yip KP, Michea L, et al. Regulation of aquaporin-2 trafficking by vasopressin in the renal collecting duct. Roles of ryanodine-sensitive Ca^{2+} stores and calmodulin[J]. J Biol Chem, 2000, 275(47): 36839-36846.

[221] Lorenz D, Krylov A, Hahm D, et al. Cyclic AMP is sufficient for triggering the exocytic recruitment of aquaporin-2 in renal epithelial cells[J]. EMBO Rep, 2003, 4(1): 88-93.

[222] Simon H, Gao Y, Franki N, et al. Vasopressin depolymerizes apical F-actin in rat inner medullary collecting duct[J]. Am J Physiol, 1993, 265(3 Pt 1): C757-C762.

[223] Noda Y, Horikawa S, Kanda E, et al. Reciprocal interaction with G-actin and tropomyosin is essential for aquaporin-2 trafficking[J]. J Cell Biol, 2008, 182(3): 587-601.

[224] Whiting JL, Ogier L, Forbush KA, et al. AKAP220 manages apical actin networks that coordinate aquaporin-2 location and renal water reabsorption[J]. Proc Natl Acad Sci USA, 2016, 113(30): E4328-E4337.

[225] Sasaki S. Aquaporin 2: from its discovery to molecular structure and medical implications[J]. Mol Aspects Med, 2012, 33(5/6): 535-546.

[226] Barile M, Pisitkun T, Yu MJ, et al. Large scale protein identification in intracellular aquaporin-2 vesicles from renal inner medullary collecting duct[J]. Mol Cell Proteomics, 2005, 4(8): 1095-1106.

[227] Mandon B, Chou CL, Nielsen S, et al. Syntaxin-4 is localized to the apical plasma membrane of rat renal collecting duct cells: possible role in aquaporin-2 trafficking[J]. J Clin Invest, 1996, 98(4): 906-913.

[228] Inoue T, Nielsen S, Mandon B, et al. SNAP-23 in rat kidney: colocalization with aquaporin-2 in collecting duct vesicles[J]. Am J Physiol, 1998, 275(5): F752-F760.

[229] Gouraud S, Laera A, Calamita G, et al. Functional involvement of VAMP/synaptobrevin-2 in cAMP-stimulated aquaporin 2 translocation in renal collecting duct cells[J]. J Cell Sci, 2002, 115(Pt 18): 3667-3674.

[230] Procino G, Barbieri C, Tamma G, et al. AQP2 exocytosis in the renal collecting duct: involvement of SNARE isoforms and the regulatory role of Munc18b[J]. J Cell Sci, 2008, 121(Pt 12): 2097-2106.

[231] Russo LM, McKee M, Brown D. Methyl-beta-cyclodextrin induces vasopressin-independent apical accumulation of aquaporin-2 in the isolated, perfused rat kidney[J]. Am J Physiol Renal Physiol, 2006, 291(1): F246-F253.

[232] Strange K, Willingham MC, Handler JS, et al. Apical membrane endocytosis via coated pits is stimulated by removal of antidiuretic hormone from isolated, perfused rabbit cortical collecting tubule[J]. J Membr Bioly, 1988, 103(1): 17-28.

[233] Lu HAJ, Sun TX, Matsuzaki T, et al. Heat shock protein 70 interacts with aquaporin-2 and regulates its trafficking[J]. J Biol Chem, 2007, 282(39): 28721-28732.

[234] Fenton RA, Pedersen CN, Moeller HB. New insights into regulated aquaporin-2 function[J]. Curr Opin Nephrol Hypertens, 2013, 22(5): 551-558.

[235] Aoki T, Suzuki T, Hagiwara H, et al. Close association of aquaporin-2 internalization with caveolin-1[J]. Acta Histochem Cytochem, 2012, 45(2): 139-146.

[236] Tajika Y, Matsuzaki T, Suzuki T, et al. Aquaporin-2 is retrieved to the apical storage compartment via early endosomes and phosphatidylinositol 3-kinase-dependent pathway[J]. Endocrinology, 2004, 145(9): 4375-4383.

[237] Street JM, Birkhoff W, Menzies RI, et al. Exosomal transmission of functional aquaporin 2 in kidney cortical collecting duct cells[J]. J Physiol, 2011, 589(Pt 24): 6119-6127.

[238] Yasui M, Zelenin SM, Celsi G, et al. Adenylate cyclase-coupled vasopressin receptor activates AQP2 promoter via a dual effect on CRE and AP1 elements[J]. Am J Physiol, 1997, 272(4 Pt 2): F443-F450.

[239] Hasler U, Leroy V, Jeon US, et al. NF-kappaB modulates aquaporin-2 transcription in renal collecting duct principal cells[J]. J Biol Chem, 2008, 283(42): 28095-28105.

[240] DiGiovanni SR, Nielsen S, Christensen EI, et al. Regulation of collecting duct water channel expression by vasopressin in Brattleboro rat[J]. Proc Natl Acad Sci USA, 1994, 91(19): 8984-8988.

[241] Rai T, Uchida S, Marumo F, et al. Cloning of rat and mouse aquaporin-2 gene promoters and identification of a negative cis-regulatory element[J]. Am J Physiol, 1997, 273(2 Pt 2): F264-F273.

[242] Hasler U, Jeon US, Kim JA, et al. Tonicity-responsive enhancer binding protein is an essential regulator of aquaporin-2 expression in renal collecting duct principal cells[J]. J Am Soc Nephrol, 2006, 17(6): 1521-1531.

[243] Yu MJ, Miller RL, Uawithya P, et al. Systems-level analysis of cell-specific AQP2 gene expression in renal collecting duct[J]. Proc Natl Acad Sci USA, 2009, 106(7): 2441-2446.

第二节 水通道蛋白家族的进化

一、引　言

AQP是位于细胞膜上的一种古老的通道蛋白，可转运水、尿素、甘油等小分子物质，以维持体液的稳态[1-7]。此外，某些AQP还可转运营养物质、代谢物前体、废物、毒素和气体分子[8-11]。除了一些原核生物外，AQP几乎分布在包括病毒在内的所有生命系统中[12]，这与它们在体液平衡和漫长进化历史中发挥的基本作用一致。基于AQP的广泛分布，并结合AQP和物种的系统发育分析，有研究者提出了AQP的进化途径[13-25]，这将有助于推测未知的水通道蛋白的功能和生理意义。

AQP的一级结构表明其整体结构相对保守，有6个跨膜结构域和5个连接环（A～E）（图1-6）[22]。它们共同组成一个四聚体，每个单体都有一个孔道，由两个高度保守的约含20个疏水氨基酸的残基组成，称为天冬氨酸-脯氨酸-丙氨酸（arginine-proline-alanine，NPA）基序，位于B环和E环中，模拟沙漏结构（表1-2）[18]。NPA基序在AQP中高度保守，是AQP家族的标志序列。研究者可以利用该氨基酸序列设计变性引物，通过PCR克隆新的AQP。

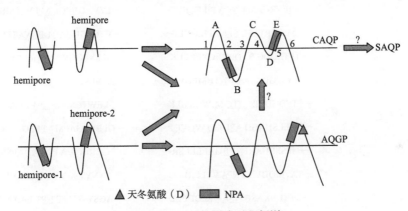

图1-6　AQP的形成过程假说

研究表明，AQP可能通过hemipore基因的复制[22]或两个相似基因hemipore-1和hemipore-2融合而成。水甘油通道蛋白（aquaglyceroporin，AQGP）可能通过缩短D环和破坏微生物中AQGP标记D残基的突变而转化为经典水通道蛋白（classical aquaporin，CAQP），而超级水通道蛋白（superaquaporin，SAQP）则可能由CAQP在多细胞生物中转化而成

表1-2　AQP在第一个和第二个NPA框中的顺序

	第一个NPA框	第二个NPA框
经典水通道蛋白（CAQP）		
AQPZ	—VGHISGGHF NPA VTIGLWAG—	—SIPVTNTSV NPA RSTAVAIFQG—
Meth	—FGRISGCHI NPA VTIALFAT—	—IGNLTGASL NPA RTFGPYLGDW—
Ch1.T	—MGTVSGAHL NPA VTLAFAMR—	—AAPVSGASM NPV RSLAPALVCG—
Ch1.P	—MGTVSGAHL NPA VTIAFAMR—	—AAPISGASM NPV RSLAPALVCG—
Cript	—FFRVSGGLF NPA VSLGMVLA—	—GVPYSGGAL NPV RSLGPAVVTH—

	第一个NPA框	第二个NPA框
Tryp1	—FGYISGGHF NPA VTMAVFLV—	—VGRISGGAF NPA AATGLQLALC—
Tryp2	—FGYISGAHF NPA ITFATFIN—	—VGGFTGGAF NPA VATGTQLVGC—
Leish	—FGYISSSHF NPA VSIAVFLV—	—AGRISGGAF NPA AASGLQVAMC—
D.disA	—VSGVSGCNL NPA VTLANLLS—	—GFNSGGAL NPV RVLGPSIISG—
D.disB	—ISGISGCQL NPA VTVGCVTT—	—LNLFTGGSL NPA RSFGPAVFSD—
D.disC	—FADVSGAHF NPA VTFATCVT—	—GGSVSGGAF NPA RVFGTALVGN—
D.disD	—CAPVSGGHL NPS ITLATFFA—	—IAPNYIFGF NIA RCLSPAIVLS—
D.disE	—CAPVSGGHL NPS ITIATFFS—	—ISPNYIFGF NMA RCLCPAIVTG—
XIP1.1	—APATSGGHV NPC ITWTEMLT—	—FSGYGGAGI NPG RCIGPAVVLG—
TIP1.1	—GANISGGHV NPA VTFGAFIG—	—GGAFSGASM NPA VAFGPAVVSW—
超级水通道蛋白（SAQP）		
CeAQP9	—IEFQRDAVA HPC PLVTNCYR—	—GINYTGMYA NPI VAWACTFN C L—
CeAQP10	—NIFNRGAMT NCA PIFEQFVF—	—LYVVGVPGL NPI VATARLYG C R—
CeAQP11	—ALCNRTAFC SPL APIEQYLF—	—VTFVGDQAL DPL VASTLFFG C R—
MtAQP11	—TFTFQDGTC DPS ECYEKFCK—	—GLFVSGGYF NPT LSFAMEYG C Q—
Dros	—GRVWGDASA CPY THMEDVVE—	—AFNFSGGYF NPV LATALKWG C R—
Urch1	—LTFDGDSTA NTC MIWQSMLK—	—GLEWTGMMF NPA LAAGITLN C G—
Urch2	—NEELSNAGD APL GQAVQVQP—	—GLEYTGAPM NPI LGFASGWG C K—
ZF11	—GFSFRGAIC NPT GALELLSR—	—GGRLTGAVF NPA LAFSIQFP C P—
ZF12	—TAVMQDVSG NPA VTLLRLLQ—	—ANNYTSGYV NPA LAYAVTLT C P—
Xenopus	—GFTFNKASG NSA VSLQDFLL—	—AGSYTGAFF NPT LAAALTFQ C S—
Chicll	—GLTLPGSTC NPC GTLQPLWG—	—GGNLTGAIF NPA LAFSLHPH C F—
Chic12	—AACANGAAS NPT VSLQEFLL—	—AAPATGAFF NPA LATASTFL C A—
AQP11	—GLTLVGTSS NPC GVMMQMML—	—GGSLTGAVF NPA LALSLHFM C F—
AQP12	—GVTLDGASA NPT VSLQEFLM—	—AGPFTSAFF NPA LAASVTFA C S—
PIP2.6	—TAGISGGHI NPA VTFGLFLA—	—TIPITGTGI NPA RSFGAAVI YN—
HIP1.1	—TGAISGGHI NPA VTLAFVVA—	—GVPYTGASM NPA RSFGPALVSG—
NIP1.2	—LGHISGAHF NPA VTIAFASC—	—AGPVSGASM NPG RSLGPAMVYS—
SIP1.1	—TVIFGSASF NPT GSAAFYVA—	—GSKYTGPAM NPA IAFGWAYMYS—
LIP	—DIVSGGSQV NPS VSVAMFVH—	—GTPYTGPAM NPM IAFGWAVQSD—
AQP1	—VGHISGAHL NPA VTLGLLLS—	—AIDYTGCGI NPA RSFGSAVLTR—
AQP8	—LGNISGGHF NPA VSLAVTVI—	—GGSISGACM NPA RAFGPAVMAG—

续表

	第一个NPA框	第二个NPA框
水甘油通道蛋白（AQGP）		
AQPV1	—FGFVS—AHL NPA MCLALFIL—	—MGGVTSIAA NPA R **D** FSPRLAHF—
GlpF	—TAGVSGAHL NPA VTIALWLF—	—MGPLTGFAM NPA R **D** FGPKVFAW—
Entero	—LFVFGGVCI NPA MALAQAIL—	—LGGTTGFAM NQA R **D** LGPRIAYQ—
Ustil	—CATTSGTQF HPA FTIAQVVF—	—CFSSSNVVA NSA R **D** IGARLVCS—
P.viy	—AAKLSGAHL NLA VTVGFATI—	—FGGNTGFAL NPS R **D** LGARLLSL—
GIP	—VGHISG—FF NPA VALAAAVV—	—GGGMTGPAL NPA R **D** LGPALVSG—
AQP3	—AGQVSGAHL NPA VTFAMCFL—	—MGENSGYAV NPA R **D** FGPRLFTA—

注：该表内数据引自参考文献[18]，有修改。

通过基因组计划积累的数据库，将有助于使用强大的生物信息学工具识别AQP，这些工具可以通过识别保守的特征性序列（NPA基序和跨膜结构域），从而发现新的AQP[17]。如果这些AQP样蛋白在适时表达之后具有水通道功能，则属于AQP家族。然而，由于它们的大部分功能尚不清楚，往往被归纳为主要内在蛋白（major intrinsic protein，MIP）。在AQP1被鉴定出来之前，最初将MIP鉴定为一种牛晶状体纤维连接膜优势蛋白，那时其通道功能尚属未知[26,27]。而AQP这一名词的运用，表明MIP具有水通道的功能[1]。

由于众多MIP蛋白已通过基因组工程鉴定，并具有转运水的功能，因此MIP家族和AQP家族通常可以互换使用。MIP作为首字母的缩略词也可来自其他一些蛋白质，如"巨噬细胞炎症蛋白"和"线粒体中间前体加工蛋白酶"。因此，在本书中，使用AQP家族而不是MIP家族，以避免混淆。

事实上，MIP是一类水转运能力较低的水通道蛋白，在发现AQP1后，MIP改名为AQP0[27]。在发现AQP1之前，被克隆和鉴定的另一个MIP家族成员是细菌甘油通道或称甘油扩散促进分子（glycerol diffusion facilitator，GlpF）。后经鉴定，哺乳动物GlpF的同源物为AQP3。然而，GlpF并没有重新命名，而是与AQP3一起被列入水甘油通道蛋白（AQGP）家族。

本节将主要从氨基酸序列而非从功能方面，来描述AQP家族的进化。文献中有许多关于AQP家族进化的详细综述，因此笔者试图找到通过进化适应环境变化的关键功能残基及其生理作用的线索[13,15,17,21,24]。本部分将概括各种AQP家族的演变和起源，以加深对AQP家族历史的基本认识。

二、水通道蛋白家族的亚家族

AQP具有一对高度保守的特征性序列，即NPA基序；NPA基序形成孔隙结构，而其转运物质的特性主要与该孔隙所带电荷的性质和大小有关[28-31]。此外，基于氨基端（hemipore-1）和羧基端（hemipore-2）约有20%的氨基酸同源性相关，彼此面对面形成一个孔道，从而推测AQP是由3个跨膜片段（three transmembrane segments，TMS）的内部串联重复序列组成。

在AQP中，氨基端相对较保守，而羧基端较不保守，这可能是为了使其孔隙具有更为独特的选择性。每一半含有两个环，其中第二个环，即B环或E环，具有高度保守的疏水残基，即形成孔隙的NPA基序（图1-6）。这一结构表明，AQP起源于AQP基因（hemipore）的串联半复制[19, 22]。另外，两个半孔可能已经独立进化，产生独特的结构（hemipore-1和hemipore-2）并融合形成AQP（图1-6）。

在鉴定AQP3为哺乳动物GlpF的同源物后，发现AQP1的细菌同源物为AQPZ[32]。随着在细菌和哺乳动物AQP家族中，发现AQPZ对应AQP1和GlpF对应AQP3，AQP家族从细菌到人类的进化途径似乎明确起来[1, 33]。此外，初级序列的差异也反映了AQP的功能两分法：水选择性和溶质渗透性。

AQP家族的两分法似乎在细菌中很早就出现了，表明AQP家族是古老的，并在进化过程中保存下来，这可能是因为其在维持水和溶质稳态中的关键作用，因为水对每一种生物都是至关重要的，尤其是GlpF，其基因位于功能基因复合体GlpFK操纵子中，同时编码甘油激酶。然而，AQPZ的作用尚不明确，因为它的缺失几乎未引起机体功能的任何缺陷，所以目前对其功能的认识尚存在争议[34]。此外，约90%的古老的细菌，甚至缺乏GlpF或AQPZ[35]。由于大多数原核生物不具有AQP，细菌中的AQP可能来源于真核生物的水平基因转移，由于对初级序列缺乏功能约束，积累的突变可产生多样化的NPA基序。

最初，根据其渗透功能的不同，AQP家族成员可分为介导水分子转运的水通道，即经典水通道蛋白（CAQP）亚家族和介导水、甘油等小分子物质转运的水甘油通道蛋白（AQGP）亚家族[2, 6]。尽管这种分类方法是基于功能的不同划分的，但本质上这种功能上的差异与通道蛋白的初级序列有关。由于对功能数据有时存在争议，且在大多数AQP中无法获得，因此，通常采用特征性序列对主序列进行分类。AQGP的特征性序列是第二个NPA基序中含有天冬氨酸残基（D），该残基可扩大孔隙，从而渗透甘油等更大的分子[31, 36]（图1-7）。然而，由于第二个NPA基序中存在一些变化，CAQP的特征性序列并不明显。因此，天冬氨酸残基（D）缺失则表示该水通道蛋白应为CAQP。当然，孔隙的大小和特征并不是简单地由天冬氨酸残基（D）决定，而是与AQGP的相关保守残基的组合有关，如较长的D环总是与D残基共存，提示AQGP亚家族通过进化实现了系统发育序列的保守性。另一方面，尽管CAQP可能来自多个系统发育来源，但是它的一些与水选择运输相关的关键保守序列被保留了下来[28, 37]。

CAQP的氨基和羧基末端部分的保守程度为30%，而AQGP的保守程度则较低，约为20%，这是由D环上的残基较长引起的[22]（图1-6）。然而，由于CAQP和AQGP在30%~40%的水平上相对保守，因此它们可能具有共同的祖先AQP。也可能是一半AQP基因的复制产生了CAQP和AQGP，且羧基端呈现多样化[30, 38]（图1-6）。由于AQGP与GlpFK操纵子在细菌中的作用比CAQP更明显，因此推测CAQP可能是通过从GlpFK操纵子中分离AQGP或通过缺失的GlpFK操纵子的剩余成分产生的。最初的AQP，很可能是AQGP，其可能含有一个更大的孔隙，允许细菌吸收营养或排泄废物。然后，第二个NPA基序中特征性Asp（D）的突变和D环的缺失可能将AQGP转化为CAQP，并具有新形成的窄孔[29]（图1-6）。然而，由于大部分原核基因来源于微生物间的水平基因转移，因此目前很难鉴定细菌中的原始AQP。

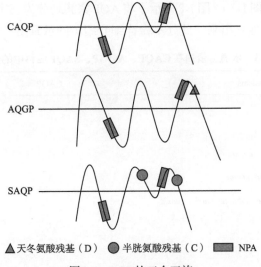

图 1-7 AQP 的三个亚族

水甘油通道蛋白（AQGP）的特征序列为第二个精氨酸-脯氨酸-丙氨酸（NPA）基序中含有天冬氨酸残基（D）。超级水通道蛋白（SAQP）的特征序列为第二个 NPA 基序下游的半胱氨酸（cysteine，Cys）残基（C）。二硫键结合的伴侣半胱氨酸残基存在于 C 环中。残基 D 和 C 的缺失表示该通道为经典水通道蛋白（CAQP）

在哺乳动物中发现 AQP11 和 AQP12，对 AQP 家族的二元分类提出了挑战，因为其具有高度退化的 NPA 基序，但总体氨基酸同源性非常低（＜15%），与超基因家族处在同一水平[39-45]。由于 NPA 基序的退化，针对该序列设计的引物则不起作用，无法通过 PCR 克隆得到 AQP11 和 AQP12。由于 AQP11 和 AQP12 的第 1 和第 6 跨膜结构域在 CAQP 和 AQGP 中相对保守，因此可通过 BLAST 搜索 EST 文库而识别。由于以往的系统发育树都是从氨基酸同源性较高的 CAQP 和 AQGP 中绘制的（＞25%），因此，AQP11 和 AQP12 的加入使得 AQP 系统发育树得到了进一步扩展。

AQP 超级基因家族的鉴定促进了研究者在基因组数据库中寻找更多成员的研究。事实上，在多细胞动物的数据库中已经发现了一类 AQP，其氨基酸序列从第一个 NPA 上游到 NPA 本身都与其他 AQP 完全不同（表 1-2）。有趣的是，这些 AQP 在细菌、单细胞真核生物和植物中均不存在（表 1-3）。由于线虫类的秀丽隐杆线虫和昆虫中的黑腹果蝇的 AQP 氨基酸序列与常规 AQP 的氨基酸序列偏离较大，因而此前未被纳入 AQP 家族[46, 47]（表 1-2）。然而，尽管它们之间的同源性很低，但它们都在缺乏 D 残基的羧末端上有一个标志性的 NPA 基序，特别是在第二个 NPA 基序的下游还存在高度保守的 Cys（C）残基——NPA××××　××××C（×是任何氨基酸残基）水通道[18, 48]（图 1-7 和图 1-8，表 1-2）。这种 Cys 可能对二硫键结合构建三维结构非常重要，并且在功能上是不可或缺的，因为它在 AQP11 中的破坏产生了与 *AQP11* 缺陷小鼠相似的表型[44, 49]。此外，在 C 环中还发现了一个用于二硫键结合的候选伙伴 Cys 残基，这

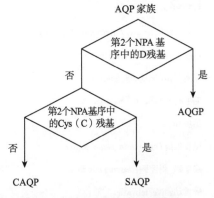

图 1-8 AQP 家族分类的简图

显示 AQP 亚家族特征，图 1-7 为本图的基础

仍有待实验加以证实（图1-7）。图1-8显示了将AQP家族划分为三个亚族（AQP8最初属于经典水通道蛋白，后被独立出来，此处不做详细介绍）的示意图[18]。

表1-3 水通道蛋白在CAQP、AQGP、SAQP亚科中的分布

有机体 Organisms	CAQP	AQGP	SAQP
微生物 Microbe			
大肠杆菌 Escherichia coli	1	1	
铜绿假单胞菌 Pseudomonas aeruginosa		1	
鼠伤寒沙门氏菌 Salmonella typhimurium	1		
马尔伯根甲醇热杆菌 Methanothermobacter marburgensis		1	
小球藻病毒 Chlorella virus	1		
真菌 Fungus			
酿酒酵母 Saccharomyces cerevisiae	2	2	
曲霉属真菌 Aspergillus nidulans	1	4	
黑粉菌属 Ustilago maydis	2	3	
稻瘟病菌 Magnaporthe grisea	3	1	
原生生物 Protist			
利什曼原虫 Leishmania major	4	1	
克氏锥虫 Trypanosoma cruzi	4		
布氏锥虫 Trypanosoma brucei		3	
弓形虫 Toxoplasma gondi	1		
恶性疟原虫 Plasmodium falciparum		1	
盘基网柄菌 Dictyostelium discoideum	5		
无脊椎动物 Invertebrate			
秀丽隐杆线虫 Caenorhabditis elegans	3	5	3
海鞘 Ciona intestinalis	4	1	1
植物 Plant			
青苔属植物 a moss Physcomitrella patens	23（8PIP、4TIP、5NIP、2SIP、2XIP、1HIP）		
橡胶树 Hevea brasiliensis	51（15PIP、17TIP、9NIF、4SIP、6XIP）		
拟南芥 Arabidopsis thaliana	35（13PIP、10TIP、9NIP、3SIP）		
毛毛杨 Populus trichocarpa	55（15PIP、17TIP、11NIP、6SIP、6XIP）		
昆虫 Insect			
黑腹果蝇 Drosophila melanogaster	7		1
采采蝇/刺舌蝇 Glossina morsitans	9		1
脊椎动物 Vertebrate			
斑马鱼 Danio rerio	11	7	2

续表

有机体 Organisms	CAQP	AQGP	SAQP
鲑鱼 Salmon	23	13	6
爪蟾 Clawed frog	12	5	2
绿安乐蜥 Green anole	9	5	2
海龟 Turtle	10	3	2
斑胸草雀 Zebra finch	7	4	2
鸭嘴兽 Platypus	8	5	2
大鼠 Rat	7	4	2
人 Human	7	4(+4)	2(+1)

注：该表由参考文献[16, 18]修改。人类有四个AQP7假基因和另一个拷贝的AQP12。

将以半胱氨酸（Cys）残基为标志性残基的亚家族归类为AQP的超基因家族，并命名为偏离NPA基序的超级水通道蛋白（SAQP）[50, 51]（表1-2）。虽然有许多AQP样蛋白已在具有偏离NPA基序的细菌基因组中发现，但它们没有这种特殊的Cys残基，而且它们的总体序列更接近CAQP。因此，这些具有不同NPA基序的细菌AQP样蛋白属于CAQP亚家族。由于SAQP在低等生物和植物中不存在，SAQP可以在多细胞动物中通过水平基因转移，从共存的具有偏离NPA基序的古老细菌中获得，这种突变可能不会对细菌产生不利影响，因为它们对AQP的依赖性较低。Cys残基的获得可能对SAQP的新功能或功能的改变至关重要，新的功能由偏离的NPA基序产生，从而形成有利于多细胞动物的独特3D结构。由于细胞质趋同，植物细胞壁的存在实际上抑制了多细胞生物的功能。因此，即使植物有机会从共生细菌中通过水平基因转移获得SAQP，也可能不需要SAQP。或者SAQP可能起源于原核生物甚至单细胞真核生物，但由于过于广泛的偏差，而无法发挥水通道的功能，进而在进化过程中丧失。事实上，AQP11是一种效率相对较低的水通道蛋白[52]。

CAQP和AQGP的三维结构分析表明，它们彼此之间具有高度的保守性[31]，而目前对SAQP还没有进行这种结构分析。如果SAQP确实具有与CAQP和AQGP相似的三维结构，那么将SAQP纳入AQP家族的有效性将变得更加确定。关于细菌中甲酸盐通道（FocA）三维结构的分析报告表明，其三维结构与AQP惊人地相似[53, 54]。虽然其一级序列与AQP家族没有同源性，但在三维结构上具有与NPA基序类似的孔形成疏水氨基酸残基，具有6个跨膜结构域。因此，FocA可能是另外一种水通道，尽管其功能研究未能显示其水通透性[54]，但其结构也由串联的重复序列组成，与AQP类似，只是其各半孔间的同源性较低，约为7%。此外，FocA可形成一种五聚体而不是四聚体，以渗透甲酸盐、亚硝酸盐和氢硫化物。目前的研究表明，FocA家族仅存在于原核生物和低等真核生物中。AQP家族目前主要由CAQP、AQGP和SAQP三个亚家族组成[18, 50]，但未来有可能发现更多的水通道家族，其三维结构与AQP家族相似，但初级序列同源性较低。

三、水通道蛋白的聚类与演化

基因的复制和结构多样化的分子进化是AQP家族系统发育框架的基础[55-59]。然而，有人提出存在从细菌到真核生物或从真核生物到细菌的水平基因转移。因此，在系统发育树中AQP和物种之间的关联缺失时，应仔细评估。水平基因转移最引人注目的例子是，植物在没有转运甘油的AQGP的情况下，却从根瘤共生细菌中获得NOD-26样固有蛋白（NOD-26 like intrinsic protein，NIP）[61]。因此，当从进化的角度比较AQP序列时，需仔细评估这种水平基因转移的风险[60, 62]。

（一）微生物AQP

第一个AQP可能起源于细菌，其支持依据是大肠杆菌（*Escherichia coli*）中存在两个AQP（AQPZ和GlpF），这两个AQP可能分别对应于CAQP和AQGP的祖先形式。然而，大多数细菌没有一套完整的AQP。一般革兰氏阴性菌只有CAQP，而革兰氏阳性菌只有AQGP。例如，革兰氏阳性胚牙乳杆菌（*Lactobacillus plantarum*）含有6个AQP，但都是AQGP。此外，只有约10%的细菌基因组含有AQP，其在细菌中的作用尚不清楚[35]。假设CAQP可能是耐冻性所必需的，通过CAQP快速挤压水分来增强细胞内的渗透性，可以防止在发生冷冻时细胞内结冰。在昆虫中，AQP也有类似作用的报道[63, 64]。AQGP可以用来运输小分子的渗透物、营养物质或毒素，而不是水分子。因此，在转移到低渗环境的情况下，AQGP可迅速转运作为渗透物的甘油，以防止细胞肿胀。AQGP的流失可能是由于微生物的寄生方式造成的。在这种寄生方式中，营养物质容易获得，环境的渗透压是稳定的。大多数细菌不存在AQP，说明在低渗环境中存在AQP可能是有害的，因为细胞中的水分通过CAQP迅速被清除，而不存在AQP则有利于防止渗透水从细胞内流失。此外，大多数细菌的渗透物质是钾或氨基酸，而不是甘油，这可能是不需要AQGP来转运的原因[8]。

基因组分析显示，在许多古细菌，如嗜热古细菌（*Thermophilic archaea*）中没有AQP，这反映出在细菌中AQP的优势有限。古细菌中一种特殊的AQP是AQPM，它是CAQP成员，存在于马尔伯根甲醇热杆菌（*Methanothermobacter marburgensis*）中[65]（表1-3）。有趣的是，AQPM具有更宽的孔隙结构，可以接受更大的分子，如硫化氢（H_2S），而不是水分子[66]。它可能是一个原始的非专门化的AQP，是更专门化的CAQP和AQGP的祖先形式。此外，尽管原核生物能在恶劣的环境中生存，但在古细菌中AQP的数量要少得多，这表明在进化过程中AQP在原核生物中的作用是很小的。

值得注意的是，小球藻病毒（chlorella virus）具有AQGP——AQPV1，这可能是在真核生物与原核生物分离时，病毒感染藻类而由细菌向病毒水平转移过程中获得的[12]（表1-3）。AQPV1可能在病毒感染和复制过程中发挥作用，其可能调节宿主细胞的水分子的运输。

（二）原生生物AQP

由于真核生物是从原核生物进化而来的，它们可能继承了原核生物的几个AQP并在质膜上表达[67, 68]。然而，原生生物中AQP基因的数量是多变的，可能在进化过程中受到环境约束而发生改变[69-71]。比如，一种囊状共生的小隐孢子虫（*Cryptosporisium parvum*）没有AQP，而恶性疟原虫（*Plasmodium falciparum*）则有。疟原虫仅含有一种AQP（PfAQP），即AQGP，用于从红细胞中摄取营养[10, 71-73]。另一方面，刚地弓形虫（*Toxoplasma gondii*）也只有一种AQP（TgAQP），即CAQP，其通过关键残基的改变，从而使其孔隙的结构和渗透性发生改变，以利于营养物质的吸收[74]。部分原生动物有多个AQP，盘基网柄菌（*Dictyostelium discoideum*）有5个AQGP，全部为CAQP[18]，布氏锥虫（*Trypanosoma brucei*）有3个AQP，全部为AQGP。3种致病性锥虫：利什曼原虫（*Leishmania major*）、克氏锥虫（*Trypanosoma cruzi*）和布氏锥虫（*Trypanosoma brucei*），含有不同的AQP亚科，这可能是由于在进化限制之外，为了适应环境，需要吸收不同的养分导致的[71]。真核生物似乎比原核生物更需要AQP，这可能与真核生物的细胞更大，需要更有效的质膜水分转运系统有关。在基因组AQP亚家族屏障之外的多个AQP可能是通过基因复制或水平基因转移产生的。

（三）真菌AQP

酿酒酵母（*Saccharomyces cerevisiae*）具有2个CAQP（ScAqy1和ScAqy2）和2个AQGP（YFL054Cp和ScFps1），表明真核生物对AQP的需求增加，可能是为了在低温下调节酵母细胞的渗透压[75-77]。高渗透压通过高渗透压甘油（high osmolality glycerol, HOG）途径诱导甘油在酵母细胞中积累，从而积累由酒精发酵产生的甘油，这是由AQGP（Fps1）的关闭来调节的[78, 79]。另一方面，对表达于真菌中的CAQP的作用尚不十分清楚。最近对38株酿酒酵母的AQP的综合分析，发现CAQP和AQGP均存在各种失活突变，这可能与在不同环境条件下的竞争优势有关。因此，在特定的环境中，AQP的存在对酵母是有害的，表明AQP即使在单细胞真核生物中也不重要。

另一种可能是，在真核生物分离之前，原核生物的水平基因转移和基因组融合[68]，以及随后原核生物内共生形成真核生物[67]，从而导致酵母中出现多个AQP。由于目前的单细胞酵母曾经是一种多细胞生物，4个AQP可能反映了在细胞壁内融合的多个细胞的组合。另一方面，面包师用的酵母菌只有2个AQGP[80, 81]。AQGP在真菌中的分布要比CAQP广泛得多，这表明甘油是酵母中的主要渗透物质，提示酵母中AQP行使摄入营养或排出毒素的功能将比水转运更重要。

（四）植物AQP

与动物相比较，由于植物为多倍体而形成了多个AQP[82, 83]。有趣的是，它们只有在高等植物中存在多样化的CAQP[84]。以拟南芥（*Arabidopsis thaliana*）为例，拟南芥有35个AQP，可进一步细分为四类：13个质膜固有蛋白（plasma membrane intrinsic protein, PIP）、10个液膜固有蛋白（tonoplast intrinsic protein, TIP）、9个NIP和3个短固有基础蛋白（short intrinsic basic protein, SIP）（表1-3）[16, 18]。植物中AQGP的缺失可以通过CAQP向AQGP的功能转化来解释，就像原生动物或昆虫一样[85, 86]。NIP可以运输除水分子外的小分子物质如甘油、硅和硼[11]，这被认为是来自细菌的水平基因转移：共生细菌在根源上可能有NIP的祖先，以吸收营养物质，也可能是负责代谢物和废物的排出[23, 61, 87]。一些NIP在内质网膜上表达，这可能让人想起以前的细胞内共生状态。植物中SAQP的缺失是一个有趣的现象，可能与细胞内TIP和SIP的存在有关，这些TIP和SIP可能使细胞内SAQP成为冗余。

在杨树（poplar tree）中发现的CAQP的另一个亚科是未分类的X固有蛋白（uncategorized X intrinsic protein，XIP）[88-90]（表1-3），但在拟南芥中缺失。与SIP类似，XIP具有较少的保守NPA框，但在第二个NPA框之后，具有保守的Cys残基（表1-2）。然而，该Cys作为NPARC，位于NPA附近，其功能意义尚不明确。因此，SIP和XIP都不属于SAQP。XIP在质膜上表达，对包括甘油[91]在内的渗透底物具有更大的选择性。有趣的是，XIP也存在于原生动物和真菌中[88, 92]。最近，在藻类中也发现了CAQP、杂交内在蛋白（hybrid intrinsic protein，HIP；类似PIP和TIP）和大内在蛋白（large intrinsic protein，LIP；类似SIP的新亚科），但它们在高等植物中并不存在[14, 93]。

在原始的陆生植物——一种青苔属植物（a moss *Physcomitrella patens*）中发现了一种AQGP——GIP，有着甘油通道的作用。然而，它可能是通过水平基因转移而来自细菌[89, 92]。在进化过程中，植物AQP是难以整合的，因为细菌水平基因转移可能受到污染（如NIP和GIP）。

根据CAQP的一级序列，植物仅有7个CAQP亚科（PIP、TIP、NIP、SIP、XIP、HIP、LIP），这似乎表明植物CAQP在功能上可与动物AQP媲美。例如，植物中AQGP的缺失实际上可以通过NIP的扩展功能来补偿[11, 87]。SAQP的缺失可以通过SIP、XIP或HIP来弥补。根据植物AQP的详细系统发育分析，这种与CAQP亚家族内微小的一级序列变化相结合的功能进化，将为包括AQP的结构-功能关系在内的进化模式提供新的见解。虽然这些结果来自CAQP亚家族，但它们将对其他AQP亚家族的功能认识产生深刻的影响。

（五）无脊椎动物AQP

SAQP首次出现在多细胞动物中，包括原口动物如昆虫及后口动物如脊椎动物等[48, 94]。秀丽隐杆线虫（*Caenorhabditis elegans*）是一种后口动物，有3个SAQP，与哺乳动物的2个SAQP高度偏离（表1-2）。它们之间的同源性也不太大，这表明线虫的SAQP可能是通过水平基因转移获得的，并来自于细菌。由于SAQP首次出现在多细胞线虫，而不是单细胞原生动物中，因此SAQP可能在多细胞生物的细胞活动中发挥作用，这些活动包括细胞分化、凋亡，细胞间通信，器官发生，交配等。如果这一推论正确，那么植物中SAQP的缺失可能反映了植物细胞壁内细胞质通信的单细胞特性，而动物细胞壁内不存在这一特性。

值得注意的是，线虫具有丰富的AQGP，具有11个AQP中的5个（45%），表明多个AQGP的保留存在一定的选择优势[46]（表1-3）。虽然甘油可能对其渗透压调节和营养很重要，但其功能研究表明，并不是所有的AQGP均可转运甘油，其中1个甚至是水选择性的。1个CAQP是水选择性的，2个CAQP却没有转运活性[46]。3个SAQP的功能尚未被研究，它们的亚细胞定位也不清楚。此外，线虫AQP的生理作用尚不清楚，即使多次敲除

AQP直至AQP缺失，也未显示任何异常[46]。有趣的是，有报道称，在低糖饮食中，AQP1可以延长寿命，这表明AQP可能在代谢而不是渗透调节中发挥作用[95]。

缓步类动物通过干燥和水溶性脱水存活，这可能受到11种AQP的调控：2个CAQP，8个AQGP和1个SAQP[48]。在其体内发现众多的AQGP，表明甘油在脱水中的重要性。缓步类动物中的一些AQGP可能来源于基因水平转移，以增加其数量。与秀丽隐杆线虫的情况一样，其SAQP也高度偏离（表1-2）。

（六）昆虫AQP

早期节肢动物，即原口动物中，3个AQP亚科都有分布[94]。一种非昆虫虱子（*Lepeophtherius salis*）具有7个AQP：2个CAQP、3个AQGP和2个SAQP[96]。血红扇头蜱（*Rhipicephalus sanguineus*）虽然有水选择性转运，但有AQGP、RsAQP1[97, 98]。由于昆虫体内的渗透物质不是甘油而是海藻糖，螨（mite）体内残留的AQGP可能已经被功能转化为水选择性AQP，或者已经获得通过细菌AQGP的水平基因转移。

有趣的是，有的昆虫失去了AQGP，而没有幼虫阶段的昆虫却保留了AQGP。如黑腹果蝇（*Drosophila melanogaster*）有8个AQP，丢失了AQGP，只含有7个CAQP和1个SAQP（表1-3）。SAQP最初并没有被纳入AQP家族，因为第一个NPA是异常的CPY，上游序列偏离[47, 94]（表1-2）。采采蝇/刺舌蝇（*Glossina morsitans*）具有10个AQP：9个CAQP和1个SAQP。在高等昆虫中，AQGP的丧失通过CAQP的功能转化来弥补，在双翅目昆虫中，如家蚕幼虫（*Bombyx mori*）[63,85,86,99,100]，CAQP变得对甘油和尿素具有渗透性。昆虫的CAQP，作为一种水通道也可能对细菌的抗冻性很重要[63, 64]。

（七）脊椎动物AQP

在脊椎动物中，AQP的3个家族都存在。大鼠共有13个AQP：CAQP（AQP0、AQP1、AQP2、AQP4、AQP5、AQP6、AQP8），AQGP（AQP3、AQP7、AQP9、AQP10）和SAQP（AQP11、AQP12）（表1-3）。

尽管在鳐鱼之前的早期脊椎动物的共同祖先中发生了两轮全基因组复制（WGD）[56,101-103]，与线虫的11个AQP相比较，哺乳动物的AQP总数相对较少。但从海鞘的7个AQP来看，这个数量分布可能是合理的，因为脊椎动物是在全基因组复制后衍生出来的。许多AQP可能在进化过程中丢失了。事实上，与线虫相比，脊椎动物中AQGP和SAQP的数量有所减少。另外，线虫可能通过水平基因转移获得了这些AQP亚科，从而增加了它们的数量。

斑马鱼作为硬骨鱼再次进行了全基因组复制，将AQP的数量增加到20个[26, 104]。一些鱼类家族，如鲑鱼，已经经历了第四轮全基因组复制，将AQP的数量进一步增加到42个[105]。有趣的是，AQP6的同源物在鸟类中缺失，但鸟类有AQP14[16]。在低等脊椎动物中，有AQP13、AQGP和AQP14～AQP16，均为CAQP，但这些AQP在哺乳动物中均已经消失。进一步分析脊椎动物各AQP同源物的组织分布和激素调节，有助于深入了解各类哺乳动物AQP的功能及其作用[106]。

虽然AQP10在脊椎动物中差异不大，但在龟[16]中已经丢失或转化为假基因[107]。不仅是啮齿动物，奶牛也有 *AQP10* 假基因，而没有任何真正的 *AQP10* 基因[108]。另一方面，人类的 AQP7 除了真实的 *AQP7* 基因[16]外，还有 4 个假基因。此外，在 AQP10 和 AQP3[40] 中发现了缺失第 6 跨膜结构域的非功能性剪接变体。这些非功能基因可能代表了假基因或缺失的过渡阶段。对于高等脊椎动物，甘油在能量代谢中的作用似乎很小，多余的 AQGP 可能并不十分必要，因此在某些哺乳动物中 AQP 可能会丢失。

虽然 SAQP 在脊椎动物中的作用尚不明确，但与其他亚科相比，仅仅 2 个 SAQP 可能还不够。事实上，秀丽隐杆线虫有 3 个 SAQP，而大西洋鲑鱼（*Atlantic salmon*）则有 6 个 SAQP[16]。有趣的是，在人类中存在 2 个具有局部复制的 *AQP12* 基因副本[16]。由于 AQP12 在胰腺腺泡中表达有限[109]，AQP12 的复制可能是在扩大 SAQP 库的过程中进行的。

四、展　　望

以进化理论为导向的 AQP 研究具有非常广阔的前景，因为它必将有利于进一步深入比较分析渗透调节和体液稳态，包括激素及其受体等[106,110,111]。由于水对于每一种生物都至关重要，因此在进化的过程中一些生物扩大了 AQP 的储备，而另一些则限制了其数量，这种适应环境变化的策略差异将成为 AQP 表达模式和功能多样性的基础。此外，AQP 可能受到渗透压和激素的调控。关于 AQP 和激素的协同进化研究值得期待，可能为探讨 AQP 的生理意义提供新的思路[106,110,111]。

虽然 AQP 通常不受开闭状态的调节，但它们的功能发挥有时可通过进出细胞膜来调节[112]。AQP 的这种移动将受到相关蛋白的调控，尽管这些蛋白可能已经与 AQP 一起进化，但目前对于这些相关蛋白却知之甚少。此外，水转运的驱动力需要有钠和钾等溶质的运输，因此与 AQP 共表达的离子通道、泵或转运体的功能状态，对于 AQP 的水转运效率具有重要影响。鉴于此，离子运动的相关调控以及它们的共同进化，也将为今后的研究提供十分广阔的领域[112]。

（许士叶）

参 考 文 献

[1] Agre P. Aquaporin water channels（Nobel Lecture）[J]. Angew Chem Int Ed Engl, 2004, 43(33): 4278-4290.

[2] Hara-Chikuma M, Verkman AS. Physiological roles of glycerol-transporting aquaporins: the aquaglyceroporins[J]. Cell Mol Life Sci, 2006, 63(12): 1386-1392.

[3] Ishibashi K, Kuwahara M, Sasaki S. Molecular biology of aquaporins[J]. Rev Physiol Biochem Pharmacol, 2000, 141: 1-32.

[4] Ishibashi K, Hara S, Kondo S. Aquaporin water channels in mammals[J]. Clin Exp Nephrol, 2009, 13(2): 107-117.

[5] Mukhopadhyay R, Bhattacharjee H, Rosen BP. Aquaglyceroporins: generalized metalloid channels[J]. Biochim Biophys Acta, 2014, 1840(5): 1583-1591.

[6] Rojek A, Praetorius J, Frøkiaer J, et al. A current view of the mammalian aquaglyceroporins[J]. Annu Rev Physiol, 2008, 70: 301-327.

[7] Zeuthen T. Water-transporting proteins[J]. J Membr Biol, 2010, 234(2): 57-73.

[8] Bienert GP, Desguin B, Chaumont F, et al. Channel-mediated lactic acid transport: a novel function for

aquaglyceroporins in bacteria[J]. Biochem J, 2013, 454（3）: 559-570.

[9] Gomes D, Agasse A, Thiébaud P, et al. Aquaporins are multifunctional water and solute transporters highly divergent in living organisms[J]. Biochim Biophys Acta, 2009, 1788（6）: 1213-1228.

[10] Liu Y, Promeneur D, Rojek A, et al. Aquaporin 9 is the major pathway for glycerol uptake by mouse erythrocytes, with implications for malarial virulence[J]. Proc Natl Acad Sci USA, 2007, 104（30）: 12560-12564.

[11] Takano J, Wada M, Ludewig U, et al. The Arabidopsis major intrinsic protein NIP5;1 is essential for efficient boron uptake and plant development under boron limitation[J]. Plant Cell, 2006, 18（6）: 1498-1509.

[12] Gazzarrini S, Kang M, Epimashko S, et al. Chlorella virus MT325 encodes water and potassium channels that interact synergistically[J]. Proc Natl Acad Sci USA, 2006, 103（14）: 5355-5360.

[13] Abascal F, Irisarri I, Zardoya R. Diversity and evolution of membrane intrinsic proteins[J]. Biochim Biophys Acta, 2014, 1840（5）: 1468-1481.

[14] Danielson JA, Johanson U. Phylogeny of major intrinsic proteins. Adv Exp Med Biol, 2010, 679: 19-31.

[15] Finn RN, Cerdà J. Evolution and functional diversity of aquaporins[J]. Biol Bul, 2015, 229（1）: 6-23.

[16] Finn RN, Chauvigné F, Hlidberg JB, et al. The lineage-specific evolution of aquaporin gene clusters facilitated tetrapod terrestrial adaptation[J]. PLoS One, 2014, 9（11）: e113686.

[17] Gupta AB, Verma RK, Agarwal V, et al. MIPModDB: a central resource for the superfamily of major intrinsic proteins[J]. Nucleic Acids Res, 2012, 40（Database issue）: D362-D369.

[18] Ishibashi K, Kondo S, Hara S, et al. The evolutionary aspects of aquaporin family[J]. Am J Physiol Regul Integr Comp Physiol, 2011, 300（3）: R566-R576.

[19] Pao GM, Wu LF, Johnson KD, et al. Evolution of the MIP family of integral membrane transport proteins[J]. Mol Microbiol, 1991, 5（1）: 33-37.

[20] Park JH, Saier MH Jr. Phylogenetic characterization of the MIP family of transmembrane channel proteins[J]. J Membr Biol, 1996, 153（3）: 171-180.

[21] Perez Di Giorgio J, Soto G, Alleva K, et al. Prediction of aquaporin function by integrating evolutionary and functional analyses[J]. J Membr Biol, 2014, 247（2）: 107-125.

[22] Reizer J, Reizer A, Saier MH Jr. The MIP family of integral membrane channel proteins: sequence comparisons, evolutionary relationships, reconstructed pathway of evolution, and proposed functional differentiation of the two repeated halves of the proteins[J]. Crit Rev Biochem Mol Biol, 1993, 28（3）: 235-257.

[23] Soto G, Alleva K, Amodeo G, et al. New insight into the evolution of aquaporins from flowering plants and vertebrates: orthologous identification and functional transfer is possible[J]. Gene, 2012, 503（1）: 165-176.

[24] Zardoya R. Phylogeny and evolution of the major intrinsic protein family[J]. Biol Cell, 2005, 97（6）: 397-414.

[25] Zardoya R, Villalba S. A phylogenetic framework for the aquaporin family in eukaryotes[J]. J Mol Evol, 2001, 52（5）: 391-404.

[26] Froger A, Clemens D, Kalman K, et al. Two distinct aquaporin 0s required for development and transparency of the zebrafish lens[J]. Invest Ophthalmol Vis Sci, 2010, 51（12）: 6582-6592.

[27] Gonen T, Sliz P, Kistler J, et al. Aquaporin-0 membrane junctions reveal the structure of a closed water pore[J]. Nature, 2004, 429（6988）: 193-197.

[28] Beitz E, Wu B, Holm LM, et al. Point mutations in the aromatic/arginine region in aquaporin 1 allow passage of urea, glycerol, ammonia, and protons[J]. Proc Natl Acad Sci USA, 2006, 103（2）: 269-274.

[29] Hub JS, de Groot BL. Mechanism of selectivity in aquaporins and aquaglyceroporins[J]. Proc Natl Acad

Sci USA, 2008, 105(4): 1198-1203.

[30] Lin X, Hong T, Mu Y, et al. Identification of residues involved in water versus glycerol selectivity in aquaporins by differential residue pair co-evolution[J]. Biochim Biophys Acta, 2012, 1818(3): 907-914.

[31] Viadiu H, Gonen T, Walz T. Projection map of aquaporin-9 at 7 A resolution[J]. J Mol Biol, 2007, 367(1): 80-88.

[32] Calamita G, Bishai WR, Preston GM, et al. Molecular cloning and characterization of AqpZ, a water channel from *Escherichia coli*[J]. J Biol Chem, 1995, 270(49): 29063-29066.

[33] Ishibashi K, Sasaki S. The dichotomy of MIP family suggests two separate origins of water channels. News Physiol Sci, 1998, 13: 137-142.

[34] Soupene E, King N, Lee H, et al. Aquaporin Z of *Escherichia coli*: reassessment of its regulation and physiological role[J]. J Bacteriol, 2002, 184(15): 4304-4307.

[35] Tanghe A, Van Dijck P, Thevelein JM. Why do microorganisms have aquaporins?[J]. Trends Microbiol, 2006, 14(2): 78-85.

[36] Fu D, Libson A, Miercke LJ, et al. Structure of a glycerol-conducting channel and the basis for its selectivity[J]. Science, 2000, 290(5491): 481-486.

[37] Fu D, Lu M. The structural basis of water permeation and proton exclusion in aquaporins[J]. Mol Membr Biol, 2007, 24(5/6): 366-374.

[38] Verma RK, Prabh ND, Sankararamakrishnan R. Intra-helical salt-bridge and helix destabilizing residues within the same helical turn: role of functionally important loop E half-helix in channel regulation of major intrinsic proteins[J]. Biochim Biophys Acta, 2015, 1848(6): 1436-1449.

[39] Ishibashi K. Aquaporin subfamily with unusual NPA boxes[J]. Biochim Biophys Acta, 2006, 1758(8): 989-993.

[40] Ishibashi K. New members of mammalian aquaporins: AQP10-AQP12[J]. Handb Exp Pharmacol, 2009, (190): 251-262.

[41] Ishibashi K, Koike S, Kondo S, et al. The role of a group III AQP, AQP11 in intracellular organelle homeostasis[J]. J Med Invest, 2009, 56 Suppl: 312-317.

[42] Itoh T, Rai T, Kuwahara M, et al. Identification of a novel aquaporin, AQP12, expressed in pancreatic acinar cells[J]. Biochem Biophys Res Commun, 2005, 330(3): 832-838.

[43] Morishita Y, Sakube Y, Sasaki S, et al. Molecular mechanisms and drug development in aquaporin water channel diseases: aquaporin superfamily (superaquaporins): expansion of aquaporins restricted to multicellular organisms[J]. J Pharmacol Sci, 2004, 96(3): 276-279.

[44] Morishita Y, Matsuzaki T, Hara-chikuma M, et al. Disruption of aquaporin-11 produces polycystic kidneys following vacuolization of the proximal tubule[J]. Mol Cell Biol, 2005, 25(17): 7770-7779.

[45] Nozaki K, Ishii D, Ishibashi K. Intracellular aquaporins: clues for intracellular water transport?[J]. Pflugers Arch, 2008, 456(4): 701-707.

[46] Huang CG, Lamitina T, Agre P, et al. Functional analysis of the aquaporin gene family in *Caenorhabditis elegans*[J]. Am J Physiol Cell Physiol, 2007, 292(5): C1867-C1873.

[47] Kaufmann N, Mathai JC, Hill WG, et al. Developmental expression and biophysical characterization of a Drosophila melanogaster aquaporin[J]. Am J Physiol Cell Physiol, 2005, 289(2): C397-C407.

[48] Grohme MA, Mali B, Wełnicz W, et al. The aquaporin channel repertoire of the tardigrade *Milnesium tardigradum*[J]. Bioinform Biol Insights, 2013, 7: 153-165.

[49] Tchekneva EE, Khuchua Z, Davis LS, et al. Single amino acid substitution in aquaporin 11 causes renal failure[J]. J Am Soc Nephrol, 2008, 19(10): 1955-1964.

[50] Benga G. On the definition, nomenclature and classification of water channel proteins (aquaporins and

relatives)[J]. Mol Aspects Med, 2012, 33(5-6): 511-513.

[51] Yakata K, Hiroaki Y, Ishibashi K, et al. Aquaporin-11 containing a divergent NPA motif has normal water channel activity[J]. Biochim Biophys Acta, 2007, 1768(3): 688-693.

[52] Yakata K, Tani K, Fujiyoshi Y. Water permeability and characterization of aquaporin-11[J]. J Struct Biol, 2011, 174(2): 315-320.

[53] Czyzewski BK, Wang DN. Identification and characterization of a bacterial hydrosulphide ion channel[J]. Nature, 2012, 483(7390): 494-497.

[54] Wang Y, Huang Y, Wang J, et al. Structure of the formate transporter FocA reveals a pentameric aquaporin-like channel[J]. Nature, 2009, 462: 467-472.

[55] Amores A, Catchen J, Ferrara A, et al. Genome evolution and meiotic maps by massively parallel DNA sequencing: spotted gar, an outgroup for the teleost genome duplication[J]. Genetics, 2011, 188(4): 799-808.

[56] Brunet FG, Roest Crollius H, Paris M, et al. Gene loss and evolutionary rates following whole-genome duplication in teleost fishes[J]. Mol Biol Evol, 2006, 23(9): 1808-1816.

[57] Cañestro C, Albalat R, Irimia M, et al. Impact of gene gains, losses and duplication modes on the origin and diversification of vertebrates[J]. Semin Cell Dev Biol, 2013, 24(2): 83-94.

[58] Donoghue PC, Purnell MA. Genome duplication, extinction and vertebrate evolution[J]. Trends Ecol Evol, 2005, 20(6): 312-319.

[59] Innan H, Kondrashov F. The evolution of gene duplications: classifying and distinguishing between models[J]. Nat Rev Genet, 2010, 11(2): 97-108.

[60] Keeling PJ, Palmer JD. Horizontal gene transfer in eukaryotic evolution[J]. Nat Rev Genet, 2008, 9(8): 605-618.

[61] Zardoya R, Ding X, Kitagawa Y, et al. Origin of plant glycerol transporters by horizontal gene transfer and functional recruitment[J]. Proc Natl Acad Sci USA, 2002, 99(23): 14893-14896.

[62] Andersson JO. Lateral gene transfer in eukaryotes[J]. Cell Mol Life Sci, 2005, 62(11): 1182-1197.

[63] Izumi Y, Sonoda S, Yoshida H, et al. Role of membrane transport of water and glycerol in the freeze tolerance of the rice stem borer, Chilo suppressalis Walker(Lepidoptera: Pyralidae)[J]. J Insect Physiol, 2006, 52(2): 215-220.

[64] Philip BN, Kiss AJ, Lee RE Jr. The protective role of aquaporins in the freeze-tolerant insect *Eurosta solidaginis*: functional characterization and tissue abundance of EsAQP1[J]. J Exp Biol, 2011, 214(Pt 5): 848-857.

[65] Lee JK, Kozono D, Remis J, et al. Structural basis for conductance by the archaeal aquaporin AqpM at 1.68 Å[J]. Proc Natl Acad Sci USA, 2005, 102(52): 18932-18937.

[66] Kozono D, Ding X, Iwasaki I, et al. Functional expression and characterization of an archaeal aquaporin. AqpM from methanothermobacter marburgensis[J]. J Biol Chem, 2003, 278(12): 10649-10656.

[67] Cavalier-Smith T. The phagotrophic origin of eukaryotes and phylogenetic classification of protozoa[J]. Int J Syst Evol Microbiol, 2002, 52(Pt 2): 297-354.

[68] Rivera MC, Lake JA. The ring of life provides evidence for a genome fusion origin of eukaryotes[J]. Nature, 2004, 431(7005): 152-155.

[69] Beitz E. Aquaporin water and solute channels from malaria parasites and other pathogenic protozoa[J]. Chem Med Chem, 2006, 1(6): 587-592.

[70] Fadiel A, Isokpehi RD, Stambouli N, et al. Protozoan parasite aquaporins[J]. Expert Rev Proteomics, 2009, 6(2): 199-211.

[71] Von Bülow J, Beitz E. Number and regulation of protozoan aquaporins reflect environmental complexity[J]. Biol Bull, 2015, 229(1): 38-46.

[72] Bahamontes-Rosa N, Wu B, Beitz E, et al. Limited genetic diversity of the *Plasmodium falciparum*

aquaglyceroporin gene[J]. Mol Biochem Parasitol, 2007, 156(2): 255-257.

[73] Promeneur D, Liu Y, Maciel J, et al. Aquaglyceroporin PbAQP during intraerythrocytic development of the malaria parasite *Plasmodium berghei*[J]. Proc Natl Acad Sci USA, 2007, 104(7): 2211-2216.

[74] Pavlovic-Djuranovic S, Schultz JE, Beitz E. A single aquaporin gene encodes a water/glycerol/urea facilitator in *Toxoplasma gondii* with similarity to plant tonoplast intrinsic proteins[J]. FEBS Lett, 2003, 555(3): 500-504.

[75] Ahmadpour D, Geijer C, Tamás MJ, et al. Yeast reveals unexpected roles and regulatory features of aquaporins and aquaglyceroporins[J]. Biochim Biophys Acta, 2014, 1840(5): 1482-1491.

[76] Laizé V, Tacnet F, Ripoche P, et al. Polymorphism of *Saccharomyces cerevisiae* aquaporins[J]. Yeast, 2000, 16(10): 897-903.

[77] Sabir F, Loureiro-Dias MC, Prista C. Comparative analysis of sequences, polymorphisms and topology of yeasts aquaporins and aquaglyceroporins[J]. FEMS Yeast Res, 2016, 16(3): fow025.

[78] Hedfalk K, Bill RM, Mullins JG, et al. A regulatory domain in the C-terminal extension of the yeast glycerol channel Fps1p[J]. J Biol Chem, 2004, 279(15): 14954-14960.

[79] Philips J, Herskowitz I. Osmotic balance regulates cell fusion during mating in *Saccharomyces cerevisiae*[J]. J Cell Biol, 1997, 138(5): 961-974.

[80] Kayingo G, Sirotkin V, Hohmann S, et al. Accumulation and release of the osmolyte glycerol is independent of the putative MIP channel Spac977. 17p in *Schizosaccharomyces pombe*[J]. Antonie Van Leeuwenhoek, 2004, 85(2): 85-92.

[81] Verma RK, Prabh ND, Sankararamakrishnan R. New subfamilies of major intrinsic proteins in fungi suggest novel transport properties in fungal channels: implications for the host-fungal interactions[J]. BMC Evol Biol, 2014, 14: 173.

[82] Adams KL, Wendel JF. Polyploidy and genome evolution in plants[J]. Curr Opin Plant Biol, 2005, 8(2): 135-141.

[83] Meyers LA, Levin DA. On the abundance of polyploids in flowering plants[J]. Evolution, 2006, 60(6): 1198.

[84] Li G, Santoni V, Maurel C. Plant aquaporins: roles in plant physiology[J]. Biochim Biophys Acta, 2014, 1840(5): 1574-1582.

[85] Finn RN, Chauvigné F, Stavang JA, et al. Insect glycerol transporters evolved by functional co-option and gene replacement[J]. Nat Commun, 2015, 6: 7814.

[86] Wallace IS, Shakesby AJ, Hwang JH, et al. Acyrthosiphon pisum AQP2: a multifunctional insect aquaglyceroporin[J]. Biochim Biophys Acta, 2012, 1818(3): 627-635.

[87] Pommerrenig B, Diehn TA, Bienert GP. Metalloido-porins: Essentiality of Nodulin 26-like intrinsic proteins in metalloid transport[J]. Plant Sci, 2015, 238: 212-227.

[88] Gupta AB, Sankararamakrishnan R. Genome-wide analysis of major intrinsic proteins in the tree plant Populus trichocarpa: characterization of XIP subfamily of aquaporins from evolutionary perspective[J]. BMC Plant Biol, 2009, 9: 134.

[89] Gustavsson S, Lebrun AS, Nordén K, et al. A novel plant major intrinsic protein in Physcomitrella patens most similar to bacterial glycerol channels[J]. Plant Physiol, 2005, 139(1): 287-295.

[90] Marjanović Z, Uehlein N, Kaldenhoff R, et al. Aquaporins in poplar: what a difference a symbiont makes![J]. Planta, 2005, 222(2): 258-268.

[91] Bienert GP, Bienert MD, Jahn TP, et al. Solanaceae XIPs are plasma membrane aquaporins that facilitate the transport of many uncharged substrates[J]. Plant J, 2011, 66(2): 306-317.

[92] Danielson JA, Johanson U. Unexpected complexity of the aquaporin gene family in the moss Physcomitrella patens[J]. BMC Plant Biol, 2008, 8: 45.

[93] Khabudaev KV, Petrova DP, Grachev MA, et al. A new subfamily LIP of the major intrinsic proteins[J].

BMC Genomics, 2014, 15(1): 173.

[94] Campbell EM, Ball A, Hoppler S, et al. Invertebrate aquaporins: a review[J]. J Comp Physiol B, 2008, 178(8): 935-955.

[95] Lee SJ, Murphy CT, Kenyon C. Glucose shortens the life span of *C. elegans* by downregulating DAF-16/FOXO activity and aquaporin gene expression[J]. Cell Metab, 2009, 10(5): 379-391.

[96] Stavang JA, Chauvigné F, Kongshaug H, et al. Phylogenomic and functional analyses of salmon lice aquaporins uncover the molecular diversity of the superfamily in Arthropoda[J]. BMC Genomics, 2015, 16(1): 618.

[97] Ball A, Campbell EM, Jacob J, et al. Identification, functional characterization and expression patterns of a water-specific aquaporin in the brown dog tick, *Rhipicephalus sanguineus*[J]. Insect Biochem Mol Biol, 2009, 39(2): 105-112.

[98] Benoit JB, Hansen IA, Szuter EM, et al. Emerging roles of aquaporins in relation to the physiology of blood-feeding arthropods[J]. J Comp Physiol B, 2014, 184(7): 811-825.

[99] Drake LL, Rodriguez SD, Hansen IA. Functional characterization of aquaporins and aquaglyceroporins of the yellow fever mosquito, *Aedes aegypti*[J]. Sci Rep, 2015, 5: 7795.

[100] Kataoka N, Miyake S, Azuma M. Aquaporin and aquaglyceroporin in silkworms, differently expressed in the hindgut and midgut of *Bombyx mori*[J]. Insect Mol Biol, 2009, 18(3): 303-314.

[101] Berthelot C, Brunet F, Chalopin D, et al. The rainbow trout genome provides novel insights into evolution after whole-genome duplication in vertebrates[J]. Nat Commun, 2014, 5: 3657.

[102] Kuraku S, Meyer A, Kuratani S. Timing of genome duplications relative to the origin of the vertebrates: did cyclostomes diverge before or after?[J]. Mol Biol Evol, 2009, 26(1): 47-59.

[103] Meyer A, Schartl M. Gene and genome duplications in vertebrates: the one-to-four(-to-eight in fish)rule and the evolution of novel gene functions[J]. Curr Opin Cell Biol, 1999, 11(6): 699-704.

[104] Tingaud-Sequeira A, Calusinska M, Finn RN, et al. The zebrafish genome encodes the largest vertebrate repertoire of functional aquaporins with dual paralogy and substrate specificities similar to mammals[J]. BMC Evol Biol, 2010, 10: 38.

[105] MacQueen DJ, Johnston IA. A well-constrained estimate for the timing of the salmonid whole genome duplication reveals major decoupling from species diversification[J]. Proc Biol Sci, 2014, 281(1778): 20132881.

[106] Suzuki M, Shibata Y, Ogushi Y, et al. Molecular machinery for vasotocin-dependent transepithelial water movement in amphibians: aquaporins and evolution[J]. Biol Bull, 2015, 229(1): 109-119.

[107] Morinaga T, Nakakoshi M, Hirao A, et al. Mouse aquaporin 10 gene(AQP10)is a pseudogene[J]. Biochem Biophys Res Commun, 2002, 294(3): 630-634.

[108] Tanaka Y, Morishita Y, Ishibashi K. Aquaporin 10 is a pseudogene in cattle and their relatives[J]. Biochem Biophys Rep, 2015, 1: 16-21.

[109] Ohta E, Itoh T, Nemoto T, et al. Pancreas-specific aquaporin 12 null mice showed increased susceptibility to caerulein-induced acute pancreatitis[J]. Am J Physiol Cell Physiol, 2009, 297(6): C1368-C1378.

[110] Konno N, Hyodo S, Yamaguchi Y, et al. Vasotocin/V2-type receptor/aquaporin axis exists in African lungfish kidney but is functional only in terrestrial condition[J]. Endocrinology, 2010, 151(3): 1089-1096.

[111] Martos-Sitcha JA, Campinho MA, Mancera JM, et al. Vasotocin and isotocin regulate aquaporin 1 function in the sea bream[J]. J Exp Biol, 2015, 218(Pt 5): 684-693.

[112] Prak S, Hem S, Boudet J, et al. Multiple phosphorylations in the C-terminal tail of plant plasma membrane aquaporins: role in subcellular trafficking of AtPIP2;1 in response to salt stress[J]. Mol Cell Proteomics, 2008, 7(6): 1019-1030.

第三节　水通道蛋白的转运特征

AQP是一组转运小分子中性溶质和某些气体分子的特异性的整合膜通道蛋白。AQP广泛分布于人体多种组织器官，与水分子的跨膜转运、细胞迁移、脑水肿、神经兴奋性调节和肿瘤的增殖生长等生理和（或）病理过程有关。近年来针对AQP介导的水、甘油、尿素、某些气体及其他分子转运等方面的研究取得了诸多进展，本节将对水通道蛋白的转运特征进行介绍。

一、水通道蛋白介导的水转运

AQP是一种内在跨膜蛋白，通过调节细胞内和细胞间的水转运，在维持组织水稳态中起着至关重要的作用。每个AQP由6个跨膜的α螺旋结构和1个位于中心的水转运孔道组成（图1-9）[1]；其独特的分子结构易受疏水效应和空间限制的影响，使之对水分子具有高度的选择性[2]。AQP通道在溶质主动转运产生的渗透梯度下开放，允许水分子沿渗透梯度快速扩散而不依靠电位驱动。作为完全由渗透梯度驱动的快速水转运通道，AQP利用渗透梯度加强膜内外水的双向转运，即水分子沿渗透梯度由低渗区向高渗区移动，从而使细胞膜两侧的渗透浓度接近等渗[3]。因此，AQP的转运并非主动过程，而依赖于跨膜渗透梯度的存在[4]。在生理情况下，AQP处于激活状态，水经AQP由低渗区向高渗区转运，通常不需要"门控"或其他方式调节。在哺乳动物中，AQP表达于需要进行大量水转运的上皮细胞、细胞膜或液泡膜中，有助于细胞内水平衡和细胞外水分子流动。

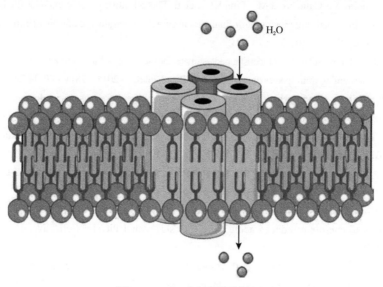

图1-9　AQP对水分子的转运

AQP具有极高的水渗透性，例如AQP1的导水可达每秒$2\times10^9 \sim 3\times10^9$个水分子/亚单位[5]。大多数AQP（AQP4和AQP6除外）的半胱氨酸残基可与$HgCl_2$等含汞试剂结合而失去水渗透性[6]。在哺乳动物中，除AQP2仅分布于肾脏集合管外，其余AQP在体内分布广泛。根据转运所介导的物质，AQP可分为仅对水具有渗透性的经典AQP（AQP0、

AQP1、AQP2、AQP4、AQP5、AQP6、AQP8，后AQP8被独立出来），对除水之外的甘油、尿素等小分子溶质（尤其是甘油）具有渗透性的水甘油通道蛋白（AQP3、AQP7、AQP9、AQP10）[7]，以及超级水通道蛋白AQP（AQP11、AQP12）[8]。此外，AQP的水渗透性差异较大，其渗透率分别为：AQP0 0.25×10^{-14} cm^3/s、AQP1 6.0×10^{-14} cm^3/s、AQP2 3.3×10^{-14} cm^3/s、AQP3 2.1×10^{-14} cm^3/s、AQP4 24×10^{-14} cm^3/s、AQP5 5.0×10^{-14} cm^3/s[9]、AQP6 93.0×10^{-14} cm^3/s、AQP8 8.2×10^{-14} cm^3/s[10]。

水稳态在神经信号转导过程中非常重要。在生理和病理条件下，星形胶质细胞在中枢神经系统的水转运中起着至关重要的作用。迄今为止，在成熟星形胶质细胞中发现5种AQP表达，分别是AQP3、AQP4、AQP5、AQP8和AQP9。AQP4在脑内含量最丰富，并且主要分布在与毛细血管、蛛网膜和软脑膜直接接触的星形胶质细胞及其终足上，这种特征性的分布方式称为极性分布，成为AQP4在中枢神经系统分布最为显著的特点。研究表明，AQP4参与脑脊液重吸收、渗透压调节等生理过程，还可能兼有细胞外渗透压感受器和水平衡调节器的功能，进而参与调控全身的水代谢[11]。渗透压是细胞肿胀的原动力，AQP4除了可受其他蛋白调节外，其表达也受渗透压的调控。研究发现，低渗环境可导致星形胶质细胞水肿和生存能力下降，伴随着星形胶质细胞AQP4的mRNA水平升高和蛋白表达增强，证实了AQP4的表达与渗透压改变直接相关[12]。由甘露醇诱导的高渗环境可以增加体外培养的星形胶质细胞中AQP4和AQP9的表达，颅内注入甘露醇可以增加大鼠大脑皮质、软脑膜和血管周围AQP4和AQP9的表达，并发现其表达调控与p38MAPK信号通路有关[13]。采用体外培养的星形胶质细胞、脑片及肝性脑病动物的实验发现，AQP4蛋白表达上调可能参与了高氨诱导的细胞肿胀。研究表明，除渗透压外，细胞氧化作用、线粒体渗透性改变也在氨诱导的星形胶质细胞水肿的病理过程中发挥作用。另外，Yang等[14]研究发现在高渗状态下，早期星形胶质细胞中所有类型AQP快速表达增加以适应外界高渗刺激，后期由于调控机制和水渗透性的差异，AQP对高渗刺激的表达模式不同，AQP3、AQP5和AQP8的表达增加持续到高渗溶液作用后6小时，而AQP4和AQP9的表达增加持续到高渗溶液作用后12小时。丝裂原活化蛋白激酶（mitogen-activated protein kinase，MAPK），特别是胞外信号调节激酶（extracellular regulated protein kinase，ERK）、c-Jun N-末端激酶（c-Jun N-terminal kinase，JNK）和p38 MAPK，是受渗透压变化而激活的重要胞内信号转导通路。高渗液诱导大鼠星形胶质细胞AQP表达时可通过不同的MAPK通路，ERK抑制剂可以抑制高渗溶液作用后星形胶质细胞AQP3、AQP5和AQP8的表达，而p38抑制剂可抑制AQP4和AQP9的表达。以上研究提示，MAPK信号通路是细胞内参与对高渗应激保护反应的重要信号途径。其余对水具有渗透性的AQP还参与细胞内其他溶质的转运，按照其参与转运的溶质在下文进行介绍。

二、水通道蛋白介导的甘油转运

水通道蛋白除了介导水分子的转运，还可以转运甘油。水甘油通道蛋白是水通道蛋白的重要亚家族，是既可以跨膜运输水分子又可以运输甘油、尿素等小分子物质（特别是甘油）的双功能水通道蛋白，其主要成员包括AQP3、AQP7、AQP9和AQP10。目前对AQP11的研究较少，但最近AQP11也被证实存在于人类和小鼠的脂肪细胞中，发挥甘油转运的功能。

与典型AQP相比，水甘油通道蛋白的孔径更宽，允许直径更大的甘油等分子通过。水甘油通道蛋白表达于身体的多个部位，包括脂肪组织、肝脏、皮肤等，发挥重要的代谢功能。

脂肪细胞为维持机体的能量平衡，不断进行生成和分解。脂肪组织在营养过剩时储存为甘油三酯，而在饥饿时，脂肪组织中的甘油三酯被水解并向能量消耗器官提供游离脂肪酸和甘油。脂肪组织是甘油的主要来源，而甘油是糖异生的重要底物之一。脂肪细胞中的水甘油通道蛋白是控制甘油进出脂肪细胞以及调控脂肪生成和分解的关键因子。人类中首个被发现的水甘油通道蛋白是AQP7，在白色脂肪和棕色脂肪等组织广泛表达。随后，AQP3和AQP9在脂肪中表达也相继被报道。在脂肪细胞亚细胞定位中，AQP3存在于细胞膜和细胞质中，AQP7主要存在于细胞质中，而AQP9分布在脂肪细胞的细胞膜上。AQP9在肝细胞中的表达更为丰富，位于朝向肝窦的肝细胞膜上，参与肝细胞对甘油的摄取，是肝脏中唯一的摄取甘油的通道。AQP3和AQP7在脂解过程中协助释放甘油，将脂肪动员产生的甘油运输至血液，而AQP9主要将甘油从血液转运至肝细胞进行糖异生（图1-10）[15, 16]。AQP7功能降低或缺失将导致甘油从脂肪细胞的释放减少而积累增加，从而促进肥胖发生。同时肾脏中AQP7在甘油重吸收中也发挥重要作用，AQP7缺乏将导致甘油尿的出现。AQP9主要在葡萄糖异生过程中转运甘油，肝脏中的AQP9在禁食时增加，在再进食时减少[17]。

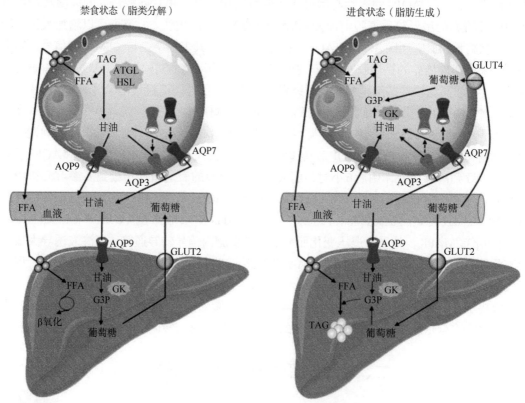

图1-10　AQP3、AQP7、AQP9介导甘油转运

TAG. triacylglycerol，甘油三酯；GLUT4. glucose transporter 4，葡萄糖转运蛋白-4；ATGL. adipose triglyceride lipase，脂肪甘油三酯脂肪酶；FFA. free fatty acid，游离脂肪酸；HSL. hormone sensitive lipase，激素敏感性脂肪酶；G3P. glycerol-3-phosphate，三磷酸甘油；GLUT2. glucose transporter 2，葡萄糖转运蛋白-2；GK. glycerol kinase，甘油激酶

扫码见彩图

AQP3是一种水转运功能相对较弱的水通道蛋白，却是一种高效的甘油转运体，其介导的甘油转运与皮肤自身的水化作用关系密切。AQP3受损引起的甘油转运异常使皮肤的水化和弹性降低并削弱了皮肤的生理屏障作用[18]。AQP10主要表达于小肠，可能参与甘油等部分小分子物质的跨膜转运过程，还可能参与脂解过程中的甘油转运[19]，但AQP10的具体生理功能及其功能异常引起的临床表现还需深入研究。

三、水通道蛋白介导的尿素转运

尿素主要由肝脏中的氨转化而来，大约90%的尿素通过肾脏排出，尿素转运过程是尿浓缩的主要机制。对AQP运输尿素的生理角色的认识相对于甘油运输要更少。水甘油通道蛋白AQP3、AQP7、AQP9和AQP10，有研究认为可能也包括AQP6，对尿素同样具有渗透性，但这些水通道蛋白在尿素转运中的生理意义尚未完全揭示[20]。表达于肾脏的AQP3除能够提高水的渗透性外，还能够增加尿素的渗透压，在尿浓缩机制中起着极其重要的作用。AQP3基因缺失虽未直接影响肾脏对尿素的浓缩功能，但却减弱了对尿中其他溶质浓缩的能力。肝脏是尿素产生的主要器官，AQP9是表达于肝细胞膜的尿素渗透性蛋白，作为摄取氨的通道，将其从血液转运到肝血窦周围肝细胞进行代谢，同时也可调节新合成尿素的流出。AQP3、AQP7和AQP9有助于角质形成细胞对尿素的吸收以改善表皮的屏障功能，在角化细胞分化、脂质合成和维持表皮稳态中起重要作用。研究发现，AQP8主要表达于肠道器官、胎盘和心脏；大鼠和人类中的AQP8可转运尿素和水，可作为汞敏感的独特性水通道蛋白，而且哺乳动物AQP8尿素渗透率高于同源植物的AQP8。尿素作为最大的氮循环池，其产量与饮食及内源性蛋白的降解有关。AQP8不仅在各个器官中发挥调节尿素转运的重要作用，也在尿浓缩机制中起到核心作用。AQP8序列与已知水通道蛋白存在的结构域差异，在尿素的渗透中起着重要的作用。当AQP8功能缺失时，尿素会堆积在其大量表达的器官，如肠道、胎盘和心脏，肠道中尿素的堆积会使肠道大量吸收NH_4^+入血，产生神经毒性，影响正常的神经生理功能[21]。AQP10在十二指肠和空肠中表达，但其参与尿素转运的作用和机制尚不清楚。

四、水通道蛋白介导的气体转运

AQP除了发挥水运输和渗透调节的作用外，某些AQP还具有转运气体的功能，如转运CO_2、O_2、NH_3和NO。其中AQP1、AQP4和AQP5可介导CO_2转运，AQP1、AQP4可介导NO和O_2转运，AQP1、AQP3、AQP4、AQP6、AQP7、AQP8和AQP9可介导NH_3转运。哺乳动物AQP对气体和水的转运具有不同的选择性。AQP1是首个被发现具有气体转运功能的水通道蛋白，气体通过AQP1四聚体的中心孔道传递到细胞膜，而不是通过水孔道（图1-11）[22]。在4个单体的中心是第5个主要由疏水氨基酸组成的孔道，它可为气体等非极性分子提供通道。AQP1

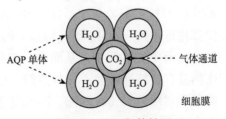

图1-11　AQP气体转运

能提高CO_2、O_2、NH_3和NO的渗透性。AQP1负责维持60%红细胞的CO_2分压，在CO_2转运中起着关键作用。CO_2和NH_3是挥发性气体，当穿过细胞膜时会引起细胞内pH的变化，AQP1介导的CO_2和NH_3跨膜转运维持了细胞内酸碱稳态[23]。在血管系统中，AQP1介导的NO从血管内皮细胞转出，以及NO转入平滑肌细胞，这有助于内皮依赖性的血管舒张[24]。除AQP1外，位于脑内的AQP4也通过其中心孔道对NO具有渗透性，但AQP4在中枢神经系统NO转运中的作用有待进一步确定。水通道蛋白对NH_3转运机制及作用尚需更多研究阐明。

五、水通道蛋白介导的其他分子转运

过氧化氢（H_2O_2）属于活性氧物种，是细胞内重要的第二信使，其必须穿过细胞膜进入细胞质才能发挥作用，同时活性氧对核酸、蛋白质和脂类具有潜在的损伤作用。H_2O_2的利用和失调，特别是内质网产生的H_2O_2，不仅影响细胞稳态，而且影响生物体的寿命。H_2O_2和水有相似的分子大小、偶极矩和氢键能力，理论上AQP具有转运H_2O_2的潜力，同时所有的AQP都是通过内质网的分泌途径发挥作用，表明每一个AQP都可能在一定程度上促进H_2O_2的跨膜转运。然而AQP对H_2O_2的渗透率不同，AQP1、AQP3、AQP5、AQP8、AQP9、AQP11的共同特征是具有一定程度的优先转运H_2O_2的能力[25]。AQP通过调节H_2O_2进出内质网来维持氧化还原稳态，而且还支持细胞信号转导。AQP1通过转运H_2O_2使得角膜细胞内活性氧含量上升，更多地参与角膜细胞中氧化应激反应。AQP3的表达通过影响下游细胞信号级联来调节细胞内H_2O_2的积累，其介导的H_2O_2摄取也是趋化因子依赖性T细胞迁移所必需的[26]。线粒体AQP8可促进线粒体H_2O_2的释放，其表达缺陷通过线粒体通透性转变机制引起线粒体去极化和细胞死亡。AQP3、AQP8介导H_2O_2跨膜转运可加速T细胞的迁移，促进肿瘤细胞的能量合成和代谢及细胞增殖。AQP11定位于内质网，部分位于线粒体或相关的细胞膜，其表达下调将严重干扰H_2O_2通过内质网的总量，但不影响线粒体或细胞膜的总量。

越来越多的证据表明，AQP还具有离子通道功能，包括AQP0、AQP1、AQP6和AQP10。AQP1中四聚体四重对称轴的中心孔道是最可能的离子传导途径，介导非选择性的单价阳离子转运[27]。AQP0、AQP6可能通过亚基内孔隙转运离子。AQP0的离子通道对电压和pH敏感，在酸性条件下开放，在中性条件下永久关闭。在pH低于5.5时，AQP6对阴离子的转运被快速可逆激活，表现出一种对硝酸盐具有显著特异性的阴离子渗透形式[28]。

某些AQP对金属亦具有转运作用，如水甘油通道蛋白（AQP3、AQP7、AQP9和AQP10）有助于类金属、硅、砷和锑的带电和非带电分子的扩散，并在类金属稳态中发挥重要作用。AQP9通过转运亚砷酸盐和亚锑酸盐参与维持类金属稳态，在肝脏亚砷酸盐排泄中发挥作用，从而在一定程度上保护动物免受砷中毒的影响[29]。AQP9对某些底物具有独特通透性，如甘露醇、山梨糖醇、嘌呤（腺嘌呤）、嘧啶（尿嘧啶、氟尿嘧啶）和单羧酸酯（乳酸酯、β-羟基丁酸），3D结构分析显示这可能是与AQP9的孔径较大有关[30]。

总之，AQP独特的分子结构赋予其对水分子高效且严格的渗透性，AQP通过利用渗透梯度加强水的膜内、外双向转运，实现细胞内外水平衡。同时，部分AQP还兼具转运甘油、尿素、某些气体、H_2O_2、离子等，甚至转运金属、类金属、嘌呤、嘧啶等的能力。随着对AQP研究的深入，AQP功能的多样性将为未来药物研发及靶向治疗提供更多的理论依据。

（李燕华）

参 考 文 献

[1] Shi ZF，Fang Q，Chen Y，et al. Methylene blue ameliorates brain edema in rats with experimental ischemic stroke via inhibiting aquaporin 4 expression[J]. Acta Pharmacol Sin，2021，42（3）：382-392.

[2] Li C，Wang W. Urea transport mediated by aquaporin water channel proteins[J]. Subcell Biochem，2014，73：227-265.

[3] Agre P. Aquaporin water channels（Nobel Lecture）[J]. Angew Chem Int Ed Engl，2004，43（33）：4278-4290.

[4] Agre P，King LS，Yasui M，et al. Aquaporin water channels：from atomic structure to clinical medicine[J]. J Physiol，2002，542（Pt 1）：3-16.

[5] Pastor-Soler N，Bagnis C，Sabolic I，et al. Aquaporin 9 expression along the male reproductive tract[J]. Biol Reprod，2001，65（2）：384-393.

[6] Benga G. The first discovered water channel protein，later called aquaporin 1：molecular characteristics，functions and medical implications[J]. Mol Aspects Med，2012，33（5/6）：518-534.

[7] Hirano Y，Okimoto N，Kadohira I，et al. Molecular mechanisms of how mercury inhibits water permeation through aquaporin-1：understanding by molecular dynamics simulation[J]. Biophys J，2010，98（8）：1512-1519.

[8] Verkman AS. More than just water channels：unexpected cellular roles of aquaporins[J]. J Cell Sci，2005，118（Pt 15）：3225-3232.

[9] Ishibashi K，Tanaka Y，Morishita Y. The role of mammalian superaquaporins inside the cell[J]. Biochim Biophys Acta，2014，1840（5）：1507-1512.

[10] Yang B，Verkman AS. Water and glycerol permeabilities of aquaporins 1-5 and MIP determined quantitatively by expression of epitope-tagged constructs in *Xenopus* oocytes[J]. J Biol Chem，1997，272（26）：16140-16146.

[11] Geng X，Yang B. Transport characteristics of aquaporins[J]. Adv Exp Med Biol，2017，969：51-62.

[12] 李燕华，孙善全. 低渗液对星形胶质细胞水通道蛋白-4表达的影响[J]. 中华医学杂志，2004，84（6）：496-501.

[13] Arima H，Yamamoto N，Sobue K，et al. Hyperosmolar mannitol simulates expression of aquaporins 4 and 9 through a p38 mitogen-activated protein kinase-dependent pathway in rat astrocytes[J]. J Biol Chem，2003，278（45）：44525-44534.

[14] Yang M，Gao F，Liu H，et al. Hyperosmotic induction of aquaporin expression in rat astrocytes through a different MAPK pathway[J]. J Cell Biochem，2013，114（1）：111-119.

[15] Lebeck J. Metabolic impact of the glycerol channels AQP7 and AQP9 in adipose tissue and liver[J]. J Mol Eedocrinol，2014，52（2）：R165-R178.

[16] Calamita G，Perret J，Delporte C. Aquaglyceroporins：drug targets for metabolic diseases?[J]. Front Physiol，2018，9：851.

[17] Hibuse T，Maeda N，Nagasawa A，et al. Aquaporins and glycerol metabolism[J]. Biochim Biophys Acta，

2006, 1758(8): 1004-1011.

[18] Nakahigashi K, Kabashima K, Ikoma A, et al. Upregulation of aquaporin-3 is involved in keratinocyte proliferation and epidermal hyperplasia[J]. J Investdermatol, 2011, 131(4): 865-873.

[19] Laforenza U, Scaffino MF, Gastaldi G. Aquaporin-10 represents an alternative pathway for glycerol efflux from human adipocytes[J]. PLoS One, 2013, 8(1): e54474.

[20] Litman T, Søgaard R, Zeuthen T. Ammonia and urea permeability of mammalian aquaporins[J]. Handb Exp Pharmacol, 2009, (190): 327-358.

[21] Liu K, Nagase H, Huang CG, et al. Purification and functional characterization of aquaporin-8[J]. Biol Cell, 2006, 98(3): 153-161.

[22] Kaldenhoff R, Kai L, Uehlein N. Aquaporins and membrane diffusion of CO_2 in living organisms[J]. Biochim Biophys Acta, 2014, 1840(5): 1592-1595.

[23] Geyer RR, Musa-Aziz R, Qin X, et al. Relative CO_2/NH_3 selectivities of mammalian aquaporins 0-9[J]. Am J Physiol Cell Physiol, 2013, 304(10): C985-C994.

[24] Herrera M, Hong NJ, Garvin JL. Aquaporin-1 transports NO across cell membranes[J]. Hypertension, 2006, 48(1): 157-164.

[25] Bienert GP, Chaumont F. Aquaporin-facilitated transmembrane diffusion of hydrogen peroxide[J]. Biochim Biophys Acta, 2014, 1840(5): 1596-1604.

[26] Hara-Chikuma M, Chikuma S, Sugiyama Y, et al. Chemokine-dependent T cell migration requires aquaporin-3-mediated hydrogen peroxide uptake[J]. J Exp Med, 2012, 209(10): 1743-1752.

[27] Yool AJ, Weinstein AM. New roles for old holes: ion channel function in aquaporin-1[J]. News Physiol Sci, 2002, 17: 68-72.

[28] Yasui M. pH regulated anion permeability of aquaporin-6[J]. Handb Exp Pharmacol, 2009, (190): 299-308.

[29] Liu Z, Shen J, Carbrey JM, et al. Arsenite transport by mammalian aquaglyceroporins AQP7 and AQP9[J]. Proc Natl Acad Sci USA, 2002, 99(9): 6053-6058.

[30] Finn RN, Cerdà J. Evolution and functional diversity of aquaporins[J]. Biol Bull, 2015, 229(1): 6-23.

第四节 哺乳动物水通道蛋白的定位分布与功能

一、心血管系统中的水通道蛋白

研究发现，心血管系统中存在某些AQP，如AQP1、AQP4、AQP7和AQP9，分别位于内皮细胞、血管平滑肌细胞和心肌细胞等部位。AQP与血管生成有关，且在脑缺血、充血性心力衰竭、高血压等疾病的病理生理过程中发挥作用。深入剖析AQP在心血管系统中的作用，有望为防治相关疾病提供新的思路。本部分将介绍AQP在心血管系统中的表达、生理功能，以及在相关疾病中的研究进展。

AQP是膜蛋白家族中的一类，其在渗透压或浓度梯度的驱动下，介导水分子及其他一些分子的跨膜转运。AQP主要分布于上皮细胞和内皮细胞，对维持机体的水、电解质平衡发挥重要作用。自1991年在红细胞中发现了第一种AQP以来，目前已鉴定出13种AQP（AQP0～AQP12）[1-3]。AQP与尿液的浓缩和腺体分泌有关，近年来借助对基因敲除小鼠模型的研究发现，AQP也与脑缺血、癌症、青光眼、肥胖和感染等疾病有关[2]。研究表明，AQP参与心血管系统的功能调节及相关疾病的发展，尤其在脑缺血、充血性心力衰竭、高血压和血管生成等方面，其作用更为明显。因此，深入研究AQP在心血管疾病中的发病机制，可为未来疾病的预防和治疗提供新的途径。

（一）心血管系统中AQP的表达和生理功能

目前，在心血管系统中已鉴定出AQP1、AQP4、AQP7和AQP9等水通道蛋白，它们分布于心脏内皮细胞和血管平滑肌细胞（smooth muscle cell，SMC）[4]，担负水、甘油和乳酸转运功能，并在血管的生理功能中发挥重要作用。同时，AQP还与某些血管疾病的病理过程存在关联。

1. AQP1 AQP1在毛细血管和小静脉内皮细胞（endothelial cell，EC）中表达较高，同时也存在于血管平滑肌细胞和角膜内皮细胞、肺和支气管内皮细胞等非血管内皮细胞。然而，在中枢神经系统的EC中尚未发现AQP1的存在[5]。目前研究仅在大鼠心肌细胞中检测到AQP1，而在小鼠心肌细胞未见其分布[6,7]。AQP1在渗透压驱动下协助水转运。同时，越来越多的证据表明，AQP1可以介导许多小分子的转运，如尿素、NH_3、H_2O_2、NO、CO_2、$Sb(OH)_3$和$As(OH)_3$[3,8-10]。AQP1协助NO进入内皮细胞，通过控制其浓度、生物利用度和扩散距离，进而调节血管张力和血压[4,11]。AQP1表达受某些血管相关疾病的影响。例如，肝硬化状态下，AQP1表达上调，增强内皮侵袭性，促进血管生成[12]。敲除小鼠*AQP1*基因可减轻胆管结扎后的肝血管生成、纤维化和门脉高压[13]。

在视网膜血管内皮细胞中，AQP1通过血管内皮生长因子（vascular endothelial growth factor，VEGF）信号通路以非依赖方式参与缺氧诱导的血管生成[14]。然而，在*AQP1*敲除新生小鼠中，缺氧诱导的视网膜病变中的微血管增殖并未受到影响[15]。此外，AQP1在恶性肿瘤的微血管内皮细胞中呈现高表达。在*AQP1*敲除小鼠的皮下或颅内植入肿瘤后，肿瘤生长和血管生成受到抑制并广泛坏死[16]。先前研究揭示，碳酸酐酶抑制剂乙酰唑胺能够抑制肿瘤组织中的AQP1蛋白表达和血管生成[17]。通过泛素-蛋白酶体系统抑制AQP1，可直接损伤小鼠黑色素瘤血管并抑制肿瘤生长[18]。因此，针对血管内皮细胞中AQP1表达的靶向调控可能在肿瘤的治疗中具有积极意义。

2. AQP4 AQP4在中枢神经系统的大脑、脊髓和视神经中呈现高表达[19-21]。该蛋白位于脑内毛细血管周围的星形胶质细胞，分布于星形胶质细胞足突、外胶质界膜、室管膜和内胶质界膜。最初大多数学者认为AQP4未表达于脑血管内皮细胞。然而，Amiry-Moghaddam于2004年通过免疫电子显微镜证实，AQP4亦存在于脑内皮细胞中，但其表达水平相较于星形胶质细胞低[22]。尽管正常内皮细胞的AQP4具有代偿作用，但实验表明，选择性敲除星形胶质细胞终足膜AQP4仍会延缓脑水肿的发生。目前，尚未明确内皮细胞中的AQP4是否参与维持脑内水平衡。

AQP4作为血脑屏障和血脑脊液（cerebrospinal fluid，CSF）屏障的关键组成部分，负责维持脑内水平衡[23]。AQP4位于与毛细血管和软脑膜直接接触的胶质细胞膜上，其高度的极性表达模式表明AQP4能介导胶质细胞、脉络丛及血管腔之间的水转运。此外，AQP4还参与促进星形胶质细胞迁移和神经信号转导。研究表明，*AQP4*缺陷的小鼠和培养的星形胶质细胞均出现星形胶质细胞迁移障碍[24]。在小鼠的心肌细胞中存在AQP4，而在大鼠心肌中尚未观察到此类现象[7]。

3. AQP7　AQP7是一种水甘油通道蛋白，主要分布于肾近端小管、睾丸、心脏、横纹肌和脂肪组织。心脏是AQP7 mRNA表达的第二大组织，仅次于脂肪组织[25]。关于心脏中AQP7的作用，目前研究较少。2009年Hibuse发现，相较于野生型小鼠，*AQP7*敲除小鼠心脏中甘油和ATP含量较低[26]。在基础条件下，*AQP7*敲除小鼠心脏具有正常的组织学和形态学。然而，当注射异丙醇或有主动脉弓缩窄（transverse aortic constriction，TAC）时，与野生型小鼠相比，*AQP7*敲除小鼠出现心肌肥厚和生存率下降，这提示甘油是心脏重要的能量底物[26]。此外，AQP7亦表达于脂肪组织的毛细血管内皮细胞中，其功能尚待研究[27]。未来研究有必要进一步明确AQP7在内皮细胞中的生理和病理意义。

4. AQP9　AQP9是一种水甘油通道蛋白，具有转运水、甘油、尿素和其他微小中性溶质的能力。该蛋白存在两种亚型，其中短亚型位于线粒体内膜，长亚型则位于细胞膜[28, 29]。AQP9广泛分布于大脑、肝、脾脏、表皮、睾丸和软脑膜血管的内皮细胞[30]。与AQP4类似，AQP9参与细胞水平衡和水肿形成[31]。此外，AQP9还与脑内能量代谢密切相关，其在神经元线粒体和葡萄糖敏感神经元中表达，并受到胰岛素负调节[30]。研究还发现，AQP9参与跨血脑屏障的乳酸和酮体转运，推测AQP9可能负责清除脑缺血中多余的乳酸和其他代谢物[32]。

（二）心血管疾病与AQP

1. 脑缺血与AQP　脑卒中是一种复杂且具有极大破坏性的神经系统疾病，目前尚无理想的治疗方法。脑水肿是脑卒中的严重并发症之一，亦是脑卒中恶化的关键因素。早期脑水肿可显著加速梗死形成，因此成为治疗的重点目标。目前在脑内已确认存在7种AQP亚型，包括AQP1、AQP3、AQP4、AQP5、AQP8、AQP9和AQP12。其中AQP1、AQP4和AQP9的含量较为丰富。在脑缺血的脑水肿期，AQP4和AQP9的表达发生变化，而AQP1的表达则保持不变[33]。脑缺血后1小时，病变核心和周边区星形胶质细胞终足的AQP4表达上升，脑缺血后48小时，损伤区周边的星形胶质细胞AQP4表达亦增加，两者变化均与脑水肿峰值出现的时间相吻合[33]。AQP4在脑缺血的早期阶段含量更为丰富[34]。多项研究表明，*AQP4*缺陷小鼠在脑缺血后脑梗死体积减小，神经行为学得到改善；阻断AQP4可部分阻止血脑屏障紊乱，缓解脑缺血引起的神经炎症[35-37]。研究发现，脑缺血后1小时AQP4缺失小鼠脑水肿程度减轻[36]。AQP4缺乏对动物行为学和病变体积的影响与缺血后第3天和第7天的水肿减轻无关[38]。

在脑缺血后24小时内，AQP9的表达随时间延长逐渐上升，但与细胞肿胀无明显关联[33]，其具体作用仍有待进一步阐明。Yang的研究发现，缺血组织周边的AQP3、AQP5和AQP8在脑缺血24小时后表达持续增强，而在核心区缺血后6小时表达下降，这表明AQP与脑缺血后水肿的发生有关[39]。因此，调控AQP表达将成为脑缺血一种有效的治疗策略。

2. 充血性心力衰竭与AQP　充血性心力衰竭（congestive heart failure，CHF）是大多数心脏病发展到终末期的典型表现，通常伴有水转运障碍。CHF急性恶化可刺激垂体，激活肾素-血管紧张素-醛固酮系统（renin-angiotensin-aldosterone system，RAAS），进而增加促肾上腺皮质激素（adrenocorticotropin, ACTH）和精氨酸血管升压素（arginine vasopressin，AVP）的释放，从而导致钠和水潴留。肾对水的重吸收及水钠潴留至关重要，

AVP增加肾集合管细胞的透水性，促使集合管更多地重吸收水分并返回血液中。

AVP能作用于肾集合管中的V2受体，并调节AQP2的表达和亚细胞定位[40, 41]。AQP2是反映肾浓缩和稀释功能的关键性标志物。AVP可促使AQP2在数分钟内从胞内囊泡转移至顶端质膜（apical plasma membrane，APM），并在数小时到数天内诱导AQP2蛋白水平升高[42, 43]。CHF大鼠肾AQP2的表达显著增加，而AQP1和AQP3表达水平保持不变[44]。同时，CHF患者尿液中AQP2排泄量亦显著增加[45]。研究发现，血浆AVP水平、肾AQP2表达和CHF严重程度之间存在密切的相关性。V2受体拮抗剂托瓦普坦能够降低CHF大鼠肾的AQP2蛋白水平[46]。美国食品药品监督管理局（FDA）在2009年批准托瓦普坦用于治疗CHF相关的低钠血症。未来，AQP2有望成为预测托瓦普坦疗效的理想指标，并为临床治疗提供指导[47]。

3. 高血压与AQP 高血压作为一种常见的心血管疾病，可引发心脏病、脑卒中、高血压视网膜病变及慢性肾疾病。血压（blood pressure，BP）的调节受多种因素的影响，如周围阻力、血管弹性、血量和心输出量等。

近来人们日益关注AQP在高血压病理生理中的作用。在自发性高血压大鼠（spontaneously hypertensive rat，SHR）模型中，肾集合管细胞的AQP2表达上调，与此同时，AVP诱导cAMP通路激活[48, 49]。醋酸去氧皮质酮（deoxycorticosterone acetate，DOCA）-盐诱导高血压大鼠模型也出现了相似的结果[50]。采用AVP V2受体拮抗剂可降低血压和尿渗透压，同时降低SHR模型及对照组中尿液AQP2水平，表明AQP2和AVP与SHR模型组中的血压增高有关[51]。

与对照WKY大鼠相比，SHR大鼠髓质中AQP2、AQP1和AQP3的表达显著增加[48]。然而，Klein Fukuoka在2006年的研究发现，血管紧张素Ⅱ或去甲肾上腺素诱导急性高血压后，肾髓质中AQP2表达降低[52]。高血压可导致脑内AQP的表达模式改变。与对照WKY大鼠相比[53]，SHR大鼠脉络丛上皮、前额皮质、纹状体和海马体中AQP4表达增高。AQP表达增加可能影响血脑屏障和血脑脊液屏障间的水转运，导致血压急性升高和血脑屏障损伤。

根据Herrera在2007年的研究，AQP1转运内皮细胞NO的中位滤过浓度（$K_{1/2}$）为0.54μmol/L，采用干扰小RNA（siRNA）抑制AQP1的表达能减少44%的NO滤过[11]。进一步研究证明，AQP1能将NO从内皮细胞中转送至血管平滑肌细胞，从而参与内皮依赖的血管舒张过程[4]。然而，AQP1缺失的人群和基因敲除小鼠并未出现高血压现象，因此需进一步收集证据来证实AQP1在高血压中的作用[54]。

4. 血管生成与AQP 血管形成过程受多种因素的调节，如VEGF、血小板衍生生长因子（platelet-derived growth factor，PDGF）、转化生长因子β（transforming growth factor-β，TGF-β）、成纤维细胞生长因子（fibroblast growth factor，FGF）、血管生成素（angiopoietin，ANG）、Notch和Wnt通路等。诸多疾病，如肿瘤、眼科疾病和伤口愈合等均与血管生成密切相关。研究已证实，AQP参与血管生成，尤其是与肿瘤有关的血管生成。肿瘤血管生成涉及数个主要步骤：基质分解、内皮细胞增殖、迁移和分化，以及内皮周围细胞的补充[55]。

在脑癌和胶质母细胞瘤中，AQP的表达与肿瘤分化度呈正相关[56, 57]。在其他癌症中，如乳腺癌、脑癌和多发性骨髓瘤，AQP的高表达可能导致局部水肿，并加剧基质崩解[58,59]。

诸多体内外多种细胞研究均证实，细胞迁移依赖AQP的参与。Saadoun的研究发现，肿瘤微血管内皮细胞中AQP1表达上调能够促进细胞迁移，且其表达水平与肿瘤微血管密度呈正相关。抑制AQP1的表达能够减少内皮细胞迁移，从而抑制肿瘤血管生成和生长[16]。表达AQP1的肿瘤细胞具有更强的转移潜能和局部浸润力。另一方面，缺乏AQP1的近曲小管上皮细胞，以及缺乏AQP3的角膜上皮细胞、肠细胞和皮肤角质形成细胞，均存在细胞迁移障碍[60]。此外，缺失AQP4可减缓激活状态星形胶质细胞的迁移，进而影响脑损伤后胶质瘢痕的形成[53]。关于AQP增强细胞迁移的机制，目前尚在研究之中。

针对AQP和肿瘤血管生成之间关系的研究，为肿瘤治疗提供了理论依据。通过药物抑制AQP表达及其介导的水转运，可能对癌细胞增殖、迁移、转移和血管生成等过程产生影响。

（三）结论和展望

鉴于AQP家族的多样性和复杂性，需要多种AQP共同协作才能发挥血管正常的生理功能。今后有必要从整合水平进一步研究AQP在血管中的表达和功能。AQP在众多疾病的发生发展和血管功能调节过程中发挥重要作用，因此，阐明AQP异常表达与血管疾病的联系具有重要的临床意义，有望为治疗血管疾病提供全新的策略和方法。

二、呼吸系统中的水通道蛋白

AQP是一种水通道转运蛋白，在呼吸系统中发挥至关重要的作用，包括促进肺泡水转运、气道加湿、胸腔积液吸收和黏膜下腺体分泌。本部分重点介绍正常和疾病状态下在肺部表达的4种AQP，即AQP1、AQP2、AQP4和AQP5，并综述它们在多种疾病模型和转基因小鼠中的功能，以期加深对AQP在呼吸系统水转运中作用的理解。目前已知AQP参与多种肺部疾病的生理和病理生理过程，但其确切作用尚不清楚，因此有必要进一步阐明AQP在肺生理和病理生理过程中的作用。

呼吸系统由呼吸道和肺组成，呼吸道包括鼻、咽、喉、气管和各级支气管。呼吸系统的主要功能是进行气体交换，并通过血液循环为外周器官提供新陈代谢所需的氧气供给。此外，呼吸系统还具有发音、嗅觉、神经内分泌、协助静脉回心等功能。在胎儿出生前，胎肺内充满水分；分娩后，新生儿肺内水分迅速被吸收，以维持肺部相对干燥，确保足够的通气和氧合作用。然而，当出现肺水肿、胸膜腔积液时，肺或呼吸道出现水转运障碍。若水分吸收过多，呼吸道会变得相对干燥，诱发浓痰和继发性气道炎症。因此，保持肺泡、组织间隙、呼吸道和胸膜腔中的水转运平衡，对于维持正常的呼吸功能至关重要。

水转运包括三种途径：由渗透梯度引起的渗透性转运，由静水压力引起的Starling机制及水分胞饮（fluid pinocytosis）。自人类发现在细胞膜上存在控制水转运的水通道蛋白已有较长时间[61]。自AQP1首次在红细胞中被发现以来，人们已在多个系统及器官检测到AQP的表达。到目前为止，肺组织中已确认有4种AQP的表达，包括位于血管内皮和胸膜中的AQP1，位于大气道上皮的AQP3，位于小气道上皮的AQP4，以及位于Ⅰ型肺泡细胞和黏膜下腺体中的AQP5。本节将总结上述AQP在正常和疾病状态下的表达，以及在各种疾病模型和转基因小鼠中的功能，旨在加深对AQP在呼吸系统水转运中作用的认识。

（一）肺和呼吸道中AQP的表达

肺组织中已发现4种AQP，包括AQP1、AQP3、AQP4和AQP5。AQP1在肺血管内皮细胞中表达[62, 63]，同时也存在于胸膜微血管内皮细胞的顶膜和基底膜[64]。AQP3位于啮齿动物气道上皮基底细胞的基底膜、黏膜下腺体细胞膜，以及人类支气管顶膜和Ⅱ型肺泡上皮细胞。AQP4位于大鼠支气管和气管中柱状细胞的基底膜，以及人类Ⅰ型肺泡上皮细胞（alveolar epithelial cell，AEC）[65-68]。AQP5的分布区域为人类Ⅰ型AEC的顶膜，以及上呼吸道黏膜下腺中浆液细胞的顶膜，同时在小鼠Ⅱ型ACE的顶膜中也检测到了AQP5[68, 69]。此外，有研究表明羊AEC的顶膜亦存在AQP5的表达[70]。

AQP的表达与肺发育及病理过程密切相关。在分娩前后，呼吸道和肺泡上皮AQP的表达具有显著差异，与水转运的适应性有关。在气道上皮和肺泡上皮中，AQP在分娩前主要负责分泌水分，而在分娩后则转变为吸收水分，清除肺内水分以实现氧合。

针对胎肺AQP的研究大多源于动物实验。绵羊胎儿是研究影响肺生理发育的关键实验动物[70, 71]。在绵羊妊娠中期，胎肺中发现AQP1、AQP3、AQP4或AQP5的mRNA和蛋白表达[72]。大鼠胎肺在出生前表达非常少量的AQP，仅在出生前能检测到AQP1和AQP4[73-75]。糖皮质激素能促进胎鼠和新生大鼠肺内AQP1的mRNA和蛋白表达，但对于出生前控制AQP1表达的因素知之甚少[67, 74]。研究者发现皮质类固醇和β肾上腺素可诱导AQP4表达增加[75]。在小鼠出生前，AQP5 mRNA表达水平极低[73]。

虽然敲除一个或多个*AQP*基因不会直接对新生小鼠的生存产生影响[76]，但这并不意味着这种情况也适用于人类等其他物种[77]。鉴于AQP在不同物种肺组织的表达和分布存在差异，关于AQP在胎肺发育和分娩时的作用难以得出一致的结论，尤其是在具有较长妊娠期的物种如人类中。

（二）肺水转运和AQP

肺在实现气体交换的同时，还承担着多项重要生理功能，其中包括水转运、代谢和免疫防御等。肺水转运过程涉及肺泡水转运、呼吸道水转运、胸膜腔水转运及黏膜下腺体分泌等环节。

1. 肺泡水转运和AQP 肺泡与毛细血管内皮之间的水转运包括渗透性转运、因气-血屏障破坏导致的水外溢和静水压驱动的水转运。AQP1和AQP5主要分布于毛细血管内皮细胞和Ⅰ型AEC的顶膜[68, 69, 78, 79]。这种细胞定位提示AQP可能在促进水分子的转运过程中发挥作用。如前所述，妊娠期AQP表达模式会发生改变，在出生后45分钟，野生型与*AQP1*、*AQP4*、*AQP5*敲除小鼠的肺湿/干重比未见明显差异[80]，这说明缓慢的水分子吸收过程并不依赖AQP，同时上述AQP在此阶段并未充分表达。

研究显示，敲除*AQP1*和*AQP5*能有效降低渗透性水转运[76, 81]。在急性肺损伤情况下，毛细血管渗透性增强导致间质和肺泡积水，然而，敲除*AQP1*或*AQP5*并不能改变肺水肿形成和吸收率[82, 83]，这可能是因为AQP介导的水转运速度相较于毛细血管渗漏引起的水转运要慢，此时通过细胞膜的水转运相当有限[83, 84]。为研究AQP5在静水压诱导肺水肿中的作用，需要设计高压灌注联合左心房外流受阻，以模拟左心衰竭引起的肺水肿，阻断AQP5并不会影响肺动脉高压引起的肺水肿[81]。进一步研究发现，AQP1和AQP5主要促进毛细血管内皮细胞与AEC的顶膜渗透性水转运，但并不参与毛细血管渗透性和静水压驱动的水转运。

*AQP1*突变患者在过量输注生理盐水后，外周支气管水肿减轻，原因是*AQP1*突变导致毛细血管网生成缺陷，因此AQP1与静水压引起的水肿之间并无显著关联[78]。相较于野生型小鼠，敲除靠近肺泡腔小气道上皮的*AQP4*并未显著影响水转运。然而，与敲除*AQP1*相比，敲除*AQP4*可显著降低渗透性水转运，表明AQP4在促进小气道上皮水转运中具有主导作用[78]。AQP1不能掩盖AQP4的潜在功能，因为在敲除*AQP1*时，AQP4的功能变得更为显著。

2. 呼吸道水转运和AQP 呼吸道必须保持高湿度，以保护气道上皮细胞，并通过黏膜下腺体分泌黏液，促进纤毛运动，以排除吸入的外来病原体。虽然纤毛上皮细胞顶膜存在AQP3和AQP4[79]，但通过对*AQP3*和*AQP4*敲除小鼠的研究发现，AQP在气道加湿、ASL水化和等渗液体吸收方面仅发挥次要作用[80]。通过计算水转运率，发现与AQP5促进唾液腺分泌相比，干燥空气刺激气道上皮的水转运相对较慢。AQP3和AQP4在气道生理中的作用甚微，提示除非受到渗透性水转运影响，缓慢的水转运不一定依赖水通道[80]。

最近研究显示敲除*AQP3*可减少呼吸道的再上皮化[85]，原因可能是水和甘油转运减少导致上皮细胞迁移减缓[86]。AQP3在呼吸道上皮生长中的作用为其在组织修复中的应用提供了可能。

3. 胸膜腔水转运和AQP 胸膜腔在调节胸膜滑液的分泌与吸收，润滑脏/壁胸膜以利于肺扩张等方面具有重要作用。脏胸膜毛细血管过滤的水分可通过壁胸膜上的淋巴导管实现重吸收。在某些恶性肿瘤中，淋巴管阻塞可能导致胸膜腔积液。AQP1表达于脏胸膜、壁胸膜的顶膜，以及内皮细胞的顶膜[64]。研究发现，AQP1有助于胸膜腔内渗透性水转运，敲除*AQP1*可显著降低渗透性水转运。然而，AQP1并不参与清除胸膜腔的等渗液体[82, 83]，且与胸腔积液的形成或清除无直接关联[64]。

4. 黏膜下腺体分泌和AQP 黏膜下腺体位于呼吸道的黏膜下层，周围富含毛细血管和神经，以保持腺体的正常分泌。研究发现，当腺体黏膜受体受到神经或化学刺激时，胞质中cAMP水平升高激活囊性纤维跨膜转导调节因子（CFTR），诱导氯化物分泌。在此过程中，Na^+顺电势差通过胞内及旁细胞途径进入细胞，同时水顺离子渗透梯度通过AQP5转运出细胞[84]。敲除小鼠*AQP5*可显著减少腺体分泌水分，使分泌液更加黏稠[87]。研究表明，照射唾液腺或干燥综合征引起的口干与AQP5的异常分布有关[88, 89]，AQP5可以通过调控唾液分泌来改善口干综合征。因此，研究调控AQP5是否有助于促进慢性阻塞性肺疾病或支气管扩张患者的气道黏液清除具有重要意义。

（三）肺癌发展和AQP

研究发现，AQP1、AQP3、AQP4和AQP5在肺癌中呈过表达状态[86, 90-92]。在肺腺癌（adenocarcinoma，ADC）和细支气管肺癌中，AQP1的表达较肺鳞状细胞癌和正常肺组织为高[90]。AQP1位于肺癌组织毛细血管的内皮细胞中，与肿瘤血管生成有关[93, 94]。此外，AQP1亦参与肺癌细胞的侵袭过程，通过shRNA敲低*AQP1*，可抑制肺癌细胞的侵袭和迁移[94]。AQP1表达与腺癌（尤其是微乳头型）术后高转移率和存活率相关[86]。以上研究表明，AQP1可作为预测肺癌分期和组织分化的重要指标。

在非小细胞肺癌（non-small cell lung carcinoma，NSCLC）中，尤其是腺癌、高分化细支气管肺泡癌和乳头状亚型，AQP3呈过表达。研究发现，AQP3在腺癌早期阶段能够调节癌细胞的生物功能[86]。例如，通过HIF-2α-VEGF通路参与肺癌的血管生成，通过AKT-MMP通路、线粒体ATP生成和甘油摄取参与肿瘤细胞浸润[95]。在NSCLC模型中，利用shRNA敲低*AQP3*的表达能够抑制癌细胞生长，这一抗癌效果已在实验中得到证实，并将在临床研究中进一步验证[95]。AQP4与肺癌细胞侵袭有关[94]。在分化良好的肺腺癌中，高水平AQP4转录和蛋白质水平具有更好的预后[91]。在肺腺癌中AQP5表达显著增加，并与NSCLC患者的不良预后相关[96]。AQP5表达细胞会出现上皮细胞标志物的缺失，同时通过SH3结合域激活c-Src以促进上皮间质转化（epithelial-mesenchymal transition，EMT），这可能是促进癌症转移的原因[96]。过表达AQP5可以激活EGFR/ERK/p38 MAPK通路，从而促进肺癌细胞的生长和侵袭[96, 97]。AQP5磷酸化受cAMP/PKA的调控，可提高肿瘤细胞的增殖能力。在肿瘤增殖和侵袭过程中，PKA中S156位点磷酸化起着重要作用[98]。因此，可以将AQP5的S156位点作为小分子抑制剂的潜在治疗靶点。此外，还可以开发针对AQP5的特异性单克隆抗体。

（四）急性肺损伤、肺部感染与AQP

研究发现，肺损伤后AQP1和AQP5表达降低[99-101]。虽然敲除*AQP1*并未导致显著的表型变化，但敲除*AQP5*却能加重铜绿假单胞菌（绿脓杆菌，*P. aeruginosa*）感染后的肺损伤[100, 102]。这可能是因为AQP1主要表达于肺毛细血管内皮细胞，敲除*AQP1*会影响渗透性水转运，渗透性变化增加，导致毛细血管泄漏期间渗透性水转运损害，但等渗液体转运不受影响。人类*AQP1*突变不会引起肺的形态学变化，但会延缓气道周围的肺积水[81]，可能是因为毛细血管网发生改变，静水压通过水通道影响等渗液体转运。在铜绿假单胞菌感染后，AQP5缺失小鼠的肺损伤加重，这可能是呼吸道表面水特性发生变化所致[100]。AQP5的缺乏导致肺部黏液产生减少，降低了铜绿假单胞菌感染前后丝裂原活化蛋白激酶（mitogen-activated protein kinase，MAPK）及NF-κB通路的活性。

AQP1和AQP5在血-气屏障中均有表达，且两者对渗透性水转运具有促进作用，因此在急性肺损伤，尤其是在肺水肿发病过程中，AQP1和AQP5发挥关键作用。研究发现，肺损伤后AQP1和AQP5表达显著降低[99, 100]。在脂多糖（LPS）诱导的急性肺损伤模型中，敲除*AQP1*后，肺水肿的形成或清除并未出现明显差异，这表明慢速水转运或从旁细胞途径泄漏的水分可能无须AQP协助进行转运。值得注意的是，AQP主要促进渗透性水转运，而非等渗性水转运。

（五）AQP与哮喘

哮喘的特征性表现是支气管收缩增加、嗜酸性粒细胞浸润、气道黏液分泌过度以及小气道上皮水肿。免疫染色实验表明，AQP1不仅在Ⅰ型和Ⅱ型肺泡上皮细胞中表达，亦存在于气道上皮。在卵清蛋白（ovalbumin，OVA）诱导的哮喘动物模型中，AQP1和AQP5的表达上升，提示AQP1和AQP5可能参与气道上皮水肿的形成[103]。支气管激发试验反映了气道对乙酰胆碱的过度活跃和高反应性[104]。小鼠实验研究表明，敲除*AQP5*会加重吸入乙酰胆碱诱导的气道反应，并伴随气道阻力增加[105]。目前尚不清楚敲除*AQP5*降低气道反应阈值的原因。此外，*AQP5*及其他哮喘基因均位于12q染色体和小鼠15号染色体的相同位点，进一步揭示了*AQP5*在哮喘发展中的潜在作用。

（六）小结

AQP在肺泡水转运、气道加湿、胸膜腔液体吸收和黏膜下腺体分泌过程中发挥重要作用。之前研究提示，AQP涉及多种肺部疾病，但其具体作用机制尚不明确。为此，有必要进一步研究AQP在肺部生理和病理生理过程中的角色，以期为疾病治疗提供有益依据。

三、消化系统中的水通道蛋白

本部分将阐述消化系统中AQP的表达特征及其生理作用。在消化道中存在AQP1、AQP3、AQP4、AQP5和AQP8的表达。在肝脏中，AQP1、AQP5、AQP8和AQP9均有表达。值得注意的是，AQP3与腹泻和炎症性肠病有关；AQP5涉及胃癌细胞增殖和迁移；AQP9则在甘油代谢、尿素转运和肝癌中发挥重要作用。然而，关于消化系统中AQP的确切分布和功能，仍需进一步深入研究。

消化系统由消化道和消化腺组成。消化道包括口腔、咽、食管、胃、小肠和大肠等，消化腺包括大消化腺和分布在消化管壁的小消化腺。大消化腺包括3对唾液腺、胰腺和肝脏，它们含有腺细胞的分泌部以及输送消化液的导管。胰腺兼有内分泌功能，其中A细胞分泌胰高血糖素，B细胞分泌胰岛素，D细胞分泌生长抑素，PP细胞分泌胰多肽，这些激素有助于调节血糖和胃肠道的运动。此外，消化道还具有吸收功能。

进食过程中，食物在唾液的作用下进行初步消化，其渗透浓度迅速从零转变为数百毫摩尔。为适应胃肠道中渗透压的快速变化，胃和肠的细胞紧密连接（以避免粪便脱水时失水），同时分泌胃液或其他消化液以平衡内容物的渗透压[106]。内容物进入小肠后，大部分水分随溶质和营养成分被吸收。随后，内容物抵达结肠，进一步脱水形成粪便。每天大约有7.5L液体分泌至消化道，包括唾液、胃液、胆汁、胰液和肠道分泌液，同时约有9L液体被吸收[107,108]。肝在物质代谢过程中起着关键作用。消化系统中的水通道蛋白在不同部位发挥着相应的生理作用[107]。

AQP的分布与其功能密切相关。在基础状态下，AQP3和AQP4主要分布于分泌性上皮（如胃）的基底侧，而顶端水通道更多地出现在吸收性上皮（如小肠）。结肠兼有吸收和分泌水的功能，因此顶膜和基底膜都存在AQP的表达[106]。AQP9在肝脏中表达，并参与脂肪代谢。本部分将阐述消化系统中的部分重要的AQP亚型[109-111]。消化系统中水转运包括旁细胞途径、跨细胞途径、扩散以及渗透压依赖的AQP途径。由于*AQP*基因敲除小鼠的水分泌表现具有稳定性，推测AQP在消化系统中的作用可能相对局限[112,113]。

（一）消化道中的AQP

1. AQP1 AQP1在胃肠道的内皮屏障中呈多样性表达，而在上皮和黏膜中则无表达。胃体、十二指肠和升结肠中的AQP1表达较幽门丰富[114]。在肛门间质中也观察到适量的AQP1，但较难确定其特异性定位[115]。

在人体内，AQP1表达于消化道黏膜下层和固有层淋巴管的内皮细胞，以及平滑肌层毛细血管内皮细胞。在小鼠中，AQP1主要表达于毛细血管和小血管内皮细胞[62,116-119]。

根据在消化道内的定位分析，推测AQP1在胃肠道黏膜和血液之间发挥吸收性水转运通道的作用。AQP1分布于小肠绒毛中的中央乳糜管内皮细胞，这些细胞在消化食物过程中产生乳糜微粒。因此，AQP1可能涉及脂肪消化过程。已有实验证明，AQP1缺失的小鼠脂肪吸收出现障碍[107,116]。

2. AQP4 AQP4特异性地表达于胃底腺（特别是胃小凹底部）壁细胞的基底膜，与调控胃酸分泌有关。为探讨AQP4在胃酸分泌中的作用，研究者构建了*AQP4*敲除小鼠模型[112]。在AQP4缺失的小鼠胃内，壁细胞在形态学上未见显著改变，敲除*AQP4*对于基础或刺激状态下的胃液分泌无明显影响，亦不会改变pH和空腹胃内血清胃泌素浓度，说明AQP4在胃酸分泌中的作用较小[113]。

AQP4还表达于小肠隐窝细胞和结肠表面上皮细胞的基底膜，提示AQP4在结肠水转运过程中发挥作用。在AQP4缺失的小鼠中，近端结肠的透水率有所降低，然而远端结肠的透水率并未受到影响，且粪便含水量与野生型小鼠无显著差异。总体看来，表面上皮细胞中的AQP4对粪便脱水和结肠水分并无显著影响[107]。

3. AQP5 AQP5最初从唾液腺中分离出来，主要在唾液腺、泪腺和胰腺等腺组织表达。在大鼠消化道，AQP5主要存在于胃和十二指肠。在胃内，AQP5的表达主要局限于幽门分泌细胞的顶膜，几乎不涉及胃底腺。而在十二指肠腺，AQP5沿分泌细胞的顶膜分布[120]。尽管免疫组化方法尚未检测到消化系统其他组织AQP5的表达，但反转录-聚合酶链反应（reverse transcriptase-polymerase chain reaction，RT-PCR）分析结果显示，肝中存在AQP5的表达。

有报道称AQP5有助于推进多种癌症的发展和侵袭[120]。在结肠癌、肺癌、慢性髓细胞性白血病、乳腺癌及胆道癌等多种癌组织中，AQP5的表达水平上调，且与临床病理特征相关。研究显示，AQP5与胃癌的发生、分化、淋巴转移和淋巴血管侵袭密切相关[106,121]。

4. AQP8 AQP8最初在胰腺、肝和睾丸中均有发现，其转录子广泛表达于消化系统器官，包括唾液腺、小肠、结肠、胰腺和肝。在消化道中，AQP8主要分布于十二指肠、空肠以及结肠中上皮细胞的顶膜[107,122]。

*AQP8*基因敲除小鼠具有正常的外观、生存状况、生长指标、器官重量和血清学表现，但睾丸体积较大。研究发现，霍乱毒素或其激动剂对*AQP8*敲除小鼠小肠液的最大分泌量无显著影响。*AQP8*基因敲除模型对结肠水分吸收或粪便脱水影响甚微，粪便含水量变化不显著。野生型小鼠与*AQP8*基因敲除小鼠之间轻微的表型差异说明，AQP8在小肠和结肠水分吸收和分泌中作用较为有限[123]。

在三硝基苯磺酸（TNBS）诱发的克罗恩病结肠炎模型中，AQP8的表达随炎症和损伤的增加而降低[124]，提示AQP8可能与炎症性肠病存在关联。

5. AQP3 AQP3在食管、近端及远端结肠中表达显著[125]。免疫组化结果显示，AQP3分布于大鼠口腔、前胃和肛门等部位的基底膜。AQP3在上述组织中的功能尚未确定。在存在粪便的恶劣环境下，AQP3可直接向上皮细胞提供水分。在口腔和肛门等皮肤延伸区域，AQP3在维持角质层水化和皮肤弹性等方面发挥关键作用[107, 126-128]。

在人类结肠中，AQP3主要分布于结肠黏膜上皮细胞[128]，与水转运密切相关。抑制结肠中AQP3的表达可能导致腹泻。研究发现，经AQP3抑制剂（$HgCl_2$和$CuSO_4$）处理1小时后，粪便中含水量增加至对照组的4倍左右，并出现严重腹泻[129]。某些泻药可通过上调AQP3引起腹泻。之前认为硫酸镁等渗透性泻药通过增加肠道中的渗透压发挥作用，然而最新研究揭示，此类药物可能是通过上调AQP3表达产生相应效果。另外，双糖酰等刺激性泻药可促进肠道蠕动，但在大鼠实验中，其作用为导致AQP3表达下调并使动物出现严重腹泻，而肠道渗透压并未改变。此外，AQP3还可能与便秘有关。吗啡是临床上常用的麻醉镇痛剂，由于其可引起肠蠕动减少而引发便秘，此时AQP3表达上调可能参与了这一过程。深入阐明AQP3水转运机制为未来开发新型泻药和抗腹泻药物提供了新的选择[129]。

在TNBS诱发的结肠炎模型中，AQP3和AQP8表达均有所降低，与此同时，肠道出现炎症和损伤。研究发现，溃疡性结肠炎（IBD）大鼠经过小肠切除和肠道功能改善后，恢复期AQP3表达出现上调[128]，这表明AQP3可能参与了炎症性肠病的发病过程[124, 125]。

研究发现，敲除小鼠*AQP3*基因会导致大肠杆菌C25在体内的转位显著上升，进而损害肠道屏障完整性[130]。

6. AQP10 关于AQP10在消化系统中的具体定位，目前尚无明确共识。有报道，小肠上皮细胞存在未剪切的AQP10片段，同时，AQP10也被发现存在于十二指肠黏膜下层毛细血管的内皮细胞[131]。此外，有研究表明，在某些物种中，AQP10可能仅充当假基因的角色[107]。

（二）消化腺中的AQP

1. AQP1 AQP1分布于胆囊、肝内胆管细胞和肝管基底膜、肝内皮屏障以及胰腺的胰管和中心细胞[115]。在肝脏，AQP1特异性表达于小叶内和小叶间导管，对细胞间的水转运起到调节作用。

2. AQP5 AQP5通常在唾液腺、泪腺和胰腺表达，参与液体分泌，并与干燥综合征（Sjögren syndrome）和糖尿病等疾病相关。在唾液腺内，AQP5存在于顶膜和腺泡细胞间的分泌小管。相较于野生型小鼠，*AQP5*敲除小鼠唾液分泌减少且出现高渗性现象，表明AQP5在唾液分泌中具有关键作用[107]。在干燥综合征模型中，AQP1表达增加，而AQP5表达减少[132]。然而，也有研究发现原发性干燥综合征患者唾液腺中AQP5的分布和密度与正常人无明显差异。因此，有必要进一步探讨AQP5在原发性干燥综合征发病机制中的作用[133]。

在胰腺中AQP5的功能可能与糖尿病和胰腺炎的发病机制有关[134]。

3. AQP8 AQP8主要分布于腮腺、唾液腺、肝脏及胰腺。在大鼠的腮腺、下颌下腺和舌下腺，AQP8存在于围绕腺泡和夹层导管周围的肌上皮细胞，但在腺泡或导管细胞中未检测到AQP8[135]。

基因敲除小鼠是研究AQP8功能的有效工具。在唾液腺中，*AQP8*转录子的表达水平较高，然而，采用免疫荧光或免疫印迹技术未能检测到其蛋白的表达。与野生型小鼠相比，敲除*AQP8*并未对唾液分泌产生影响。同时，*AQP8/AQP5*双敲除小鼠和*AQP5*敲除小鼠的唾液分泌也未受到明显影响。在肝脏中，AQP8主要分布在肝细胞的胞内囊泡中。在高脂肪喂养下，*AQP8*敲除小鼠未出现脂肪泻、血脂、肝功能或胰酶异常，仅血浆甘油三酯和胆固醇略有上升，提示AQP8与肝脏、胆或胰腺功能无直接关联[128]。

AQP8仅在人类胰腺腺泡细胞的顶膜中表达，可能与胰液分泌功能有关。

4. AQP7　　在人类小肠和结肠表面上皮细胞均存在AQP7。在大鼠，AQP7分布于绒毛中肠上皮细胞的顶区，以及结肠和盲肠上皮细胞，这提示AQP7可能参与绒毛上皮的快速水转运过程[136]。

5. AQP9　　在肝内，AQP9位于肝细胞的基底膜（窦状小管）[137, 138]，这是亚砷酸盐进入哺乳动物细胞的主要途径，其积累可导致肝细胞损伤，诱发肝细胞癌。胆管结扎诱发的肝外胆汁淤积可导致肝细胞基底膜中AQP9表达下调，表明其可能与胆汁转运相关。

AQP9与肝脏的甘油代谢密切相关。甘油是脂质溶解过程中甘油三酯的产物，经门静脉进入肝脏，随后参与葡萄糖合成。AQP9特异性地表达于门静脉窦状小管，是肝脏中甘油吸收的唯一通道。

*AQP9*敲除模型有助于研究AQP9在甘油代谢中的作用。与*AQP9*杂合子小鼠相比，*AQP9*敲除小鼠出现明显的高血糖和高甘油三酯血症。当*AQP9*敲除小鼠与Leprdb/Leprdb肥胖和2型糖尿病模型小鼠交配后，Leprdb/Leprdb *AQP9*敲除小鼠的血糖水平较Leprdb/Leprdb *AQP9*杂合子小鼠为低。同时，Leprdb/Leprdb *AQP9*敲除小鼠的血浆甘油水平亦低于Leprdb/Leprdb *AQP9*杂合子小鼠。以上结果提示，AQP9可能在肝脏甘油吸收及葡萄糖代谢中发挥作用[138]。

6. AQP12　　AQP12选择性地表达于胰腺腺泡细胞，然而，*AQP12*基因敲除小鼠胰腺相关功能未出现明显异常。因此，有必要进一步研究以揭示AQP12的功能[131]。

四、泌尿系统中的水通道蛋白

肾存在多种AQP，其中AQP1分布于近曲小管、髓袢降支细支段和肾直小血管，AQP2～AQP6位于集合管，AQP7位于近曲小管，AQP8位于近曲小管和集合管，AQP11则位于近曲小管的内质网。AQP2受血管升压素调控，在影响尿液浓缩能力的肾病中具有重要意义。运用*AQP*敲除小鼠模型，研究人员明确了AQP在肾跨上皮水转运中的作用，并揭示了AQP在肾生理及在基础和临床研究中的应用潜力。

泌尿系统包括肾、输尿管、膀胱和尿道。肾作为泌尿系统的核心器官，负责重新吸收水分和浓缩尿液，产生的尿液流经肾盂、输尿管和膀胱，最终经尿道排出。

肾调节水的分泌和重吸收，实现尿液的浓缩和稀释。在此过程中，近曲小管、髓袢细段、远端小管和集合管等透水组织负责水的分泌和重吸收，而髓袢升支等非透水组织则有助于建立肾皮质到髓质的渗透梯度。抗利尿激素（antidiuretic hormone，ADH）作用于远端小管和集合管，改变其透水性，进而调节尿液浓度。水通道蛋白在水转运过程中起到关键作用，促使肾小管、集合管和肾直小血管中上皮和内皮的水分顺利通过。

（一）AQP在泌尿系统中的定位

肾内存在至少9种AQP：AQP1分布于近曲小管、髓袢降支细段和肾直小血管；AQP2～AQP6位于集合管；AQP7位于近端小管；AQP8位于近端小管和集合管；AQP11位于近端小管细胞的内质网[139-143]。

哺乳动物的尿道上皮通常是不透水的，但在特定条件下，尿道上皮也能够表达AQP2和AQP3，承担水分子和溶质的有效转运[144, 145]。

1. AQP1 AQP1分布于近曲小管和髓袢降支细段（thin descending limb of Henle，TDLH）的基底膜和顶膜，以及髓质直小血管降支（descending vasa recta，DVR）的血管内皮[114, 146-154]。AQP1分布于近端小管除起始部分之外的S1～S3段[151, 155]。此外，AQP1在输尿管和膀胱中的毛细血管和动脉的内皮细胞亦有表达[115, 144]。

2. AQP2 AQP2表达于从连接管开始的肾集合管主细胞。在血管升压素的调控下，AQP2可以通过融合方式从胞内囊泡转运至顶膜[156-160]。在输尿管和膀胱，AQP2分布于除了与内腔相接触之外的上皮细胞膜周边区域[161]。

3. AQP3 AQP3最初因为具有甘油转运功能而被命名为甘油固有蛋白（glycerol intrinsic protein，GLIP），在皮质基侧质膜和外髓集合管上皮表达[79, 162-167]。在尿道上皮基底层和中间层的周边细胞上，AQP3表达显著[168]。

4. AQP4 AQP4表达于内髓质集合管上皮和近端小管S3区的基底膜[169-172]。冷冻电子断层成像技术显示，AQP4单体在细胞膜上以四聚体形式存在，四聚体进一步聚集形成更高级的结构，即粒子正交阵列（orthogonal arrays of particles，OAP）[173, 174]。

5. AQP5 Procino等研究者首次证实，肾集合管中的B型闰细胞表达AQP5[175]。

6. AQP6 AQP6主要分布在集合管A型闰细胞内的囊泡[176, 177]，与H$^+$-ATPase共表达[178]。

7. AQP7 AQP7是一种甘油水通道蛋白，表达于近直小管（S3段）。与此同时，AQP1亦可在此部位表达[179, 180]。

8. AQP8 关于AQP8是否在肾中表达，目前尚有争议。Elkjaer等首次报道AQP8位于近曲小管。但尚不清楚AQP8的超微定位[181]。

9. AQP9 AQP9是一种特有的甘油水通道蛋白，仅在鸟类的尿浓缩系统中存在[182]。在哺乳动物中，AQP9分布于白细胞，能在白细胞流经内髓质的高渗环境时，有效防止白细胞发生变形[183]。

10. AQP11 AQP11并不属于典型的水通道蛋白。在肾内，AQP11主要在近端小管的细胞内表达[184, 185]。荧光标记发现，AQP11分布于AQP11转染细胞的内质网，以及*AQP11*转基因小鼠的肾组织中[186-188]。

此外，肾内可能还存在其他有待确认的AQP。

（二）泌尿系统中AQP的功能

1. AQP1 为了研究AQP1在近端小管中的作用，Verkman采用靶向基因敲除技术构建了AQP1敲除小鼠模型[189-194]。这些小鼠除了生长较慢，在生存能力、外部特征及器官形态等方面与野生型小鼠无显著区别。

经测定，AQP1敲除小鼠中近端小管S2段的跨上皮渗透水通透性（permeability，Pf）较野生型小鼠降低，这说明驱动跨细胞水转运的渗透浓度是通过AQP1完成的。相比野生型小鼠，AQP1敲除小鼠近曲小管顶端质膜囊泡的Pf（10℃）明显降低，汞剂无法抑制敲除小鼠近端小管中的低透水性。此外，AQP1敲除小鼠固有膜中Pf约为0.006cm/s（37℃），与通过膜脂质的水转运相似，表明其他AQP亚型和非AQP转运蛋白对近端小管透水作用较小。在近端小管中，细胞旁途径完成的渗透驱动跨上皮水转运占比不到20%[195]。

AQP1敲除小鼠尿量增加，表明集合管中对水的吸收减少。在TDLH和DVR中敲除AQP1可能引发逆流机制障碍，进而阻碍髓间质高渗性状态的形成。对AQP1敲除小鼠进行断水，dDAVP刺激集合管透水率（几乎可以平衡尿液和髓质间渗透压）并没有增加尿液渗透压[189, 196]。鉴于NaCl转运通道功能正常，且集合管具有透水性能，即使AQP1敲除小鼠，仍可完成适度的尿液浓缩。

有学者采用TDLH片段研究了AQP1在体外环境下对TDLH透水率的影响[197]。结果显示，AQP1敲除小鼠TDLH的Pf显著下降，证实了AQP1是TDLH的主要水转运通道。同时，这一结果也说明水转运介导的TDLH渗透平衡对于肾逆流浓缩的实现具有积极作用。

关于水的重吸收和溶质进入对TDLH渗透平衡的影响，长期以来存在争议[198]。研究发现，AQP1基因敲除小鼠表现出尿液浓缩缺陷，TDLH的透水率下降，表明TDLH中的高透水性在尿液浓缩过程中具有重要作用。敲除AQP1不会对TDLH的NaCl和尿素渗透性产生影响，因此，流出TDLH管腔外的渗透水转运在逆流倍增机制中具有重要作用，而溶质本身并不足以形成高浓度的尿液[195]。

在经历断水36小时后，AQP1敲除小鼠的平均体重下降了35%，血液渗透压上升至517mOsm/kg H_2O，而野生型小鼠的平均体重降幅为20%～22%，血液渗透性增加到311～325mOsm/kg H_2O。值得注意的是，几乎所有严重高渗缺水的AQP1敲除小鼠都可通过口服水的方式得到恢复，且无并发症发生[189]。

对AQP1敲除小鼠进行尿液渗透压的测量，发现在断水前后每8小时一次的检测中，其数值普遍低于650mOsm/kg H_2O。这一现象是由于缺失AQP1导致无法形成渗透梯度，从而使尿液渗透压保持在较低水平。相比之下，断水后野生型小鼠的尿液渗透压从1400mOsm/kg H_2O显著增加到3000mOsm/kg H_2O。在大多数断水的AQP1敲除小鼠中，尿钠含量低于10mmol/L。V2受体激动剂dDAVP并不能升高AQP1敲除小鼠的尿液渗透压，表明AQP1敲除小鼠的尿浓缩缺陷不是由中央渗透感应器感应[189]。AQP1基因敲除小鼠所出现的尿液浓缩缺陷，主要可能源于近端小管等渗性水重吸收障碍以及髓质逆流倍增机制受到了破坏[189, 199, 200]。

2. AQP2 在基础状态下，AQP2位于肾集合管上皮细胞内的囊泡。当受到ADH的刺激时，AQP2通过囊泡-胞外融合的方式从胞内转移到顶膜[201]。顶膜的透水率受AQP2的转运调节[202-209]。由于AQP3和AQP4在基底膜持续存在，当AQP2出现在顶膜时，水分子能轻易通过主细胞进行跨细胞重吸收[210]。AQP2的突变可导致肾性尿崩症（nephrogenic diabetes insipidus，NDI）[211, 212]，因此在尿浓缩中具有重要的作用。值得注意的是，选择性敲除小鼠集合管 AQP2 会导致严重的尿浓缩缺陷，而 AQP2 全基因敲除小鼠会在2周内因严重脱水而死亡[188, 213]。以上结果均证实了 AQP2 在肾水转运中的关键作用。

脱水或高钠血症可导致AVP水平升高，AVP与集合管上皮细胞基底膜中的V2受体结合，进而通过升高胞内cAMP水平激活蛋白激酶A（protein kinase A，PKA）[214-218]。激活的PKA可使AQP2 C端Ser256位点发生磷酸化，并促使AQP2从胞内囊泡转移至顶膜[3, 219-229]。

AQP2的Ser256位点磷酸化对其转位至细胞表面至关重要[210, 230, 231]。此外，其他潜在的磷酸化位点，如Ser261、Ser264和Ser269，也可能对AQP2的转位产生影响[232-235]。对大鼠内髓集合管细胞的磷酸化蛋白分析显示，AQP2磷酸化主要发生在Ser261位点[236, 237]。ADH引起Ser256的单磷酸化和Ser256和Ser261的双磷酸化增加，表明这两个位点的磷酸化与AQP2转位密切相关[238]。免疫荧光检测显示，AQP2的Ser261磷酸化主要发生在细胞内，而非内质网、高尔基体和溶酶体[239]。Ser261点突变分析表明，Ser261磷酸化对AQP2转位无明显影响[240]。

AQP2的细胞内转位使跨细胞透水率增加，促进尿液浓缩。当恢复正常状态后，AQP2通过泛素介导的胞饮作用，存储在胞内囊泡或者进行降解[216]。

3. AQP3 在 AQP3 敲除小鼠中，尽管围产期存活率和出生后生长状况均保持正常，但显著表现出多尿症状，其消耗和排泄的液体量比野生型小鼠高出10倍。值得注意的是，AQP3敲除小鼠的平均尿液渗透压（262mOsm/kg H_2O）显著低于野生型小鼠（1270mOsm/kg H_2O）。在使用dDAVP和断水36小时后，AQP3敲除小鼠的尿液渗透浓度显著增加，但仍比野生型小鼠显著降低。这些结果揭示了AQP3敲除小鼠中NDI的独特表现，尽管由于利尿冲洗作用，导致髓间渗透压低于野生型小鼠，但AQP3敲除小鼠的逆流交换基本保持正常。进一步的研究表明，AQP1/AQP3双敲除小鼠中，AQP1引起的逆流交换缺陷或AQP3引起的集合管功能异常，将导致NDI出现不同症状[241]。

采用空间滤波显微镜对皮质集合管基底膜的Pf进行测量，结果显示，AQP3敲除小鼠皮质集合管的中位渗透平衡时间为2.7秒，显著长于野生型小鼠的1.1秒。这一现象可能是由于敲除AQP3，导致了皮质集合管基底膜的水通透性降低[242]，从而使得水分子的交换速率低于野生型小鼠。实验证据表明，AQP3在尿液浓缩中起重要作用，通过将水分子转运至集合管上皮基底膜，从而有助于尿液的浓缩。

4. AQP4 基因敲除小鼠被广泛应用于评估AQP4在尿液浓缩中的作用[243, 244]。在水合小鼠（hydrated mice）中，尿液渗透浓度未出现显著改变。野生型与AQP4敲除小鼠的血清钠浓度及渗透压相似。然而，在连续断水36小时后，AQP4缺失小鼠的最大尿液渗透压低于野生型小鼠，表明AQP4缺失小鼠出现了中度尿浓缩障碍[243, 245, 246]。在断水18～48小时后，利用血管升压素并进行灌注内髓集合管（IMCD），野生型和AQP4敲除小鼠基底膜跨上皮Pf分别为0.056cm/s和0.013cm/s，这提示AQP4在IMCD条件下负责基底膜中大部分的水转运。

尽管AQP4敲除小鼠的IMCD透水率显著下降，但其尿液浓缩能力仅受到轻微影响[195]，这与基于正常沿集合管水转运通道分布的预期相符。对抗糖尿病条件下的啮齿动物进行的穿刺研究发现，水分主要在皮质而非髓质集合管进行重吸收[247]。

5. AQP5 根据文献报道，*AQP5*敲除小鼠的肾功能与野生型小鼠无异[248]。然而，关于AQP5在肾中的具体作用，尚需要进一步研究和确定。

6. AQP6 AQP6是一种典型的AQP，但其特性与其他AQP有所不同。在应用AQP抑制剂Hg^{2+}或在酸性环境下，AQP6对水和离子的渗透性呈现增加趋势[146, 249-251]，提示AQP6可能与集合管的泌酸功能存在关联。

关于AQP6在肾小管中的转运作用，目前尚无明确认识。闰细胞富含线粒体，为细胞正常功能运作提供所需能量[252]。闰细胞中存在运输质子（H^+）-ATPase的胞内囊泡[251]，而AQP6即分布于此类囊泡中。研究表明，H^+-ATPase根据酸碱度变化从胞内囊泡转移到顶膜，但闰细胞细胞膜中并未发现AQP6[176, 253]，AQP6在细胞膜中的缺乏提示其在细胞内发挥作用。越来越多的研究表明，AQP6可能并未直接参与简单水转运（simple fluid transport），但可能在维持胞内酸碱平衡中具有一定的作用[254, 255]，具体机制有待深入阐明。

7. AQP7 相较于野生型小鼠外髓质囊泡的Pf（20×10^{-3}cm/s），*AQP7*敲除小鼠Pf（18×10^{-3}cm/s）未出现显著减少[256]，这提示AQP7对近直小管的透水性影响较小。根据对*AQP1/AQP7*双敲除小鼠的研究，AQP7对近直小管透水率仅为AQP1的1/8[189]。

在实验中，*AQP7*基因敲除小鼠并未出现尿液浓缩缺陷。然而，与*AQP1*敲除小鼠5.7ml的尿量相比，*AQP1/AQP7*双敲除小鼠的尿量（7.3ml）显著增加，说明在近直小管中，通过AQP7重吸收的水量相当可观。

*AQP7*敲除小鼠的血清甘油水平（0.036mg/ml）略低于野生型小鼠（0.04mg/ml）。但*AQP7*敲除小鼠尿液中的甘油含量（1.7mg/ml）显著高于野生型小鼠（0.005mg/ml），这表明AQP7在近直小管中负责甘油的重吸收[256-259]。尽管在肾水转运中作用较小，但AQP7仍然是肾吸收甘油的主要途径。

AQP7具有氨转运功能，但部分转运机制尚无法解释[9, 260]。在近端小管中，谷氨酰胺代谢产生的HCO_3^-和NH_4^+被释放至管液中。NH_4^+从近端小管细胞向外转运，进入管腔体转换为NH_3[261]。AQP7与NH_3或（和）NH_4^+分泌有关，在短时间内可调节膜两侧的氨浓度。

8. AQP8 经RT-PCR分析，在肾组织中检测到AQP8的转录子。在基础水平或断水36小时之后[123]，*AQP8*敲除小鼠与野生型小鼠的尿液渗透压无明显差异。无论是*AQP1*敲除小鼠或者*AQP8/AQP1*双敲除小鼠，其尿液渗透压均无显著差异，据此推测AQP8在尿浓缩过程中并未发挥重要作用。有研究发现，AQP8有助于促进肾中氨的转运[260, 262-264]。Molinas等通过敲除人肾近端小管HK-2细胞的*AQP8*，发现氨的释放减少，同时酸性环境下AQP8的表达有所上调[265]。

9. AQP11 *AQP11*敲除小鼠死于严重肾衰竭，说明AQP11在生物中具有至关重要的作用[185]。

（三）肾疾病与AQP

1. AQP1 常染色体显性遗传性多囊肾（autosomal dominant polycystic kidney disease，ADPKD）是人类常见的单基因疾病，全世界每4000~1000人中就有一人患病，其特征性表现是进行性增大的肾小管源性液性囊肿[266]。ADPKD的囊肿发生与囊性上皮细胞的异常增殖及囊性液体分泌密切相关，受到一系列复杂的机制调控。在人类ADPKD中，71%的肾囊肿内皮表达AQP1，其中44%位于近端小管[267]。约2/3的囊肿表达AQP1，或者肾集合管表达AQP2[268, 269]。

AQP1能够延缓肾囊肿的进展[266]。在敲除*AQP1*的多囊肾病（PKD）小鼠中，其肾大小和囊肿数量显著大于表达AQP1的PKD小鼠，这一现象源于大量的近端小管囊肿的存在。在福斯科林（forskolin）的作用下，犬肾细胞系（MDCK）细胞形成包围中央腔的囊肿。值得注意的是，稳定过表达AQP1的MDCK细胞形成的囊肿状细胞群并未出现明显的管腔。AQP1过表达能够降低β-catenin和周期蛋白D1的水平，从而抑制Wnt信号通路。β-catenin的磷酸化是其失活形式，AQP1过表达的MDCK细胞中β-catenin磷酸化水平上调，表明AQP1可促进β-catenin降解。

免疫共沉淀实验揭示了AQP1与β-catenin、GSK3β、LRP6及Axin1的相互作用。亚细胞分离实验表明，β-catenin、GSK3β和Axin1同时存于细胞质和细胞膜，而LRP6和AQP1仅分布于细胞膜。细胞膜上的"破坏信号复合体（destruction signaling complex）"与AQP1相互作用并稳定细胞膜上的"复合体"。AQP1缺失，会降低破坏复合物的稳定性，并阻止β-catenin泛素化，从而导致β-catenin募集并转移至细胞核，与TCF（一种转录因子）结合，β-catenin/TCF复合物可上调Wnt靶基因的表达。因此，AQP1可通过AQP1-巨细胞信号复合体抑制Wnt信号通路，从而在肾囊肿的发生发展过程中发挥关键作用。

研究表明，AQP1和肾、膀胱等与透水密切相关器官的肿瘤有关[270-272]。AQP1具有促进细胞生长和迁移的作用，并在肿瘤血管生成中发挥关键作用[16, 273]。尽管AQP1不能作为疾病预后的评估指标，但由于其分布和表达的独特性，因而被视为一种极好的生物标志物[274, 275]。

2. AQP2　锂治疗能降低AQP2表达，进而导致NDI的发生，其特征是肾集合管对ADH无法产生反应[159, 276]。*AQP2*突变可能引起先天性NDI[173, 237, 277-289]。已经发现与*AQP2*突变有关的常染色体隐性和常染色体显性疾病。在隐性疾病中，AQP2失去水通道功能，或因突变被错误转送至内质网[290-293]。在显性疾病中，变异的AQP2出现在异常的细胞器中，如高尔基复合体、晚期内涵体、溶酶体或基底膜等部位。

AQP2-T126M基因突变小鼠常可作为人类隐性NDI模型[294-296]。通过免疫印迹分析，在野生型小鼠的全肾组织中检测到34～40kDa的蛋白条带，证实AQP2具有复杂糖基化片段，同时也检测到29kDa非糖基化的AQP2片段。*AQP2*-T126M的核心糖基化在31kDa，大部分可被糖苷内切酶H降解[297]。*AQP2*-T126M突变小鼠出现多尿，其排出的尿液比同窝野生型小鼠增多7倍。在断水18小时后，野生型小鼠的尿液渗透浓度从1840mOsmol·L^{-1}显著增加到2872mOsmol·L^{-1}，而*AQP2*敲除小鼠的尿渗透浓度并未增加，*AQP2*-T126M突变小鼠的尿渗透浓度增加到1027mOsmol·L^{-1}[297]。17-AAG作为Hsp90抑制剂能够部分保护具有缺陷的*AQP2*-T126M细胞，这一现象表明该类似物或其他Hsp90抑制剂具备作为治疗NDI的药物的潜力[297]。

在结扎左冠状动脉引起的充血性心力衰竭（congestive heart failure，CHF）模型中，大鼠肾中AQP2表达呈上调趋势[43]。随着水渗透压的升高，质膜中AQP2的含量也显著增加[298]。肝硬化是一种与水潴留有关的临床综合征，AQP2在不同肝硬化动物模型中的表达方式存在差异[174, 299-301]。

3. AQP4 在对45名单侧输尿管梗阻以及15名肾功能损伤的儿童进行的研究中发现[302]，AQP4表达随肾功能损伤程度的加重而降低。在盐敏感的高血压患者中，AQP4表达上升，然而AQP1和AQP2的表达却呈下调趋势[303]。

4. AQP5 在对糖尿病肾病患者的肾活检中，均发现了AQP5，而在正常对照组中却未检测到该物质，提示AQP5与糖尿病肾病有关[304]。AQP5的表达上调可能导致多尿，这可能与AQP2的膜定位破坏有关。尿液AQP5被视为糖尿病肾病的一种潜在的新型标志物[305]。

5. AQP6 AQP6的表达随肾细胞癌和嗜酸细胞瘤的发展过程发生改变，有望成为肾癌诊断的潜在标志物[306, 307]。

6. AQP11 在 *AQP11* 敲除小鼠中观察到了与人类多囊肾病（polycystic kidney disease，PKD）相似的肾肿大、贫血及多囊等表现。然而，与人类PKD不同的是，*AQP11* 敲除小鼠的肾髓质中未出现囊肿，而是在皮质中出现大量囊肿，同时AQP11表达水平升高。

研究表明，*AQP11* 敲除小鼠可出现近曲小管细胞空泡化和内质网管腔扩张[185]。在此过程中，内质网应激反应基因（如 *Hsppa5* 和 *Hsp90b1*）的表达增加，同时在空泡化的近曲小管中检测到TUNEL阳性细胞和caspase-3阳性细胞。此外，空泡细胞Ki-67呈阳性且有EGFR表达，提示近曲小管出现内质网应激，进而导致凋亡和细胞增殖[188]。以上结果表明，敲除内质网中 *AQP11* 会影响内质网功能，可能导致内质网应激和肾损伤。

Inoue等研究者发现，与肾囊肿的情况类似，*AQP11* 敲除小鼠中多囊蛋白-1的表达升高，多囊素-2的表达下降，近曲小管的初级纤毛出现拉长现象[187]。内质网功能障碍导致多囊蛋白-1的N-糖基化异常，进而影响其转位至初级纤毛，表明 *AQP11* 的缺失将导致多囊素-1损伤，间接引发PKD[188]。尽管肾囊肿和AQP11之间的关系仍存在诸多悬而未决的问题，但通过进一步分析，有望最终确定AQP11的生理功能。

（陆蔚天）

参 考 文 献

[1] Verkman AS, Anderson MO, Papadopoulos MC. Aquaporins: important but elusive drug targets[J]. Nat Rev Drug Discov, 2014, 13（4）: 259-277.

[2] Gomes D, Agasse A, Thiébaud P, et al. Aquaporins are multifunctional water and solute transporters highly divergent in living organisms[J]. Biochim Biophys Acta, 2009, 1788（6）: 1213-1228.

[3] Day RE, Kitchen P, Owen DS, et al. Human aquaporins: regulators of transcellular water flow[J]. Biochim Biophys Acta, 2014, 1840（5）: 1492-1506.

[4] Herrera M, Garvin JL. Novel role of AQP-1 in NO-dependent vasorelaxation[J]. Am J Physiol Renal Physiol, 2007, 292（5）: F1443-F1451.

[5] Verkman AS. Aquaporins in endothelia[J]. Kidney Int, 2006, 69（7）: 1120-1123.

[6] Au CG, Cooper ST, Lo HP, et al. Expression of aquaporin 1 in human cardiac and skeletal muscle[J]. J Mol Cell Cardiol, 2004, 36（5）: 655-662.

[7] Rutkovskiy A, Valen G, Vaage J. Cardiac aquaporins[J]. Basic Res Cardiol, 2013, 108（6）: 393.

[8] Nakhoul NL, Davis BA, Romero MF, et al. Effect of expressing the water channel aquaporin-1 on the CO_2 permeability of *Xenopus* oocytes[J]. Am J Physiol, 1998, 274（2）: C543-C548.

[9] Geyer RR, Musa-Aziz R, Qin X, et al. Relative CO_2/NH_3 selectivities of mammalian aquaporins 0-9[J].

Am J Physiol Cell Physiol, 2013, 304(10): C985-C994.

[10] Almasalmeh A, Krenc D, Wu B, et al. Structural determinants of the hydrogen peroxide permeability of aquaporins[J]. FEBS J, 2014, 281(3): 647-656.

[11] Herrera M, Hong NJ, Garvin JL. Aquaporin-1 transports NO across cell membranes[J]. Hypertension, 2006, 48(1): 157-164.

[12] Huebert RC, Vasdev MM, Shergill U, et al. Aquaporin-1 facilitates angiogenic invasion in the pathological neovasculature that accompanies cirrhosis[J]. Hepatology, 2010, 52(1): 238-248.

[13] Huebert RC, Jagavelu K, Hendrickson HI, et al. Aquaporin-1 promotes angiogenesis, fibrosis, and portal hypertension through mechanisms dependent on osmotically sensitive microRNAs[J]. Am J Pathol, 2011, 179(4): 1851-1860.

[14] Kaneko K, Yagui K, Tanaka A, et al. Aquaporin 1 is required for hypoxia-inducible angiogenesis in human retinal vascular endothelial cells[J]. Microvasc Res, 2008, 75(3): 297-301.

[15] Ruiz-Ederra J, Verkman AS. Aquaporin-1 independent microvessel proliferation in a neonatal mouse model of oxygen-induced retinopathy[J]. Invest Ophthalmol Vis Sci, 2007, 48(10): 4802-4810.

[16] Saadoun S, Papadopoulos MC, Hara-Chikuma M, et al. Impairment of angiogenesis and cell migration by targeted aquaporin-1 gene disruption[J]. Nature, 2005, 434(7034): 786-792.

[17] Xiang Y, Ma B, Li T, et al. Acetazolamide inhibits aquaporin-1 protein expression and angiogenesis[J]. Acta Pharmacol Sin, 2004, 25(6): 812-816.

[18] Chou B, Hiromatsu K, Okano S, et al. Antiangiogenic tumor therapy by DNA vaccine inducing aquaporin-1-specific CTL based on ubiquitin-proteasome system in mice[J]. J Immunol, 2012, 189(4): 1618-1626.

[19] Papadopoulos MC, Manley GT, Krishna S, et al. Aquaporin-4 facilitates reabsorption of excess fluid in vasogenic brain edema[J]. FASEB J, 2004, 18(11): 1291-1293.

[20] Oshio K, Binder DK, Yang B, et al. Expression of aquaporin water channels in mouse spinal cord[J]. Neuroscience, 2004, 127(3): 685-693.

[21] Nagelhus EA, Veruki ML, Torp R, et al. Aquaporin-4 water channel protein in the rat retina and optic nerve: polarized expression in Müller cells and fibrous astrocytes[J]. J Neurosci, 1998, 18(7): 2506-2519.

[22] Amiry-Moghaddam M, Frydenlund DS, Ottersen OP. Anchoring of aquaporin-4 in brain: molecular mechanisms and implications for the physiology and pathophysiology of water transport[J]. Neuroscience, 2004, 129(4): 999-1010.

[23] Benfenati V, Ferroni S. Water transport between CNS compartments: functional and molecular interactions between aquaporins and ion channels[J]. Neuroscience, 2010, 168(4): 926-940.

[24] Saadoun S, Papadopoulos MC, Watanabe H, et al. Involvement of aquaporin-4 in astroglial cell migration and glial scar formation[J]. J Cell Sci, 2005, 118(Pt 24): 5691-5698.

[25] Sjöholm K, Palming J, Olofsson LE, et al. A microarray search for genes predominantly expressed in human omental adipocytes: adipose tissue as a major production site of serum amyloid A[J]. J Clin Endocrinol Metab, 2005, 90(4): 2233-2239.

[26] Hibuse T, Maeda N, Nakatsuji H, et al. The heart requires glycerol as an energy substrate through aquaporin 7, a glycerol facilitator[J]. Cardiovasc Res, 2009, 83(1): 34-41.

[27] Skowronski MT, Lebeck J, Rojek A, et al. AQP7 is localized in capillaries of adipose tissue, cardiac and striated muscle: implications in glycerol metabolism[J]. Am J Physiol Renal Physiol, 2007, 292(3): F956-F965.

[28] Amiry-Moghaddam M, Lindland H, Zelenin S, et al. Brain mitochondria contain aquaporin water channels: evidence for the expression of a short AQP9 isoform in the inner mitochondrial membrane[J].

FASEB J, 2005, 19(11): 1459-1467.

[29] Badaut J, Fukuda AM, Jullienne A, et al. Aquaporin and brain diseases[J]. Biochim Biophys Acta, 2014, 1840(5): 1554-1565.

[30] Badaut J, Petit JM, Brunet JF, et al. Distribution of aquaporin 9 in the adult rat brain: preferential expression in catecholaminergic neurons and in glial cells[J]. Neuroscience, 2004, 128(1): 27-38.

[31] Badaut J, Regli L. Distribution and possible roles of aquaporin 9 in the brain[J]. Neuroscience, 2004, 129(4): 971-981.

[32] Badaut J. Aquaglyceroporin 9 in brain pathologies[J]. Neuroscience, 2010, 168(4): 1047-1057.

[33] Ribeiro Mde C, Hirt L, Bogousslavsky J, et al. Time course of aquaporin expression after transient focal cerebral ischemia in mice[J]. J Neurosci Res, 2006, 83(7): 1231-1240.

[34] Hirt L, Ternon B, Price M, et al. Protective role of early aquaporin 4 induction against postischemic edema formation[J]. J Cereb Blood Flow Metab, 2009, 29(2): 423-433.

[35] Hirt L, Fukuda AM, Ambadipudi K, et al. Improved long-term outcome after transient cerebral ischemia in aquaporin-4 knockout mice[J]. J Cereb Blood Flow Metab, 2017, 37(1): 277-290.

[36] Akdemir G, Ratelade J, Asavapanumas N, et al. Neuroprotective effect of aquaporin-4 deficiency in a mouse model of severe global cerebral ischemia produced by transient 4-vessel occlusion[J]. Neurosci Lett, 2014, 574: 70-75.

[37] Katada R, Akdemir G, Asavapanumas N, et al. Greatly improved survival and neuroprotection in aquaporin-4-knockout mice following global cerebral ischemia[J]. FASEB J, 2014, 28(2): 705-714.

[38] Yao X, Derugin N, Manley GT, et al. Reduced brain edema and infarct volume in aquaporin-4 deficient mice after transient focal cerebral ischemia[J]. Neurosci Lett, 2015, 584: 368-372.

[39] Yang M, Gao F, Liu H, et al. Temporal changes in expression of aquaporin-3, -4, -5 and -8 in rat brains after permanent focal cerebral ischemia[J]. Brain Res, 2009, 1290: 121-132.

[40] Hasler U, Mordasini D, Bens M, et al. Long term regulation of aquaporin-2 expression in vasopressin-responsive renal collecting duct principal cells[J]. J Biol Chem, 2002, 277(12): 10379-10386.

[41] Zelenina M, Christensen BM, Palmér J, et al. Prostaglandin E_2 interaction with AVP: effects on AQP2 phosphorylation and distribution[J]. Am J Physiol Renal Physiol, 2000, 278(3): F388-F394.

[42] Nielsen S, Chou CL, Marples D, et al. Vasopressin increases water permeability of kidney collecting duct by inducing translocation of aquaporin-CD water channels to plasma membrane[J]. Proc Natl Acad Sci USA, 1995, 92(4): 1013-1017.

[43] Xu DL, Martin PY, Ohara M, et al. Upregulation of aquaporin-2 water channel expression in chronic heart failure rat[J]. J Clin Invest, 1997, 99(7): 1500-1505.

[44] Yu CM, Wing-Hon Lai K, Li PS, et al. Normalization of renal aquaporin-2 water channel expression by fosinopril, valsartan, and combination therapy in congestive heart failure: a new mechanism of action[J]. J Mol Cell Cardiol, 2004, 36(3): 445-453.

[45] Funayama H, Nakamura T, Saito T, et al. Urinary excretion of aquaporin-2 water channel exaggerated dependent upon vasopressin in congestive heart failure[J]. Kidney Int, 2004, 66(4): 1387-1392.

[46] Veeraveedu PT, Watanabe K, Ma M, et al. Effects of nonpeptide vasopressin V2 antagonist tolvaptan in rats with heart failure[J]. Biochem Pharmacol, 2007, 74(10): 1466-1475.

[47] Imamura T, Kinugawa K. Urine aquaporin-2: a promising marker of response to the arginine vasopressin type-2 antagonist, tolvaptan in patients with congestive heart failure[J]. Int J Mol Sci, 2016, 17(1): 105.

[48] Lee J, Kim S, Kim J, et al. Increased expression of renal aquaporin water channels in spontaneously hypertensive rats[J]. Kidney Blood Press Res, 2006, 29(1): 18-23.

[49] Kim SW, Wang W, Kwon TH, et al. Increased expression of ENaC subunits and increased apical

targeting of AQP2 in the kidneys of spontaneously hypertensive rats[J]. Am J Physiol Renal Physiol, 2005, 289(5): F957-F968.

[50] Lee J, Kang DG, Kim Y. Increased expression and shuttling of aquaporin-2 water channels in the kidney in DOCA-salt hypertensive rats[J]. Clin Exp Hypertens, 2000, 22(5): 531-541.

[51] Buemi M, Nostro L, Di Pasquale G, et al. Aquaporin-2 water channels in spontaneously hypertensive rats[J]. Am J Hypertens, 2004, 17(12 Pt 1): 1170-1178.

[52] Klein JD, Murrell BP, Tucker S, et al. Urea transporter UT-A1 and aquaporin-2 proteins decrease in response to angiotensin Ⅱ or norepinephrine-induced acute hypertension[J]. Am J Physiol Renal Physiol, 2006, 291(5): F952-F959.

[53] Tomassoni D, Bramanti V, Amenta F. Expression of aquaporins 1 and 4 in the brain of spontaneously hypertensive rats[J]. Brain Res, 2010, 1325: 155-163.

[54] Kone BC. NO break-ins at water gate[J]. Hypertension, 2006, 48(1): 29-30.

[55] Clapp C, Martínez de la Escalera G. Aquaporin-1: a novel promoter of tumor angiogenesis[J]. Trends Endocrinol Metab, 2006, 17(1): 1-2.

[56] Warth A, Kröger S, Wolburg H. Redistribution of aquaporin-4 in human glioblastoma correlates with loss of agrin immunoreactivity from brain capillary basal laminae[J]. Acta Neuropathol, 2004, 107(4): 311-318.

[57] Saadoun S, Papadopoulos MC, Davies DC, et al. Aquaporin-4 expression is increased in oedematous human brain tumours[J]. J Neurol Neurosurg, 2002, 72(2): 262-265.

[58] Endo M, Jain RK, Witwer B, et al. Water channel (aquaporin 1) expression and distribution in mammary carcinomas and glioblastomas[J]. Microvasc Res, 1999, 58(2): 89-98.

[59] Saadoun S, Papadopoulos MC, Davies DC, et al. Increased aquaporin 1 water channel expression in human brain tumours[J]. Br J Cancer, 2002, 87(6): 621-623.

[60] Papadopoulos MC, Saadoun S, Verkman AS. Aquaporins and cell migration[J]. Pflugers Arch, 2008, 456(4): 693-700.

[61] Krane CM, Fortner CN, Hand AR, et al. Aquaporin 5-deficient mouse lungs are hyperresponsive to cholinergic stimulation[J]. Proc Natl Acad Sci USA, 2001, 98(24): 14114-14119.

[62] Nielsen S, Smith BL, Christensen EI, et al. Distribution of the aquaporin CHIP in secretory and resorptive epithelia and capillary endothelia[J]. Proc Natl Acad Sci USA, 1993, 90(15): 7275-7279.

[63] Folkesson HG, Matthay MA, Hasegawa H, et al. Transcellular water transport in lung alveolar epithelium through mercury-sensitive water channels[J]. Proc Natl Acad Sci USA, 1994, 91(11): 4970-4974.

[64] Song Y, Yang B, Matthay MA, et al. Role of aquaporin water channels in pleural fluid dynamics[J]. Am J Physiol Cell Physiol, 2000, 279(6): C1744-C1750.

[65] Kreda SM, Gynn MC, Fenstermacher DA, et al. Expression and localization of epithelial aquaporins in the adult human lung[J]. Am J Respir Cell Mol Biol, 2001, 24(3): 224-234.

[66] Frigeri A, Gropper MA, Umenishi F, et al. Localization of MIWC and GLIP water channel homologs in neuromuscular, epithelial and glandular tissues[J]. J Cell Sci, 1995, 108(Pt 9): 2993-3002.

[67] King LS, Nielsen S, Agre P. Aquaporins in complex tissues. Ⅰ. Developmental patterns in respiratory and glandular tissues of rat[J]. Am J Physiol, 1997, 273(5): C1541-C1548.

[68] Nielsen S, King LS, Christensen BM, et al. Aquaporins in complex tissues. Ⅱ. Subcellular distribution in respiratory and glandular tissues of rat[J]. Am J Physiol, 1997, 273(5): C1549-C1561.

[69] Funaki H, Yamamoto T, Koyama Y, et al. Localization and expression of AQP5 in cornea, serous salivary glands, and pulmonary epithelial cells[J]. Am J Physiol, 1998, 275(4): C1151-C1157.

[70] Lipsett J, Cool JC, Runciman SI, et al. Effect of antenatal tracheal occlusion on lung development in the sheep model of congenital diaphragmatic hernia: a morphometric analysis of pulmonary structure and

maturity[J]. Pediatr Pulmonol, 1998, 25(4): 257-269.

[71] Hooper SB, Harding R. Fetal lung liquid: a major determinant of the growth and functional development of the fetal lung[J]. Clin Exp Pharmacol Physiol, 1995, 22(4): 235-247.

[72] Liu H, Hooper SB, Armugam A, et al. Aquaporin gene expression and regulation in the ovine fetal lung[J]. J Physiol, 2003, 551(Pt 2): 503-514.

[73] Ruddy MK, Drazen JM, Pitkanen OM, et al. Modulation of aquaporin 4 and the amiloride-inhibitable sodium channel in perinatal rat lung epithelial cells[J]. Am J Physiol, 1998, 274(6): L1066-L1072.

[74] Yasui M, Serlachius E, Löfgren M, et al. Perinatal changes in expression of aquaporin-4 and other water and ion transporters in rat lung[J]. J Physiol, 1997, 505(Pt 1)(Pt 1): 3-11.

[75] Umenishi F, Carter EP, Yang B, et al. Sharp increase in rat lung water channel expression in the perinatal period[J]. Am J Respir Cell Mol Biol, 1996, 15(5): 673-679.

[76] Verkman AS, Yang B, Song Y, et al. Role of water channels in fluid transport studied by phenotype analysis of aquaporin knockout mice[J]. Exp Physiol, 2000, 85 Spec No: 233S-241S.

[77] King LS, Yasui M. Aquaporins and disease: lessons from mice to humans[J]. Trends Endocrinol Metab, 2002, 13(8): 355-360.

[78] Bai C, Fukuda N, Song Y, et al. Lung fluid transport in aquaporin-1 and aquaporin-4 knockout mice[J]. J Clin Invest, 1999, 103(4): 555-561.

[79] Frigeri A, Gropper MA, Turck CW, et al. Immunolocalization of the mercurial-insensitive water channel and glycerol intrinsic protein in epithelial cell plasma membranes[J]. Proc Natl Acad Sci USA, 1995, 92(10): 4328-4331.

[80] Song Y, Jayaraman S, Yang B, et al. Role of aquaporin water channels in airway fluid transport, humidification, and surface liquid hydration[J]. J Gen Physiol, 2001, 117(6): 573-582.

[81] Ma T, Fukuda N, Song Y, et al. Lung fluid transport in aquaporin-5 knockout mice[J]. J Clin Invest, 2000, 105(1): 93-100.

[82] 蒋进军,白春学,洪群英,等.水通道与钠通道在小鼠胸腔液体转运中的作用[J].中华结核和呼吸杂志, 2003, 26(1): 26-29.

[83] Jiang JJ, Bai CX, Hong QY, et al. Effect of aquaporin-1 deletion on pleural fluid transport[J]. Acta Pharmacol Sin, 2003, 24(4): 301-305.

[84] Ma T, Song Y, Gillespie A, et al. Defective secretion of saliva in transgenic mice lacking aquaporin-5 water channels[J]. J Biol Chem, 1999, 274(29): 20071-20074.

[85] Levin MH, Verkman AS. Aquaporin-3-dependent cell migration and proliferation during corneal re-epithelialization[J]. Invest Ophthalmol Vis Sci, 2006, 47(10): 4365-4372.

[86] Machida Y, Ueda Y, Shimasaki M, et al. Relationship of aquaporin 1, 3, and 5 expression in lung cancer cells to cellular differentiation, invasive growth, and metastasis potential[J]. Hum Pathol, 2011, 42(5): 669-678.

[87] Song Y, Verkman AS. Aquaporin-5 dependent fluid secretion in airway submucosal glands[J]. J Biol Chem, 2001, 276(44): 41288-41292.

[88] Li Z, Zhao D, Gong B, et al. Decreased saliva secretion and down-regulation of AQP5 in submandibular gland in irradiated rats[J]. Radiat Res, 2006, 165(6): 678-687.

[89] Wang D, Iwata F, Muraguchi M, et al. Correlation between salivary secretion and salivary AQP5 levels in health and disease[J]. J Med Invest, 2009, 56(Suppl): 350-353.

[90] Hoque MO, Soria JC, Woo J, et al. Aquaporin 1 is overexpressed in lung cancer and stimulates NIH-3T3 cell proliferation and anchorage-independent growth[J]. Am J Pathol, 2006, 168(4): 1345-1353.

[91] Warth A, Muley T, Meister M, et al. Loss of aquaporin-4 expression and putative function in non-small

cell lung cancer[J]. BMC Cancer, 2011, 11: 161.

[92] Chae YK, Woo J, Kim MJ, et al. Expression of aquaporin 5 (AQP5) promotes tumor invasion in human non small cell lung cancer[J]. PLoS One, 2008, 3(5): e2162.

[93] López-Campos JL, Sánchez Silva R, Gómez Izquierdo L, et al. Overexpression of Aquaporin-1 in lung adenocarcinomas and pleural mesotheliomas[J]. Histol Histopathol, 2011, 26(4): 451-459.

[94] Xie Y, Wen X, Jiang Z, et al. Aquaporin 1 and aquaporin 4 are involved in invasion of lung cancer cells[J]. Clin Lab, 2012, 58(1/2): 75-80.

[95] Xia H, Ma YF, Yu CH, et al. Aquaporin 3 knockdown suppresses tumour growth and angiogenesis in experimental non-small cell lung cancer[J]. Exp Physiol, 2014, 99(7): 974-984.

[96] Zhang Z, Chen Z, Song Y, et al. Expression of aquaporin 5 increases proliferation and metastasis potential of lung cancer[J]. J Pathol, 2010, 221(2): 210-220.

[97] Xu JL, Xia R. The emerging role of aquaporin 5 (AQP5) expression in systemic malignancies[J]. Tumour Biol, 2014, 35(7): 6191-6192.

[98] Woo J, Lee J, Chae YK, et al. Overexpression of AQP5, a putative oncogene, promotes cell growth and transformation[J]. Cancer Lett, 2008, 264(1): 54-62.

[99] Su X, Song Y, Jiang J, et al. The role of aquaporin-1 (AQP1) expression in a murine model of lipopolysaccharide-induced acute lung injury[J]. Respir Physiol Neurobiol, 2004, 142(1): 1-11.

[100] Zhang ZQ, Song YL, Chen ZH, et al. Deletion of aquaporin 5 aggravates acute lung injury induced by Pseudomonas aeruginosa[J]. J Trauma, 2011, 71(5): 1305-1311.

[101] Wang F, Huang H, Lu F, et al. Acute lung injury and change in expression of aquaporins 1 and 5 in a rat model of acute pancreatitis[J]. Hepatogastroenterology, 2010, 57(104): 1553-1562.

[102] Song Y, Fukuda N, Bai C, et al. Role of aquaporins in alveolar fluid clearance in neonatal and adult lung, and in oedema formation following acute lung injury: studies in transgenic aquaporin null mice[J]. J Physiol, 2000, 525(Pt 3): 771-779.

[103] Ablimit A, Hasan B, Lu W, et al. Changes in water channel aquaporin 1 and aquaporin 5 in the small airways and the alveoli in a rat asthma model[J]. Micron, 2013, 45: 68-73.

[104] Holgate ST. Pathogenesis of asthma[J]. Clin Expl Allergy, 2008, 38(6): 872-897.

[105] Krane CM, Goldstein DL. Comparative functional analysis of aquaporins/glyceroporins in mammals and anurans[J]. Mamm Genome, 2007, 18(6/7): 452-462.

[106] Laforenza U. Water channel proteins in the gastrointestinal tract[J]. Mol Aspects Med, 2012, 33(5/6): 642-650.

[107] Matsuzaki T, Tajika Y, Ablimit A, et al. Aquaporins in the digestive system[J]. Med Electron Microsc, 2004, 37(2): 71-80.

[108] Masyuk AI, Marinelli RA, LaRusso NF. Water transport by epithelia of the digestive tract[J]. Gastroenterology, 2002, 122(2): 545-562.

[109] Ma T, Verkman AS. Aquaporin water channels in gastrointestinal physiology[J]. J Physiol, 1999, 517(Pt 2)(Pt 2): 317-326.

[110] Jiang L, Li J, Liu X, et al. Expression of aquaporin-4 water channels in the digestive tract of the guinea pig[J]. J Mol Histol, 2014, 45(2): 229-241.

[111] Pelagalli A, Squillacioti C, Mirabella N, et al. Aquaporins in health and disease: an overview focusing on the gut of different species[J]. Int J Mol Sci, 2016, 17(8): 1213.

[112] Wang KS, Komar AR, Ma T, et al. Gastric acid secretion in aquaporin-4 knockout mice[J]. Am J Physiol Gastrointest Liver Physiol, 2000, 279(2): G448-G453.

[113] Wang KS, Ma T, Filiz F, et al. Colon water transport in transgenic mice lacking aquaporin-4 water

channels[J]. Am J Physiol Gastrointest Liver Physiol, 2000, 279(2): G463-G470.

[114] Benga G. The first discovered water channel protein, later called aquaporin 1: molecular characteristics, functions and medical implications[J]. Mol Aspects Med, 2012, 33(5-6): 518-534.

[115] Mobasheri A, Marples D. Expression of the AQP-1 water channel in normal human tissues: a semiquantitative study using tissue microarray technology[J]. Am J Physiol Cell Physiol, 2004, 286(3): C529-C537.

[116] Ma T, Jayaraman S, Wang KS, et al. Defective dietary fat processing in transgenic mice lacking aquaporin-1 water channels[J]. Am J Physiol Cell Physiol, 2001, 280(1): C126-C134.

[117] De Luca A, Vassalotti G, Pelagalli A, et al. Expression and localization of aquaporin-1 along the intestine of colostrum suckling buffalo calves[J]. Anat Histol Embryol, 2015, 44(5): 391-400.

[118] Talbot NC, Garrett WM, Caperna TJ. Analysis of the expression of aquaporin-1 and aquaporin-9 in pig liver tissue: comparison with rat liver tissue[J]. Cells Tissues Organs, 2003, 174(3): 117-128.

[119] Casotti G, Waldron T, Misquith G, et al. Expression and localization of an aquaporin-1 homologue in the avian kidney and lower intestinal tract[J]. Comp Biochem Physiol A Mol Integr Physiol, 2007, 147(2): 355-362.

[120] Parvin MN, Kurabuchi S, Murdiastuti K, et al. Subcellular redistribution of AQP5 by vasoactive intestinal polypeptide in the Brunner's gland of the rat duodenum[J]. Am J Physiol Gastrointest Liver Physiol, 2005, 288(6): G1283-G1291.

[121] Huang YH, Zhou XY, Wang HM, et al. Aquaporin 5 promotes the proliferation and migration of human gastric carcinoma cells[J]. Tumour Biol, 2013, 34(3): 1743-1751.

[122] Fischer H, Stenling R, Rubio C, et al. Differential expression of aquaporin 8 in human colonic epithelial cells and colorectal tumors[J]. BMC Physiol, 2001, 1: 1.

[123] Yang B, Song Y, Zhao D, et al. Phenotype analysis of aquaporin-8 null mice[J]. Am J Physiol Cell Physiol, 2005, 288(5): C1161-C1170.

[124] Zhao G, Li J, Wang J, et al. Aquaporin 3 and 8 are down-regulated in TNBS-induced rat colitis[J]. Biochem Biophys Res Commun, 2014, 443(1): 161-166.

[125] Zhao GX, Dong PP, Peng R, et al. Expression, localization and possible functions of aquaporins 3 and 8 in rat digestive system[J]. Biotech Histochem, 2016, 91(4): 269-276.

[126] Silberstein C, Kierbel A, Amodeo G, et al. Functional characterization and localization of AQP3 in the human colon[J]. Braz J Med Biol Res, 1999, 32(10): 1303-1313.

[127] Koyama Y, Yamamoto T, Tani T, et al. Expression and localization of aquaporins in rat gastrointestinal tract[J]. Am J Physiol, 1999, 276(3): C621-C627.

[128] Ikarashi N, Kon R, Sugiyama K. Aquaporins in the colon as a new therapeutic target in diarrhea and constipation[J]. Int J Mol Sci, 2016, 17(7): 1172.

[129] Ikarashi N, Kon R, Iizasa T, et al. Inhibition of aquaporin-3 water channel in the colon induces diarrhea[J]. Biol Pharm Bull, 2012, 35(6): 957-962.

[130] Zhang W, Xu Y, Chen Z, et al. Knockdown of aquaporin 3 is involved in intestinal barrier integrity impairment[J]. FEBS Lett, 2011, 585(19): 3113-3119.

[131] Ishibashi K. New members of mammalian aquaporins: AQP10-AQP12[J]. Handb Exp Pharmacol, 2009, (190): 251-262.

[132] Wu X, Ren C, Zhou H, et al. Therapeutic effect of Zeng Ye decoction on primary Sjögren's syndrome via upregulation of aquaporin-1 and aquaporin-5 expression levels[J]. Mol Med Rep, 2014, 10(1): 429-434.

[133] Beroukas D, Hiscock J, Jonsson R, et al. Subcellular distribution of aquaporin 5 in salivary glands in

primary Sjögren's syndrome[J]. Lancet, 2001, 358（9296）：1875-1876.

[134] Delporte C. Aquaporins in secretory glands and their role in Sjögren's syndrome[J]. Handb Exp Pharmacol, 2009, （190）：185-201.

[135] Delporte C, Steinfeld S. Distribution and roles of aquaporins in salivary glands[J]. Biochim Biophys Acta, 2006, 1758（8）：1061-1070.

[136] Laforenza U, Gastaldi G, Grazioli M, et al. Expression and immunolocalization of aquaporin-7 in rat gastrointestinal tract[J]. Biol Cell, 2005, 97（8）：605-613.

[137] Okada S, Misaka T, Matsumoto I, et al. Aquaporin-9 is expressed in a mucus-secreting goblet cell subset in the small intestine[J]. FEBS Lett, 2003, 540（1-3）：157-162.

[138] Maeda N. Implications of aquaglyceroporins 7 and 9 in glycerol metabolism and metabolic syndrome[J]. Mol Aspects Med, 2012, 33（5-6）：665-675.

[139] Kortenoeven ML, Fenton RA. Renal aquaporins and water balance disorders[J]. Biochim Biophys Acta, 2014, 1840（5）：1533-1549.

[140] Michalek K. Aquaglyceroporins in the kidney：present state of knowledge and prospects[J]. J Physiol Pharmacol, 2016, 67（2）：185-193.

[141] Nielsen S, Frør J, Knepper MA. Renal aquaporins：key roles in water balance and water balance disorders[J]. Curr Opin Nephrol Hypertens, 1998, 7（5）：509-516.

[142] Verkman AS. Role of aquaporin water channels in kidney and lung[J]. Am J Med Sci, 1998, 316（5）：310-320.

[143] Yamamoto T, Sasaki S. Aquaporins in the kidney：emerging new aspects[J]. Kidney Int, 1998, 54（4）：1041-1051.

[144] Spector DA, Wade JB, Dillow R, et al. Expression, localization, and regulation of aquaporin-1 to-3 in rat urothelia[J]. Am J Physiol Renal Physiol, 2002, 282（6）：F1034-F1042.

[145] Spector DA, Yang Q, Liu J, et al. Expression, localization, and regulation of urea transporter B in rat urothelia[J]. Am J Physiol Renal Physiol, 2004, 287（1）：F102-F108.

[146] Agre P, Nielsen S. The aquaporin family of water channels in kidney[J]. Nephrologie, 1996, 17（7）：409-415.

[147] Bedford JJ, Leader JP, Walker RJ. Aquaporin expression in normal human kidney and in renal disease[J]. J Am Soc Nephro, 2003, 14（10）：2581-2587.

[148] Ma T, Frigeri A, Tsai ST, et al. Localization and functional analysis of CHIP28k water channels in stably transfected Chinese hamster ovary cells[J]. J Biol Chem, 1993, 268（30）：22756-22764.

[149] Maunsbach AB, Marples D, Chin E, et al. Aquaporin-1 water channel expression in human kidney[J]. J Am Soc Nephrol, 1997, 8（1）：1-14.

[150] Nielsen S, Agre P. The aquaporin family of water channels in kidney[J]. Kidney Int, 1995, 48（4）：1057-1068.

[151] Nielsen S, Pallone T, Smith BL, et al. Aquaporin-1 water channels in short and long loop descending thin limbs and in descending vasa recta in rat kidney[J]. Am J Physiol, 1995, 268（6 Pt 2）：F1023-F1037.

[152] Nielsen S, Smith BL, Christensen EI, et al. CHIP28 water channels are localized in constitutively water-permeable segments of the nephron[J]. J Cell Biol, 1993, 120（2）：371-383.

[153] Verkman AS, Shi LB, Frigeri A, et al. Structure and function of kidney water channels[J]. Kidney Int, 1995, 48（4）：1069-1081.

[154] Zhang R, Skach W, Hasegawa H, et al. Cloning, functional analysis and cell localization of a kidney proximal tubule water transporter homologous to CHIP28[J]. J Cell Biol, 1993, 120（2）：359-369.

[155] Maeda Y, Smith BL, Agre P, et al. Quantification of Aquaporin-CHIP water channel protein in microdissected renal tubules by fluorescence-based ELISA[J]. J Clin Invest, 1995, 95（1）: 422-428.

[156] Christensen BM, Marples D, Jensen UB, et al. Acute effects of vasopressin V2-receptor antagonist on kidney AQP2 expression and subcellular distribution[J]. Am J Physiol, 1998, 275（2）: F285-F297.

[157] Coleman RA, Wu DC, Liu J, et al. Expression of aquaporins in the renal connecting tubule[J]. Am J Physiol Renal Physiol, 2000, 279（5）: F874-F883.

[158] Kishore BK, Terris JM, Knepper MA. Quantitation of aquaporin-2 abundance in microdissected collecting ducts: axial distribution and control by AVP[J]. Am J Physiol, 1996, 271（1 Pt 2）: F62-F70.

[159] Marples D, Christensen S, Christensen EI, et al. Lithium-induced downregulation of aquaporin-2 water channel expression in rat kidney medulla[J]. J Clin Invest, 1995, 95（4）: 1838-1845.

[160] Sabolić I, Katsura T, Verbavatz JM, et al. The AQP2 water channel: effect of vasopressin treatment, microtubule disruption, and distribution in neonatal rats[J]. J Membr Biol, 1995, 143（3）: 165-175.

[161] Mobasheri A, Wray S, Marples D. Distribution of AQP2 and AQP3 water channels in human tissue microarrays[J]. J Mol Histol, 2005, 36（1-2）: 1-14.

[162] Baum MA, Ruddy MK, Hosselet CA, et al. The perinatal expression of aquaporin-2 and aquaporin-3 in developing kidney[J]. Pediatr Res, 1998, 43（6）: 783-790.

[163] Ecelbarger CA, Terris J, Frindt G, et al. Aquaporin-3 water channel localization and regulation in rat kidney[J]. Am J Physiol, 1995, 269（5 Pt 2）: F663-F672.

[164] Echevarria M, Windhager EE, Tate SS, et al. Cloning and expression of AQP3, a water channel from the medullary collecting duct of rat kidney[J]. Proc Natl Acad Sci USA, 1994, 91（23）: 10997-11001.

[165] Hara-Chikuma M, Verkman AS. Physiological roles of glycerol-transporting aquaporins: the aquaglyceroporins[J]. Cell Mol Life Sci, 2006, 63（12）: 1386-1392.

[166] Ishibashi K, Sasaki S, Fushimi K, et al. Immunolocalization and effect of dehydration on AQP3, a basolateral water channel of kidney collecting ducts[J]. Am J Physiol, 1997, 272（2 Pt 2）: F235-F241.

[167] Kwon TH, Nielsen J, Masilamani S, et al. Regulation of collecting duct AQP3 expression: response to mineralocorticoid[J]. Am J Physiol Renal Physiol, 2002, 283（6）: F1403-F1421.

[168] Rubenwolf PC, Georgopoulos NT, Clements LA, et al. Expression and localisation of aquaporin water channels in human urothelium in situ and in vitro[J]. Eur Urol, 2009, 56（6）: 1013-1023.

[169] Deen PM, van Os CH. Epithelial aquaporins[J]. Curr Opin Cell Biol, 1998, 10（4）: 435-442.

[170] Kim YH, Earm JH, Ma T, et al. Aquaporin-4 expression in adult and developing mouse and rat kidney[J]. J Am Soc Nephrol, 2001, 12（9）: 1795-1804.

[171] Terris J, Ecelbarger CA, Marples D, et al. Distribution of aquaporin-4 water channel expression within rat kidney[J]. Am J Physiol, 1995, 269（6 Pt 2）: F775-F785.

[172] van Hoek AN, Ma T, Yang B, et al. Aquaporin-4 is expressed in basolateral membranes of proximal tubule S3 segments in mouse kidney[J]. Am J Physiol Renal Physiol, 2000, 278（2）: F310-F316.

[173] Schrier RW, Fassett RG, Ohara M, et al. Pathophysiology of renal fluid retention[J]. Kidney Int Suppl, 1998, 67: S127-S132.

[174] Fernández-Llama P, Turner R, Dibona G, et al. Renal expression of aquaporins in liver cirrhosis induced by chronic common bile duct ligation in rats[J]. J Am Soc Nephrol, 1999, 10（9）: 1950-1957.

[175] Procino G, Mastrofrancesco L, Sallustio F, et al. AQP5 is expressed in type-B intercalated cells in the collecting duct system of the rat, mouse and human kidney[J]. Cell Physiol Biochem, 2011, 28（4）: 683-692.

[176] Rabaud NE, Song L, Wang Y, et al. Aquaporin 6 binds calmodulin in a calcium-dependent manner[J]. Biochem Biophys Res Commun, 2009, 383（1）: 54-57.

[177] Yasui M, Kwon TH, Knepper MA, et al. Aquaporin-6: an intracellular vesicle water channel protein in renal epithelia[J]. Proc Natl Acad Sci USA, 1999, 96(10): 5808-5813.

[178] Yasui M, Hazama A, Kwon TH, et al. Rapid gating and anion permeability of an intracellular aquaporin[J]. Nature, 1999, 402(6758): 184-187.

[179] Ishibashi K, Imai M, Sasaki S. Cellular localization of aquaporin 7 in the rat kidney[J]. Exp Nephrol, 2000, 8(4-5): 252-257.

[180] Nejsum LN, Elkjaer M, Hager H, et al. Localization of aquaporin-7 in rat and mouse kidney using RT-PCR, immunoblotting, and immunocytochemistry[J]. Biochem Biophys Res Commun, 2000, 277(1): 164-170.

[181] Elkjaer ML, Nejsum LN, Gresz V, et al. Immunolocalization of aquaporin-8 in rat kidney, gastrointestinal tract, testis, and airways[J]. Am J Physiol Renal Physiol, 2001, 281(6): F1047-F1057.

[182] Nishimura H, Yang Y. Aquaporins in avian kidneys: function and perspectives[J]. Am J Physiol Regul Integr Comp Physiol, 2013, 305(11): R1201-R1214.

[183] Takata K, Matsuzaki T, Tajika Y. Aquaporins: water channel proteins of the cell membrane[J]. Prog Histochem Cytochem, 2004, 39(1): 1-83.

[184] Atochina-Vasserman EN, Biktasova A, Abramova E, et al. Aquaporin 11 insufficiency modulates kidney susceptibility to oxidative stress[J]. Am J Physiol Renal Physiol, 2013, 304(10): F1295-F1307.

[185] Morishita Y, Matsuzaki T, Hara-chikuma M, et al. Disruption of aquaporin-11 produces polycystic kidneys following vacuolization of the proximal tubule[J]. Mol Cell Biol, 2005, 25(17): 7770-7779.

[186] Ikeda M, Andoo A, Shimono M, et al. The NPC motif of aquaporin-11, unlike the NPA motif of known aquaporins, is essential for full expression of molecular function[J]. J Biol Chem, 2011, 286(5): 3342-3350.

[187] Inoue Y, Sohara E, Kobayashi K, et al. Aberrant glycosylation and localization of polycystin-1 cause polycystic kidney in an AQP11 knockout model[J]. J Am Soc Nephrol, 2014, 25(12): 2789-2799.

[188] Matsuzaki T, Yaguchi T, Shimizu K, et al. The distribution and function of aquaporins in the kidney: resolved and unresolved questions[J]. Anat Sci Int, 2017, 92(2): 187-199.

[189] Ma T, Yang B, Gillespie A, et al. Severely impaired urinary concentrating ability in transgenic mice lacking aquaporin-1 water channels[J]. J Biol Chem, 1998, 273(8): 4296-4299.

[190] Schnermann J, Chou CL, Ma T, et al. Defective proximal tubular fluid reabsorption in transgenic aquaporin-1 null mice[J]. Proc Natl Acad Sci USA, 1998, 95(16): 9660-9664.

[191] Verkman AS. Dissecting the roles of aquaporins in renal pathophysiology using transgenic mice[J]. Semin Nephrol, 2008, 28(3): 217-226.

[192] Verkman AS. Knock-out models reveal new aquaporin functions[J]. Handb Exp Pharmacol, 2009, (190): 359-381.

[193] Yang B, Folkesson HG, Yang J, et al. Reduced osmotic water permeability of the peritoneal barrier in aquaporin-1 knockout mice[J]. Am J Physiol, 1999, 276(1): C76-C81.

[194] Yang B, Ma T, Dong JY, et al. Partial correction of the urinary concentrating defect in aquaporin-1 null mice by adenovirus-mediated gene delivery[J]. Hum Gene Ther, 2000, 11(4): 567-575.

[195] Verkman AS. Lessons on renal physiology from transgenic mice lacking aquaporin water channels[J]. J Am Soc Nephrol, 1999, 10(5): 1126-1135.

[196] Cai Q, McReynolds MR, Keck M, et al. Vasopressin receptor subtype 2 activation increases cell proliferation in the renal medulla of AQP1 null mice[J]. Am J Physiol Renal Physiol, 2007, 293(6): F1858-F1864.

[197] Chou CL, Knepper MA, Hoek AN, et al. Reduced water permeability and altered ultrastructure in thin descending limb of Henle in aquaporin-1 null mice[J]. J Clin Invest, 1999, 103(4): 491-496.

[198] Moore LC, Marsh DJ. How descending limb of Henle's loop permeability affects hypertonic urine formation[J]. Am J Physiol, 1980, 239(1): F57-F71.

[199] Jen JF, Stephenson JL. Externally driven countercurrent multiplication in a mathematical model of the urinary concentrating mechanism of the renal inner medulla[J]. Bull Math Biol, 1994, 56(3): 491-514.

[200] Knepper MA. Molecular physiology of urinary concentrating mechanism: regulation of aquaporin water channels by vasopressin[J]. Am J Physiol, 1997, 272(1Pt 2): F3-F12.

[201] Tajika Y, Matsuzaki T, Suzuki T, et al. Immunohistochemical characterization of the intracellular pool of water channel aquaporin-2 in the rat kidney[J]. Anat Sci Int, 2002, 77(3): 189-195.

[202] Boone M, Deen PM T. Physiology and pathophysiology of the vasopressin-regulated renal water reabsorption[J]. Pflugers Arch, 2008, 456(6): 1005-1024.

[203] Brown D. The ins and outs of aquaporin-2 trafficking[J]. Am J Physiol Renal Physiol, 2003, 284(5): F893-F901.

[204] Eto K, Noda Y, Horikawa S, et al. Phosphorylation of aquaporin-2 regulates its water permeability[J]. J Biol Chem, 2010, 285(52): 40777-40784.

[205] Katsura T, Verbavatz JM, Farinas J, et al. Constitutive and regulated membrane expression of aquaporin 1 and aquaporin 2 water channels in stably transfected LLC-PK1 epithelial cells[J]. Proc Natl Acad Sci USA, 1995, 92(16): 7212-7216.

[206] Noda Y, Sasaki S. Regulation of aquaporin-2 trafficking and its binding protein complex[J]. Biochim Biophys Acta, 2006, 1758(8): 1117-1125.

[207] Valenti G, Procino G, Tamma G, et al. Minireview: aquaporin 2 trafficking[J]. Endocrinology, 2005, 146(12): 5063-5070.

[208] Yamamoto T, Sasaki S, Fushimi K, et al. Vasopressin increases AQP-CD water channel in apical membrane of collecting duct cells in Brattleboro rats[J]. Am J Physiol, 1995, 268(6 Pt 1): C1546-C1551.

[209] Zhang XY, Wang B, Guan YF. Nuclear receptor regulation of aquaporin-2 in the kidney[J]. Int J Mol Sci, 2016, 17(7): 1105.

[210] Takata K, Matsuzaki T, Tajika Y, et al. Localization and trafficking of aquaporin 2 in the kidney[J]. Histochem Cell Biol, 2008, 130(2): 197-209.

[211] Bichet DG, Bockenhauer D. Genetic forms of nephrogenic diabetes insipidus(NDI): vasopressin receptor defect(X-linked)and aquaporin defect(autosomal recessive and dominant)[J]. Best Pract Res Clin Endocrinol Metab, 2016, 30(2): 263-276.

[212] Deen PM, Verdijk MA, Knoers NV, et al. Requirement of human renal water channel aquaporin-2 for vasopressin-dependent concentration of urine[J]. Science, 1994, 264(5155): 92-95.

[213] Rojek A, Füchtbauer EM, Kwon TH, et al. Severe urinary concentrating defect in renal collecting duct-selective AQP2 conditional-knockout mice[J]. Proc Natl Acad Sci USA, 2006, 103(15): 6037-6042.

[214] Bichet DG. Vasopressin receptors in health and disease[J]. Kidney Int, 1996, 49(6): 1706-1711.

[215] Bichet DG. Vasopressin receptor mutations in nephrogenic diabetes insipidus[J]. Semin Nephrol, 2008, 28(3): 245-251.

[216] Frick A, Eriksson UK, de Mattia F, et al. X-ray structure of human aquaporin 2 and its implications for nephrogenic diabetes insipidus and trafficking[J]. Proc Natl Acad Sci USA, 2014, 111(17): 6305-6310.

[217] Kuwahara M, Fushimi K, Terada Y, et al. cAMP-dependent phosphorylation stimulates water permeability of aquaporin-collecting duct water channel protein expressed in *Xenopus* oocytes[J]. J Biol

Chem, 1995, 270(18): 10384-10387.

[218] Robben JH, Sze M, Knoers NV, et al. Functional rescue of vasopressin V2 receptor mutants in MDCK cells by pharmacochaperones: relevance to therapy of nephrogenic diabetes insipidus[J]. Am J Physiol Renal Physiol, 2007, 292(1): F253-F260.

[219] Katsura T, Gustafson CE, Ausiello DA, et al. Protein kinase A phosphorylation is involved in regulated exocytosis of aquaporin-2 in transfected LLC-PK1 cells[J]. Am J Physiol, 1997, 272(6 Pt 2): F817-F822.

[220] Moeller HB, Fenton RA. Cell biology of vasopressin-regulated aquaporin-2 trafficking[J]. Pflugers Arch, 2012, 464(2): 133-144.

[221] Moeller HB, Olesen ET, Fenton RA. Regulation of the water channel aquaporin-2 by posttranslational modification[J]. Am J Physiol Renal Physiol, 2011, 300(5): F1062-F1073.

[222] Nedvetsky PI, Tamma G, Beulshausen S, et al. Regulation of aquaporin-2 trafficking[J]. Handb Exp Pharmacol, 2009, (190): 133-157.

[223] Nejsum LN, Zelenina M, Aperia A, et al. Bidirectional regulation of AQP2 trafficking and recycling: involvement of AQP2-S256 phosphorylation[J]. Am J Physiol Renal Physiol, 2005, 288(5): F930-F938.

[224] Procino G, Carmosino M, Tamma G, et al. Extracellular calcium antagonizes forskolin-induced aquaporin 2 trafficking in collecting duct cells[J]. Kidney Int, 2004, 66(6): 2245-2255.

[225] Sasaki S, Yui N, Noda Y. Actin directly interacts with different membrane channel proteins and influences channel activities: AQP2 as a model[J]. Biochim Biophys Acta, 2014, 1838(2): 514-520.

[226] van Balkom BW, Savelkoul PJ, Markovich D, et al. The role of putative phosphorylation sites in the targeting and shuttling of the aquaporin-2 water channel[J]. J Biol Chem, 2002, 277(44): 41473-41479.

[227] Vukićević T, Schulz M, Faust D, et al. The trafficking of the water channel aquaporin-2 in renal principal cells—a potential target for pharmacological intervention in cardiovascular diseases[J]. Front Pharmacol, 2016, 7: 23.

[228] Whiting JL, Ogier L, Forbush KA, et al. AKAP220 manages apical actin networks that coordinate aquaporin-2 location and renal water reabsorption[J]. Proc Natl Acad Sci USA, 2016, 113(30): E4328-E4337.

[229] Yui N, Lu HA, Chen Y, et al. Basolateral targeting and microtubule-dependent transcytosis of the aquaporin-2 water channel[J]. Am J Physiol Cell Physiol, 2013, 304(1): C38-C48.

[230] Carney EF. Cell biology: vasopressin-independent AQP2 trafficking[J]. Nat Rev Nephrol, 2016, 12(9): 509.

[231] Fushimi K, Sasaki S, Marumo F. Phosphorylation of serine 256 is required for cAMP-dependent regulatory exocytosis of the aquaporin-2 water channel[J]. J Biol Chem, 1997, 272(23): 14800-14804.

[232] Cheung PW, Nomura N, Nair A V, et al. EGF receptor inhibition by erlotinib increases aquaporin 2-mediated renal water reabsorption[J]. J Am Soc Nephrol, 2016, 27(10): 3105-3116.

[233] Fenton RA, Moeller HB, Hoffert JD, et al. Acute regulation of aquaporin-2 phosphorylation at Ser-264 by vasopressin[J]. Proc Natl Acad Sci USA, 2008, 105(8): 3134-3139.

[234] Hoffert JD, Fenton RA, Moeller HB, et al. Vasopressin-stimulated increase in phosphorylation at Ser269 potentiates plasma membrane retention of aquaporin-2[J]. J Biol Chem, 2008, 283(36): 24617-24627.

[235] Park EJ, Kwon TH. A minireview on vasopressin-regulated aquaporin-2 in kidney collecting duct cells[J]. Electrolyte Blood Press, 2015, 13(1): 1-6.

[236] Hoffert JD, Pisitkun T, Saeed F, et al. Dynamics of the G protein-coupled vasopressin V2 receptor signaling network revealed by quantitative phosphoproteomics[J]. Mol Cell Proteomics, 2012, 11(2): M111.014613.

[237] Robben JH, Knoers NV, Deen PM. Cell biological aspects of the vasopressin type-2 receptor and aquaporin 2 water channel in nephrogenic diabetes insipidus[J]. Am J Physiol Renal Physiol, 2006, 291(2): F257-F270.

[238] Hoffert JD, Pisitkun T, Wang G, et al. Quantitative phosphoproteomics of vasopressin-sensitive renal cells: regulation of aquaporin-2 phosphorylation at two sites[J]. Proc Natl Acad Sci USA, 2006, 103(18): 7159-7164.

[239] Hoffert JD, Nielsen J, Yu MJ, et al. Dynamics of aquaporin-2 serine-261 phosphorylation in response to short-term vasopressin treatment in collecting duct[J]. Am J Physiol Renal Physiol, 2007, 292(2): F691-F700.

[240] Lu HJ, Matsuzaki T, Bouley R, et al. The phosphorylation state of serine 256 is dominant over that of serine 261 in the regulation of AQP2 trafficking in renal epithelial cells[J]. Am J Physiol Renal Physiol, 2008, 295(1): F290-F294.

[241] Yang B, Ma T, Verkman AS. Erythrocyte water permeability and renal function in double knockout mice lacking aquaporin-1 and aquaporin-3[J]. J Biol Chem, 2001, 276(1): 624-628.

[242] Ma T, Song Y, Yang B, et al. Nephrogenic diabetes insipidus in mice lacking aquaporin-3 water channels[J]. Proc Natl Acad Sci USA, 2000, 97(8): 4386-4391.

[243] Ma T, Yang B, Gillespie A, et al. Generation and phenotype of a transgenic knockout mouse lacking the mercurial-insensitive water channel aquaporin-4[J]. J Clin Invest, 1997, 100(5): 957-962.

[244] Verbavatz JM, Ma T, Gobin R, et al. Absence of orthogonal arrays in kidney, brain and muscle from transgenic knockout mice lacking water channel aquaporin-4[J]. J Cell Sci, 1997, 110(Pt22): 2855-2860.

[245] Chou CL, Ma T, Yang B, et al. Fourfold reduction of water permeability in inner medullary collecting duct of aquaporin-4 knockout mice[J]. Am J Physiol, 1998, 274(2): C549-C554.

[246] Yang B, van Hoek AN, Verkman AS. Very high single channel water permeability of aquaporin-4 in baculovirus-infected insect cells and liposomes reconstituted with purified aquaporin-4[J]. Biochemistry, 1997, 36(24): 7625-7632.

[247] Karlberg L, Källskog O, Ojteg G, et al. Renal medullary blood flow studied with the 86-Rb extraction method. Methodological considerations[J]. Acta Physiol Scand, 1982, 115(1): 11-18.

[248] Tamma G, Procino G, Svelto M, et al. Cell culture models and animal models for studying the pathophysiological role of renal aquaporins[J]. Cell Mol Life Sci, 2012, 69(12): 1931-1946.

[249] Hazama A, Kozono D, Guggino WB, et al. Ion permeation of AQP6 water channel protein. Single channel recordings after Hg^{2+} activation[J]. J Biol Chem, 2002, 277(32): 29224-29230.

[250] Holm LM, Klaerke DA, Zeuthen T. Aquaporin 6 is permeable to glycerol and urea[J]. Pflugers Arch, 2004, 448(2): 181-186.

[251] Promeneur D, Kwon TH, Yasui M, et al. Regulation of AQP6 mRNA and protein expression in rats in response to altered acid-base or water balance[J]. Am J Physiol Renal Physiol, 2000, 279(6): F1014-F1026.

[252] Ohshiro K, Yaoita E, Yoshida Y, et al. Expression and immunolocalization of AQP6 in intercalated cells of the rat kidney collecting duct[J]. Arch Histol Cytol, 2001, 64(3): 329-338.

[253] Madsen KM, Tisher CC. Response of intercalated cells of rat outer medullary collecting duct to chronic metabolic acidosis[J]. Lab Invest, 1984, 51(3): 268-276.

[254] Ikeda M, Beitz E, Kozono D, et al. Characterization of aquaporin-6 as a nitrate channel in mammalian cells. Requirement of pore-lining residue threonine 63[J]. J Biol Chem, 2002, 277(42): 39873-39879.

[255] Jun JG, Maeda S, Kuwahara-Otani S, et al. Expression of adrenergic and cholinergic receptors in murine renal intercalated cells[J]. J Vet Med Sci, 2014, 76(11): 1493-1500.

[256] Sohara E, Rai T, Miyazaki J, et al. Defective water and glycerol transport in the proximal tubules of

AQP7 knockout mice[J]. Am J Physiol Renal Physiol, 2005, 289（6）: F1195-F1200.
[257] Lin EC. Glycerol utilization and its regulation in mammals[J]. Annu Rev Biochem, 1977, 46: 765-795.
[258] Maeda N, Funahashi T, Hibuse T, et al. Adaptation to fasting by glycerol transport through aquaporin 7 in adipose tissue[J]. Proc Natl Acad Sci USA, 2004, 101（51）: 17801-17806.
[259] Sohara E, Rai T, Sasaki S, et al. Physiological roles of AQP7 in the kidney: lessons from AQP7 knockout mice[J]. Biochim Biophys Acta, 2006, 1758（8）: 1106-1110.
[260] Litman T, Søgaard R, Zeuthen T. Ammonia and urea permeability of mammalian aquaporins[J]. Handb Exp Pharmacol, 2009, （190）: 327-358.
[261] Koeppen BM. The kidney and acid-base regulation[J]. Adv Physiol Educ, 2009, 33（4）: 275-281.
[262] Jahn TP, Møller AL, Zeuthen T, et al. Aquaporin homologues in plants and mammals transport ammonia[J]. FEBS Lett, 2004, 574（1/2/3）: 31-36.
[263] Saparov SM, Liu K, Agre P, et al. Fast and selective ammonia transport by aquaporin-8[J]. J Biol Chem, 2007, 282（8）: 5296-5301.
[264] Soria LR, Fanelli E, Altamura N, et al. Aquaporin-8-facilitated mitochondrial ammonia transport[J]. Biochem Biophys Res Commun, 2010, 393（2）: 217-221.
[265] Molinas SM, Trumper L, Marinelli RA. Mitochondrial aquaporin-8 in renal proximal tubule cells: evidence for a role in the response to metabolic acidosis[J]. Am J Physiol Renal Physiol, 2012, 303（3）: F458-F466.
[266] Wang W, Li F, Sun Y, et al. Aquaporin-1 retards renal cyst development in polycystic kidney disease by inhibition of Wnt signaling[J]. FASEB J, 2015, 29（4）: 1551-1563.
[267] Bachinsky DR, Sabolic I, Emmanouel DS, et al. Water channel expression in human ADPKD kidneys[J]. Am J Physiol, 1995, 268（3 Pt 2）: F398.
[268] Devuyst O. The expression of water channels AQP1 and AQP2 in a large series of ADPKD kidneys[J]. Nephron, 1998, 78（1）: 116-117.
[269] Devuyst O, Burrow CR, Smith BL, et al. Expression of aquaporins-1 and-2 during nephrogenesis and in autosomal dominant polycystic kidney disease[J]. Am J Physiol, 1996, 271（1 Pt 2）: F169-F183.
[270] King LS, Agre P. Pathophysiology of the aquaporin water channels[J]. Annu Rev Physiol, 1996, 58: 619-648.
[271] Liu J, Zhang WY, Ding DG. Expression of aquaporin 1 in bladder uroepithelial cell carcinoma and its relevance to recurrence[J]. Asian Pac J Cancer Prev, 2015, 16（9）: 3973-3976.
[272] Morrissey JJ, Mobley J, Figenshau RS, et al. Urine aquaporin 1 and perilipin 2 differentiate renal carcinomas from other imaged renal masses and bladder and prostate cancer[J]. Mayo Clin Proc, 2015, 90（1）: 35-42.
[273] McCoy E, Sontheimer H. Expression and function of water channels（aquaporins）in migrating malignant astrocytes[J]. Glia, 2007, 55（10）: 1034-1043.
[274] Mazal PR, Exner M, Haitel A, et al. Expression of kidney-specific cadherin distinguishes chromophobe renal cell carcinoma from renal oncocytoma[J]. Hum Pathol, 2005, 36（1）: 22-28.
[275] Mobasheri A, Airley R, Hewitt SM, et al. Heterogeneous expression of the aquaporin 1（AQP1）water channel in tumors of the prostate, breast, ovary, colon and lung: a study using high density multiple human tumor tissue microarrays[J]. Int J Oncol, 2005, 26（5）: 1149-1158.
[276] Bichet DG. Nephrogenic diabetes insipidus[J]. Am J Med, 1998, 105（5）: 431-442.
[277] Bockenhauer D, Bichet DG. Inherited secondary nephrogenic diabetes insipidus: concentrating on humans[J]. Am J Physiol Renal Physiol, 2013, 304（8）: F1037-F1042.
[278] Chen YC, Cadnapaphornchai MA, Schrier RW. Clinical update on renal aquaporins[J]. Biol Cell, 2005,

97（6）：357-371.

[279] Dollerup P, Thomsen TM, Nejsum LN, et al. Partial nephrogenic diabetes insipidus caused by a novel AQP2 variation impairing trafficking of the aquaporin-2 water channel[J]. BMC Nephrol, 2015, 16: 217.

[280] Frøkiaer J, Marples D, Knepper MA, et al. Pathophysiology of aquaporin-2 in water balance disorders[J]. Am J Med Sci, 1998, 316（5）：291-299.

[281] Holmes RP. The role of renal water channels in health and disease[J]. Mol Aspects Med, 2012, 33（5-6）：547-552.

[282] Klein N, Kümmerer N, Hobernik D, et al. The AQP2 mutation V71M causes nephrogenic diabetes insipidus in humans but does not impair the function of a bacterial homolog[J]. FEBS Open Bio, 2015, 5: 640-646.

[283] Kwon TH, Frøkiær J, Nielsen S. Regulation of aquaporin-2 in the kidney: a molecular mechanism of body-water homeostasis[J]. Kidney Res Clin Pract, 2013, 32（3）：96-102.

[284] Lee MD, King LS, Agre P. The aquaporin family of water channel proteins in clinical medicine[J]. Medicine（Baltimore）, 1997, 76（3）：141-156.

[285] Moeller HB, Rittig S, Fenton RA. Nephrogenic diabetes insipidus: essential insights into the molecular background and potential therapies for treatment[J]. Endocr Rev, 2013, 34（2）：278-301.

[286] Nielsen S, Frøkiaer J, Marples D, et al. Aquaporins in the kidney: from molecules to medicine[J]. Physiol Rev, 2002, 82（1）：205-244.

[287] Procino G, Mastrofrancesco L, Mira A, et al. Aquaporin 2 and apical calcium-sensing receptor: new players in polyuric disorders associated with hypercalciuria[J]. Semin Nephrol, 2008, 28（3）：297-305.

[288] Tamarappoo BK, Verkman AS. Defective aquaporin-2 trafficking in nephrogenic diabetes insipidus and correction by chemical chaperones[J]. J Clin Invest, 1998, 101（10）：2257-2267.

[289] Yang B, Ma T, Xu Z, et al. cDNA and genomic cloning of mouse aquaporin-2: functional analysis of an orthologous mutant causing nephrogenic diabetes insipidus[J]. Genomics, 1999, 57（1）：79-83.

[290] Deen PM, Croes H, van Aubel RA, et al. Water channels encoded by mutant aquaporin-2 genes in nephrogenic diabetes insipidus are impaired in their cellular routing[J]. J Clin Invest, 1995, 95（5）：2291-2296.

[291] Kamsteeg EJ, Hendriks G, Boone M, et al. Short-chain ubiquitination mediates the regulated endocytosis of the aquaporin-2 water channel[J]. Proc Natl Acad Sci USA, 2006, 103（48）：18344-18349.

[292] Noda Y, Sohara E, Ohta E, et al. Aquaporins in kidney pathophysiology[J]. Nat Rev Nephrol, 2010, 6（3）：168-178.

[293] Valenti G, Frigeri A, Ronco PM, et al. Expression and functional analysis of water channels in a stably AQP2-transfected human collecting duct cell line[J]. J Biol Chem, 1996, 271（40）：24365-24370.

[294] Sauer B. Inducible gene targeting in mice using the Cre/lox system[J]. Methods, 1998, 14（4）：381-392.

[295] Yang B, Gillespie A, Carlson EJ, et al. Neonatal mortality in an aquaporin-2 knock-in mouse model of recessive nephrogenic diabetes insipidus[J]. J Biol Chem, 2001, 276（4）：2775-2779.

[296] Yang B, Zhao D, Qian L, et al. Mouse model of inducible nephrogenic diabetes insipidus produced by floxed aquaporin-2 gene deletion[J]. Am J Physiol Renal Physiol, 2006, 291（2）：F465-F472.

[297] Yang B, Zhao D, Verkman AS. Hsp90 inhibitor partially corrects nephrogenic diabetes insipidus in a conditional knock-in mouse model of aquaporin-2 mutation[J]. FASEB J, 2009, 23（2）：503-512.

[298] Nielsen S, Terris J, Andersen D, et al. Congestive heart failure in rats is associated with increased expression and targeting of aquaporin-2 water channel in collecting duct[J]. Proc Natl Acad Sci USA, 1997, 94（10）：5450-5455.

[299] Fernández-Llama P, Jimenez W, Bosch-Marcé M, et al. Dysregulation of renal aquaporins and Na-Cl cotransporter in CCl$_4$-induced cirrhosis[J]. Kidney Int, 2000, 58(1): 216-228.

[300] Fujita N, Ishikawa SE, Sasaki S, et al. Role of water channel AQP-CD in water retention in SIADH and cirrhotic rats[J]. Am J Physiol, 1995, 269(6): F926-F931.

[301] Jonassen TE, Nielsen S, Christensen S, et al. Decreased vasopressin-mediated renal water reabsorption in rats with compensated liver cirrhosis[J]. Am J Physiol, 1998, 275(2): F216-F225.

[302] Li ZZ, Xing L, Zhao ZZ, et al. Decrease of renal aquaporins 1-4 is associated with renal function impairment in pediatric congenital hydronephrosis[J]. World J Pediatr, 2012, 8(4): 335-341.

[303] Procino G, Romano F, Torielli L, et al. Altered expression of renal aquaporins and α-adducin polymorphisms may contribute to the establishment of salt-sensitive hypertension[J]. Am J Hypertens, 2011, 24(7): 822-828.

[304] Wu H, Chen L, Zhang X, et al. Aqp5 is a new transcriptional target of Dot1a and a regulator of Aqp2[J]. PLoS One, 2013, 8(1): e53342.

[305] Lu Y, Chen L, Zhao B, et al. Urine AQP5 is a potential novel biomarker of diabetic nephropathy[J]. J Diabetes Complications, 2016, 30(5): 819-825.

[306] Tan MH, Wong CF, Tan HL, et al. Genomic expression and single-nucleotide polymorphism profiling discriminates chromophobe renal cell carcinoma and oncocytoma[J]. BMC Cancer, 2010, 10: 196.

[307] Yusenko MV, Zubakov D, Kovacs G. Gene expression profiling of chromophobe renal cell carcinomas and renal oncocytomas by Affymetrix GeneChip using pooled and individual tumours[J]. Int J Biol Sci, 2009, 5(6): 517-527.

第五节 水通道蛋白的表达调控机制

AQP是对水与部分溶质（甘油、尿素等）具有跨细胞膜转运能力的小分子蛋白。迄今为止，在人体内共发现13种水通道蛋白（AQP0～AQP12）。AQP家族在维持人体多个系统器官水、电解质的运输平衡中发挥着重要作用。在缺血缺氧、创伤、感染、肿瘤等病理条件下，AQP的表达量和（或）表达部位发生异常，将介导水肿的发生；部分人体器官（如脑、脊髓、视网膜）的水肿可导致严重后果。因此，对AQP表达调控机制的研究，将有助于减轻或阻止相关疾病所造成的损伤，并对研究新的治疗策略及开发新的治疗药物有重要意义。

如前所述，在不同的损伤应激或者药物处理条件下，机体内AQP可发生磷酸化、泛素化、类泛素化、谷胱甘肽化、糖基化及其他翻译后修饰（见本章第一节），在AQP修饰中（后）将激活体内多条信号通路，最终调控AQP的表达量及表达部位，影响其功能。

本节将针对人体内各种AQP，分别介绍其表达调控的信号通路。

一、AQP1的表达调控机制

AQP1是AQP家族中发现的第一种AQP，在体内广泛分布，除了促进水分子转运外，AQP1还对于二氧化碳（CO_2）和氨（NH_3）具有极高的渗透性，是AQP家族中发现的第一种具有气体转运功能的AQP。在脑内，AQP1表达于脉络丛上皮细胞的顶膜中，在调节其所分布组织的水转运中起着重要作用，并且AQP1在细胞迁移和生长中也发挥着重要作用。

研究表明，AQP1的功能和表达可受到多种因素的影响。首先，AQP1的水通透性可被激素所调控：在爪蟾卵母细胞中发现，血管升压素处理可增加AQP1的水通透性，而心房钠尿肽（atrial natriuretic peptide，ANP）处理则可降低其水通透性[1]。其次，AQP1可发生磷酸化或泛素化修饰而影响其表达定位：在体及离体研究均表明，AQP1被蛋白激酶A（protein kinase A，PKA）磷酸化将导致其从细胞内小室运输到顶膜[2,3]；而且，蛋白激酶C（protein kinase C，PKC）通过磷酸化AQP1的Thr157和Thr239等位点，将调节AQP1通道的水渗透性和离子电导[4]；最近的一项研究表明，信号分子环磷酸腺苷（cyclic adenosine monophosphate，cAMP）和环磷酸鸟苷（cyclic guanosine monophosphate，cGMP）可促进小鼠肾近端小管细胞表达的AQP1从完整的内涵体腔（intact endosomal compartment）运输到刷状缘膜域（brush border membrane），又因AQP1氨基酸序列中显示出了Lys243和Lys267两个潜在的泛素化位点，该研究还发现cAMP和cGMP也可降低AQP1的泛素化水平并提高其稳定性[5]。此外，不同的渗透条件也将影响AQP1的表达：在低渗条件下，AQP1与Ca^{2+}/钙调蛋白、PKC和微管相互作用，导致AQP1迅速发生易位[6,7]；而在高渗条件下，通过促进启动子介导的AQP1合成[8]和抑制AQP1蛋白降解[9]，增加AQP1的表达[10,11]。

二、AQP2的表达调控机制

关于AQP2的调控，详见本章第一节。

三、AQP3的表达调控机制

AQP3分布于多种组织中，其对水、甘油、尿素及H_2O_2均有通透性，对以上物质的通透功能可能与细胞迁移、炎症和肿瘤的发生密切相关。在中枢神经系统中，AQP3的表达首先在脑膜细胞（meningeal cell）中被发现，进一步的研究发现其在梨状皮质、海马和背侧丘脑的星形胶质细胞和神经元中也有表达[12,13]，但AQP3在中枢神经系统中的表达可能与物种类别有关，例如在猪脑中则未发现AQP3的表达[14]。迄今为止，AQP3在中枢神经系统中的作用尚不明确，仅有的研究表明，在脑缺血6小时内AQP3的表达即可上调，提示在脑水肿早期AQP3参与了神经元水肿的形成过程[15]。对于AQP3的调控机制目前所知甚少，少量的研究提示AQP3可能通过cAMP-PKA信号通路进行短期调节[16-18]。

四、AQP4的表达调控机制

AQP4是神经系统中表达最多的水通道蛋白，在维持血脑屏障的完整性及脑的正常功能中发挥着重要作用。在鼠脑缺血模型中发现，AQP4表达呈现短暂降低后持续上升的双向调节过程[19]；而在颅内压增高时，则发现AQP4表达上升[20]，而敲除*AQP4*后将加重脑的损伤[21]。上述研究表明，AQP4表达的变化在脑损伤后的多种病理生理过程中，尤其是中枢神经系统水肿的形成中发挥着重要作用，AQP4 逐渐成为中枢神经系统水肿治疗的药物靶点。之前所有将 AQP4 作为干预靶点的研究策略均集中于对 AQP4 抑制剂的筛选及鉴定上，以阻断 AQP4 的通道功能。目前已经发现的 AQP4 抑制剂包括血管升压素、褪黑素、PKC、汞（Hg^{2+}）、凝血酶、多巴胺、缺氧、四乙基铵（tetraethylammonium，TEA）、布美他尼、乙酰唑胺（AZA）、siAQP4、姜黄素和 H_2S 等，以上因素均可抑制 AQP4 的表达，并对细胞毒性脑肿胀、癫痫、胶质瘢痕等损伤具有潜在的治疗效果[22-27]；部分促进 AQP4 表达的因素，包括谷氨酸、营养因子、营养不良蛋白、连接蛋白 43（Cx43）、K^+ 通道蛋白（Na^+-K^+-ATPase；NKCC1）、Kir4.1、铅（Pb^{2+}）、cAMP 和乳酸等[23, 28-30]，可能在加快血管源性脑水肿中过量液体的清除方面具有治疗潜力。

然而，任何针对机体中 AQP4 的非靶向治疗都可能对内耳、眼和肾脏的功能产生副作用，因为上述器官也表达 AQP4，并与这些器官的生理功能密切相关。在 *AQP4* 敲除小鼠中发现，其听觉和视觉信号转导受损，表现为听觉脑干反应阈值增加[31, 32]和视网膜电位降低[33]；同时还需要考虑到在神经系统水肿的消退阶段，AQP4 介导的向血管转运水的功能是有益的，因此长期抑制 AQP4 表达的治疗方法已逐渐被放弃。当前，对于 AQP4 表达部位及表达量的调控成为人们关注的焦点，并发现 AQP4 存在多种调控机制。

（一）蛋白激酶磷酸化修饰对 AQP4 的调节

细胞内囊泡运输的调节方式对于膜蛋白功能的调控发挥着重要的作用[34]，蛋白激酶的磷酸化修饰是囊泡运输的关键调控子，膜蛋白在蛋白激酶作用下可逆性的磷酸化，是对其功能进行动态调节最常见的方式。在信号转导过程中，通过蛋白的磷酸化修饰来传递信息，可快速、简便、可逆地调节蛋白的定位及其活性[35]。通过对 AQP4 相应基团进行可逆性的磷酸化修饰，可使 AQP4 的表达从细胞膜转移至细胞内，而不是直接干预其活性，从而调控 AQP4 的水转运功能，以达到促进水肿清除和延缓水肿形成的效果。AQP4 具有能够被 PKA、PKC、PKG、酪蛋白激酶（CK）和 Ca^{2+}/钙调蛋白依赖性蛋白激酶（CaMK）等多种酶类磷酸化的潜在位点。大量研究显示磷酸化修饰参与了 AQP4 蛋白[36-38]及其 mRNA[39]的调节，并可调控其亚细胞定位。

1. 钙调蛋白及蛋白激酶 A 在 AQP4 调控中的作用　AQP4 的第 111 位丝氨酸（Ser111）是 PKA 磷酸化和钙依赖 CaMK Ⅱ 磷酸化的潜在位点。PKA 对 Ser111 的磷酸化增加了 AQP4 的水通透能力[38, 40, 41]。据报道，毛喉萜（forskolin）、血管升压素（arginine vasopressin，AVP）和 V2 受体激动剂等促进 cAMP 产生的药物，均可增加转染了 AQP4 的肾细胞系的水通透性[40]。转染了 AQP4 cDNA 的星形胶质细胞在 AQP4 的 Ser111 位点被磷酸化后，细胞膜水通透性增加，而 Ca^{2+}/CaMK Ⅱ 抑制剂则可逆转这一现象，提示 CaMK Ⅱ 介导的 Ser111 磷酸化可增加 AQP4 的水通透性[42]。

星形胶质细胞表面 AQP4 的丰度在应激刺激下可发生迅速和可逆的变化，其变化与 CaM 或 PKA 的磷酸化作用有关。研究表明，AQP4 和 CaM 之间的直接相互作用导致 AQP4 羧基端的特异性构象变化，并促使 AQP4 定位于细胞膜。研究表明，CaM 在星形胶质细胞 AQP4 的易位中至少有两种不同的作用：首先，瞬时受体电位阳离子通道亚家族成员 4（transient receptor potential cation channel subfamily V member 4，TRPV4，为对 Ca^{2+} 具有选择通透性的阳离子通道）开放后，Ca^{2+}内流增加，激活 CaM，进而导致腺苷酸环化酶的激活、cAMP 的产生和 PKA 的激活；其次，CaM 优先与经过 PKA 磷酸化的 AQP4 结合，结合的强度是通过 PKA 对 AQP4 蛋白第 276 位丝氨酸（Ser276）的磷酸化作用来调节的；此外，CaM 可直接与 AQP4 相结合，其间的相互作用导致 AQP4 羧基端的特异性构象变化，并促使 AQP4 定位于细胞膜[43]。由此可见，CaM 和 PKA 参与了损伤后 AQP4 向细胞膜的转位。临床所使用的抗精神病药物三氟拉嗪即通过抑制 CaM 的功能，从而抑制 AQP4 在血-脊髓屏障（blood-spinal cord barrier）中的表达，以减轻中枢神经系统水肿，有助于其功能的恢复[43]。

2. 蛋白激酶 C 在 AQP4 调控中的作用　AQP4 第 180 位的丝氨酸（Ser180）是公认的 PKC 作用位点。多项研究表明[44,45]，Ser180 在 PKC 对 AQP4 的调节中发挥了重要作用。在转染 AQP4 的胶质瘤细胞系中发现，PKC 对 AQP4 的 Ser180 位点具有直接作用并能够改变 AQP4 的水通透能力[45]。PKC 的高活性可抑制 AQP4 的水通透性和细胞的迁移能力，Ser180 除了作为 PKC 磷酸化的靶点，目前也被认为是 Ca^{2+}/CaMK Ⅱ 磷酸化的位点。

在猪肾细胞系（LLC-PK1 cell）中使用 PKC 激动剂佛波醇 12,13-二丁酸酯（phorbol 12,13-dibutyrate，PDBu）短期孵育，未见 AQP4 的改变，但是延长 PKC 激动剂的孵育时间，可导致 AQP4 对水的通透能力迅速下降[38]，出现这种情况的原因可能是 PKC 长时间的孵育过程，使 AQP4 其他位点被磷酸化（如 Ser267 和 Ser273）。使用 PKC 的激动剂佛波酯（phorbol ester）将 AQP4 磷酸化，将导致 AQP4 水通透能力降低，并且该效应呈现剂量依赖性的特点[36]。PKC 的激活也可抑制 AQP4 的表达，这与其降低星形胶质细胞水通透性的影响是一致的。此外，PKC 对 AQP4 的磷酸化过程可能参与了多巴胺对视网膜[46]的调节作用。

3. 其他种类的蛋白激酶在 AQP4 调控中的作用　有研究证实，AQP4 蛋白中第 111 位的丝氨酸（Ser111）参与 CaMK Ⅱ 对 AQP4 的调节。CaMK Ⅱ 并不能直接催化 Ser111 的磷酸化，但是通过一氧化氮合酶（nitric oxide synthetase，NOS）和产生的 NO 及蛋白激酶 G（protein kinase G，PKG）而发挥作用[41]，即 CaMK Ⅱ-NO-cGMP-PKG 信号通路在 AQP4 的 Ser111 磷酸化过程中发挥着作用。除了 PKG 外，Ser111 也依赖于 cAMP 的 PKA 磷酸化位点，有研究显示其可上调肾上皮细胞和卵母细胞中 AQP4 的水通透性[47]；而且，Ser111 还被证实是谷氨酸和铅对 AQP4 发挥作用的位点，可以作为开发提高 AQP4 水通透性药物的潜在靶点。

AQP4 中还存在其他可被磷酸化的位点，它们在 AQP4 的合成、加工和降解中具有重要的功能。在肾上皮细胞中，Ser276 被酪蛋白激酶 Ⅱ（casein kinase 2，CK Ⅱ）磷酸化，促进内化的 AQP4 被运送到溶酶体，使 AQP4 的降解增加[48]。在小鼠原代星形胶质细胞中，Ser276 和其他 3 种末端含有羧基的序列 Ser285、Thr289、Ser316，可能参与了 AQP4 向高尔基体的转运。以上研究都未提及这些氨基酸序列对 AQP4 水通透性的影响，但是它们的磷酸化可能调节细胞表面 AQP4 的表达，因而可影响整个细胞膜的水通透性。在组胺处理的胃壁细胞中，AQP4 内化并进入晚期内涵体后，PKA 可将内化的 AQP4 磷酸化，帮助 AQP4 定位于细胞内的循环小泡中[49]，从而减少质膜中水通过 AQP4 的跨膜转运。

由于不同磷酸化位点对AQP4的水通透功能可能具有相反的效应，提示星形胶质细胞AQP4的功能可被不同的生理和病理刺激因素所影响。因此，可将AQP4作为潜在的治疗靶点，以研发快速上调或者下调星形胶质细胞水通透性的药物。

（二）细胞内运输对AQP4的调控

细胞膜蛋白通过细胞内囊泡内吞进入内涵体（endosome）的过程称为内化，研究表明，针对内化过程的调节可达到对通道蛋白进行调控的目的。AQP4在细胞膜与内涵体之间存在着动态运输，使内涵体可作为已内化AQP4的储存场所。体外实验发现，在转染AQP4的人胃肿瘤细胞系中，组胺可引起AQP4的内化，并伴随着细胞水通透性的降低；在蟾蜍卵母细胞中使用PKC激动剂后，也可观察到AQP4的内化。内化后的AQP4可返回细胞膜[49]或者分选至溶酶体内降解[48]，但是与AQP2的转运相比，AQP4的内化和再循环的发生相对缓慢[49]。CKⅡ磷酸化已被证实参与了AQP4在细胞内腔及高尔基体和晚期溶酶体的分布[48]，但CKⅡ是否可导致AQP4的内化仍不清楚。

（三）金属离子对AQP4的调节

AQP4一直被认为是汞不敏感性的AQP，因为它在与AQP1相对应的Cys189位点缺乏半胱氨酸[50, 51]。然而，停流分析（stopped-flow analysis）证实汞可抑制脂蛋白体（proteoliposome）中AQP4 M23的水通透性[25]。AQP4和AQP1在水通透功能被抑制时所需汞的剂量和应答时间不同，提示汞对AQP4水通透性抑制效应的分子机制与AQP1不一样。通过AQP4中6个半胱氨酸残基定点突变的方法，证实其178位半胱氨酸（Cys178）和253位半胱氨酸（Cys253）是汞作用的位点。

Zn^{2+}和Cu^{2+}对卵巢癌细胞中AQP4的水通透性没有影响[52]，但可抑制脂蛋白体中AQP4 M23的水通透性[53]，人体内的Zn^{2+}主要是由星形胶质细胞摄入，在培养的星形胶质细胞中发现，低渗透压环境刺激将使细胞内的Zn^{2+}浓度在NOS作用下增加[54]。氧化还原复合物并不直接影响AQP4的水通透功能，但是这些复合物作用所致Zn^{2+}的释放，可能在AQP4的调节中发挥重要作用。

（四）蛋白间相互作用对AQP4进行调节

近年来的研究表明，AQP4的极性分布与其发挥锚定作用的蛋白分子有关[28, 55-57]。AQP4的锚定机制是指与其锚定有关的多种蛋白（或复合物）及其相互作用。抗肌萎缩蛋白聚糖复合物（dystroglycan complex，DGC）是与AQP4的锚定相关的重要复合物，为连接细胞外基质（extracellular matrix，ECM）与细胞骨架的一组相互作用蛋白，包括抗肌萎缩蛋白聚糖（dystroglycan，DG）、抗肌萎缩蛋白（dystrophin，DYS）同源体Dp71、α-syntrophin以及$α_1$-dystrobrevin等，该复合物能将多种膜蛋白分子定位到相应细胞膜域，它们的异常与多种肌营养不良性疾病的脑、眼病变有关。

α-syntrophin中含有的PDZ结构域与突触后致密区95（post-synaptic density-95，PSD-95）、大盘（discs-large）或闭锁小带1（zona occludens 1，ZO-1）蛋白中的特征性结合基序一致，其可与AQP4的羧基端发生相互作用，以此将AQP4锚定于细胞膜的特定膜域[58]。

在缺乏α-syntrophin表达的小鼠中，星形胶质细胞AQP4出现了异常定位，其在大脑皮质和小脑邻近血管周围的终足膜上减少，但在朝着神经纤维网的膜部其表达量增加。相似的定位改变也发生于dystrophin敲除的小鼠[59]。在人胶质母细胞瘤中，AQP4伴随着α-syntrophin及α-dystroglycan从终足分布至整个细胞表面[60]，但dystrophin的分布则未发生改变[29, 60]。在中央颞叶癫痫的患者，硬化区血管周围AQP4的丢失与脑周围特异的dystrophin亚型的减少是呈比例发生的[61]。α-dystroglycan是与细胞外基质类肝素硫酸蛋白聚糖（agrin）相结合的蛋白，可参与血脑屏障的形成和维持[30]。在培养的星形胶质细胞中，agrin的表达可以增加细胞膜中AQP4的表达量[62]。在体实验提示蛋白聚糖在星形胶质细胞终足膜域（endfoot membranes of astrocytes）AQP4/粒子正交阵列（orthogonal arrays of particle，OAP）的成簇中起着关键作用[63]。

在视网膜中，laminin与DG复合物均分布于Müller细胞靠近玻璃体和视网膜血管周围的终足膜域[28, 56]。特别有意义的是，这些膜域也是AQP4的高表达区。AQP4与DGC的多个分子，如α-DG、β-DG、Dp71以及α-syntrophin之间存在相互作用[64]，提示AQP4可能与DGC形成更大的复合物而定位于Müller细胞特殊膜域。DGC的完整性对维持AQP4的正确定位至关重要。在Dp71缺乏的Mdx3cv小鼠视网膜内，AQP4在Müller细胞终足膜域的极性分布消失，呈弥散状分布于其所有突起上，这种分布异常可能与Mdx3cv小鼠ERG上b波振幅下降有关。同时，视网膜血管周围AQP4的分布明显减少（约75%）[64]。表达低糖基化型α-DG的Largemyd小鼠[患有先天性肌营养不良症（dystroglycanopathies）]，表现出ERG上b波振幅下降和α-DG与ECM分子如层粘连蛋白（laminin）或蛋白聚糖的相互作用严重受损，并伴随相应的α-syntrophin缺失[65]。上述结果提示，ECM分子与DG复合物通过相互作用，参与视网膜电流图（electroretinogram，ERG）的形成，它们之间的作用可能与AQP4对神经元活动所致局部积聚水分子的快速清除有关。

AQP4与内向整流性钾通道（Kir4.1）之间也可发生相互作用[66]，在AQP4定位异常的小鼠模型中发现Kir4.1和AQP4存在功能上的偶联[28]。Kir4.1和AQP4在星形胶质细胞中的亚细胞分布模式非常相似[67]，它们在血管周围的星形胶质细胞终足上丰富表达，均由血管周围的层粘连蛋白所锚定。在含有DGC的膜域Kir4.1和AQP4关联，并且Kir4.1能够直接结合到该复合物中的α-syntrophin上[55]。然而，目前并没有Kir4.1和AQP4相互作用的直接证据。在人星形胶质细胞中这两种蛋白再分布（redistribution）的特点不相同[29]，并且对其表达水平的调节也有区别。*AQP4*的敲除未能影响到星形胶质细胞中Kir4.1的表达及功能[68, 69]；在原代培养的胶质细胞中，抑制Kir4.1的功能也没有显著改变AQP4的水通透性[69]。由此可见，Kir4.1和AQP4之间的相互作用需要进一步的研究。

在肾脏和小脑中发现，AQP4可与Na^+-K^+-ATP酶产生相互作用[70]。由于星形胶质细胞能够参与细胞外K^+的清除和神经递质的摄取，Na^+-K^+-ATP酶在星形胶质细胞的活性中起着重要作用。Na^+-K^+-ATP酶产生了跨膜Na^+浓度梯度，这一浓度梯度有利于对K^+的继发主动转运及对神经递质的摄取，也在神经元受刺激后从细胞外空间清除K^+的过程中发挥主动作用。Na^+-K^+-ATP酶是否能够影响AQP4的定位及其水通透性仍不清楚。

小脑中的AQP4可与代谢型谷氨酸受体（mGluR，Ⅰ型mGluR的一种）发生相互作用[70]，此相互作用可使谷氨酸增强AQP4对水的通透性；而且，小脑中mGluR的激活也可增加

Na^+-K^+-ATP酶的活性，Na^+-K^+-ATP酶和mGluR在星形胶质细胞膜中可能形成蛋白复合体。既然AQP4的氨基末端存在着与其他蛋白相互作用的区域，这表明该区域也存在着连接AQP4和Na^+-K^+-ATP酶、mGluR的锚定蛋白。

在星形胶质细胞中，AQP4被siRNA抑制后可下调连接蛋白43（connexin 43，Cx43），提示AQP4和Cx43有功能上的联系[71]。在急性高氨血症患者中也可见AQP4和Cx43的表达同时下降[72]，但在神经胶质瘤[73]和自闭症[74]患者中这两种蛋白表达量的变化是不同的。以OAP形式存在的AQP4与Cx43表达部位相邻[75]，但是没有实验结果证实脑中AQP4和Cx43存在着直接相互作用。

有学者提出星形胶质细胞中AQP4与纤维状肌动蛋白（F-actin）存在着相互作用，并且可影响星形胶质细胞肌动蛋白微丝的构成[71]。然而，培养的星形胶质细胞中，*AQP4*敲除后的效应存在着物种特异性。在*AQP4*敲除小鼠中，未见其星形胶质细胞形态的改变[76]，肌动蛋白的调节没有影响星形胶质细胞膜中AQP4的移动性和OAP的稳定性[77]。

（五）激素对神经系统AQP4的调控

糖皮质激素可上调视网膜Müller细胞中AQP4的表达量，并可调节视网膜中AQP4的表达部位，从而维持视网膜神经上皮层中水的平衡[78]。另有研究显示[79]，曲安奈德（triamcinolone acetonide，糖皮质激素的一种）可下调正常视网膜中AQP4的表达，但上调葡萄膜炎状态下视网膜中AQP4的表达。有研究显示，血管升压素也可通过介导AQP4的内化以调节AQP4的水通透性，其机制与Ser180被PKC磷酸化有关。

（六）AQP4的降解在其表达调控中的作用

中枢神经系统中AQP4的表达还可受到溶酶体或者泛素-蛋白酶体系统的影响。高眼压模型大鼠在腹腔注射内皮素后视网膜AQP4表达的下降，即与泛素-蛋白酶体系统降解的AQP4的增加有关[80]。针对人视网膜缺血及糖尿病所致视网膜病变及相关动物模型[46, 80-82]的研究发现，视网膜AQP4的表达量或者亚细胞定位可发生改变。因此，研究AQP4的调节机制，对于深入阐明疾病状态下视网膜损伤的分子机制及AQP4在其中发挥的作用，以及治疗手段的更新和相关的药物开发，具有重要意义。目前，关于视网膜AQP4的调节机制已经取得一些进展，但与临床应用还有一定的距离，需要付出更多的努力。

（七）AQP4表达调控的信号通路

1. ERK1/2信号转导通路对AQP4的调控作用在神经系统疾病中的应用　在癫痫持续状态下发现，星形胶质细胞中的AQP4受到67kDa的层粘连蛋白受体（laminin receptor，LR）-ERK1/2-PI3K-AKT信号通路调节，此调节作用有望成为神经系统疾病中血管源性水肿的重要治疗靶点之一[83]，但目前细胞外信号调节激酶1/2（extracellular regulated protein kinase1/2，ERK1/2）对于AQP4的调控作用仍存在争议。ERK1/2激活降低了星形胶质细胞损伤后AQP4的表达，而增加了氧-葡萄糖剥夺诱导的AQP4的表达[84]。丝氨酸/苏氨酸激酶（AKT）是磷脂酰肌醇-3-羟激酶（phosphoinositide 3 kinase，PI3K）和ERK1/2介导的AQP4调控的常见下游效应器[85, 86]。

2. p38-MAPK信号转导通路对AQP4调控作用在神经系统疾病中的应用 丝裂原活化蛋白激酶（mitogen-activated protein kinase，MAPK）是经典的细胞信号转导途径，在响应各种内在和外在的细胞刺激时被激活。哺乳动物的MAPK主要包括JNK（c-Jun氨基端激酶）、ERK1/2和p38激酶。MAPK信号通路参与调控细胞增殖、分化、存活和凋亡等多种功能，并被证实参与细胞发生、侵袭、转移，以及胶质瘤的发展和预后等，是调节神经胶质瘤形成和进展的最重要途径。p38丝裂原活化蛋白激酶（p38-MAPK）是MAPK超家族的一员，在外界刺激下被激活。使用p38激活剂茴香霉素、替莫唑胺（TMZ）抑制U87和U251胶质瘤细胞的增殖、迁移和侵袭，并诱导细胞阻滞在G_2/M期，以上变化与p38磷酸化水平的增加并激活p38-MAPK通路，导致AQP4表达水平的降低有关[87]。

五、AQP5的表达调控机制

AQP5表达于小鼠和人类子宫中，对于H_2O和CO_2均有通透性。在低氧条件下，AQP5的表达可受到PKA的调控[13]。AQP5中的两个位点已确认能够被PKA磷酸化，分别是胞质D袢中的Ser156位点和羧基末端的Thr259位点。然而，在上述磷酸化位点突变后细胞中AQP5的表达量与野生型细胞比较没有差别，表明AQP5磷酸化在生理条件下可能并未发生。相比之下，cAMP-PKA的AQP5 Thr259位点磷酸化最近被证明与AQP5的侧向扩散有关，可能调节唾液腺等腺体分泌物中的水含量。但另有观点认为，导致AQP5从胞质转运到质膜的原因是细胞内Ca^{2+}的增加，而不是PKA所致AQP5的磷酸化。

AQP5启动子中均存在雌激素反应元件，通过激活磷脂酰肌醇-3-羟激酶和蛋白激酶B（PI3K/AKT）通路，AQP5可促进小鼠体内子宫内膜样细胞的异位植入，可能与促雌激素分泌的作用有关。此外，由于子宫内AQP5的表达可能会受黄体酮等其他因素的影响，导致子宫在增殖后期和分泌期的AQP5表达频率较低[88]。

六、AQP6的表达调控机制

AQP6 mRNA表达于小鼠大脑后部、小脑和脊髓中[89,90]，而AQP6蛋白则存在于大鼠脑神经细胞的突触囊泡中，可能参与了突触小泡的形成和分泌[89,91]。AQP6在中枢神经系统中的作用尚不清楚。AQP6 mRNA的表达受组织特异性和年龄相关性的调控，可能在小鼠的发育中发挥作用[89]。AQP6在中枢神经系统中的作用及其调控机制有待进一步研究。

七、AQP7的表达调控机制

AQP7可促进水、甘油、尿素、氨、亚砷酸盐和NH_3的运输，作为甘油通道蛋白，其主要在脂肪代谢中发挥作用。使用Northern印迹分析首次在大鼠脑中检测到AQP7的弱表达[92]，组织学研究发现，其主要分布于小鼠脑脉络膜丛上皮细胞（CPEC）和内皮祖细胞（EPC）的顶膜中[93]，提示AQP7可能参与脑脊液（CSF）分泌。目前在猪脑中检测到AQP7 mRNA 的表达[14]。关于 AQP7 的调控机制，尚需更多的研究。

八、AQP8 的表达调控机制

AQP8 是一种水通道蛋白，首次发现于肾脏近端小管和集合小管细胞的胞内区域。在中枢神经系统中，AQP8 主要表达于梨状皮质、海马和丘脑背侧星形胶质细胞、神经元和少突胶质细胞；在脊髓中央管内室管膜细胞中也有表达，在室管膜和脉络丛则为弱阳性[12, 94]。多项研究表明，AQP8 可输送水[95, 96]和氨[25, 96]，而输送氨可能与其维持酸碱平衡的功能有关。

在正常组织中，AQP8 在细胞膜的表达水平非常低[67, 97]，但在疾病条件下其表达量可增加。在星形胶质细胞瘤中，AQP8 的表达水平随着肿瘤级别的增加而上调[94]；反之，在胶质瘤细胞中下调 AQP8 的表达，则对细胞增殖和迁移具有明显的抑制作用[94]。另外，AQP8 在脑缺血后表达上调，提示 AQP8 参与水肿早期的形成过程[15]。由此可见，AQP8 可能在脑疾病（水肿和肿瘤）的发生发展中发挥重要作用，并可作为星形细胞瘤和其他种类胶质瘤的潜在治疗药物。

关于 AQP8 的调控机制研究表明，胰高血糖素或其第二信使 cAMP 强烈诱导 AQP8 从细胞内囊泡重新分布到肝细胞膜[98]，从而提高细胞膜的水渗透性，促进水转运和胆汁形成。这些研究表明，PKA 和 PI3K 通路都参与了胰高血糖素诱导的 AQP8 转运[98, 99]。

九、AQP9 的表达调控机制

AQP9 是水甘油通道蛋白家族的成员，对水、甘油、尿素和一元羧酸盐均有通透性，在啮齿动物和灵长类动物的脑中均有表达[100]。AQP9 主要表达于室管膜细胞、下丘脑细胞[101]、星形胶质细胞、软脑膜血管内皮细胞、儿茶酚胺能神经元细胞等[102, 103]。研究表明，在恶性星形胶质细胞瘤中 AQP9 的表达量增加[104]。在体研究发现，没有给予应激刺激时，*AQP9* 敲除小鼠并无明显的异常表现[105]；然而，在体外培养的星形胶质细胞中抑制 AQP9 的表达将减少甘油摄取，增加葡萄糖摄取和氧化代谢[106]，另一项研究也表明，在培养的星形胶质细胞中，AQP9 在缺氧时表达降低，在复氧时表达恢复[13]。

AQP9 的表达受到多种信号通路的调控。有研究表明，PKA 信号转导途径可能通过二丁酰环腺苷酸（dbcAMP）诱导的相关因子调控 AQP9 的表达[107]；通过 PKC 信号通路则可以直接下调 AQP9 mRNA 及其蛋白的表达，该过程不需要新合成蛋白的参与[108]。p38-MAPK 信号途径可调节高渗条件下星形胶质细胞 AQP9 的表达[109, 110]。在体研究表明，在大脑中动脉缺血条件下，AQP9 的表达也可通过 p38 MAPK 信号转导途径参与调控脑缺血后脑水肿的发生[111]。

十、AQP10 的表达调控机制

AQP10 是一种水甘油通道蛋白，存在于小肠、结肠上皮细胞和脂肪细胞中[112-114]，同时也在胎牙、咀嚼肌和肿胀的牙龈中发现了AQP10 mRNA的表达。目前关于AQP10的表达及其调控作用的研究不多。有课题组报道，可将阵列技术应用于AQP10表达和修饰的识别，以深入揭示其功能[115]。

十一、AQP11的表达调控机制

AQP11表达于大鼠和小鼠中枢神经系统的多种细胞[116]，如大脑皮质和海马的神经元、浦肯野细胞[117]及脉络丛上皮和脑毛细血管内皮[118, 119]。AQP11具有传统的N端Asn-Pro-Ala（NPA）特征基序和独特的氨基酸序列模式，其中包括Asn-Pro-Cys（NPC）基序，AQP11在其第一个NPA基序中含有半胱氨酸残基而不是丙氨酸，在第二个基序中含有亮氨酸而不是精氨酸[120]，上述结构对其功能的充分发挥至关重要。

AQP11缺失小鼠的大脑表现正常，没有任何形态和功能异常[118]，然而AQP11缺失小鼠的血脑屏障（blood-brain barrier，BBB）中AQP4表达减少了一半，表明AQP11可能与AQP4在功能上具有相关性[118]。另外，在AQP11缺失小鼠中还发现其肾脏近端小管内质网中出现空泡化及发生多囊性肾病[121]，并在出生后两个月内死亡；这提示AQP11是多囊素-1和多囊素-2正确糖基化所必需的[122]。有研究提示，当渗透环境发生改变时，AQP11可能降低其表达以保护脑组织[119]。由于AQP11具有独特的高亲和力的汞离子结合位点，即三半胱氨酸基位点（tri-cysteine motif site），分布在浦肯野细胞中的AQP11可能与自闭症患者中的汞等阳离子相互作用。因此，AQP11有望成为自闭症的治疗靶点[123]。AQP11在中枢神经系统中的生理作用及其调控机制有待进一步研究。

十二、AQP12的表达调控机制

AQP12属于超级水通道蛋白家族成员，其结构与AQP11有32%的相似性。AQP12表达于胰腺腺泡细胞的胞质中[124]。迄今为止，关于AQP12的功能研究还不够深入。在胰腺 *AQP12* 敲除小鼠中发现其异常改变是有限的，可能是因为其他水通道蛋白亚型的代偿作用。在生理条件及病理条件下对AQP12进行调控，将有助于阐明其在胰腺功能中的作用。

在过去三十年里，研究人员对AQP的认识取得了重大进展。AQP通过协调水和溶质在不同室腔之间的运输，参与了机体内许多重要的生理和病理过程，是治疗机体多种功能障碍的重要靶点。然而，对于完全理解各器官系统中AQP的生理功能和病理意义还需要进行更多的工作。本节重点介绍了当前AQP可能存在的各种调控机制。随着新实验技术的产生和研究理论的进步，未来对AQP的超微结构、生理功能及其病理机制的认识将不断提高。

（甘胜伟）

参 考 文 献

[1] Patil RV, Han Z, Wax MB. Regulation of water channel activity of aquaporin 1 by arginine vasopressin and atrial natriuretic peptide[J]. Biochem Biophys Res Commun, 1997, 238(2): 392-396.

[2] Han Z, Patil RV. Protein kinase A-dependent phosphorylation of aquaporin-1[J]. Biochem Biophys Res Commun, 2000, 273(1): 328-332.

[3] Marinelli RA, Pham L, Agre P, et al. Secretin promotes osmotic water transport in rat cholangiocytes by increasing aquaporin-1 water channels in plasma membrane. Evidence for a secretin-induced vesicular translocation of aquaporin-1[J]. J Biol Chem, 1997, 272(20): 12984-12988.

[4] Zhang W, Zitron E, Hömme M, et al. Aquaporin-1 channel function is positively regulated by protein kinase C[J]. J Biol Chem, 2007, 282(29): 20933-20940.

[5] Pohl M, Shan Q, Petsch T, et al. Short-term functional adaptation of aquaporin-1 surface expression in the proximal tubule, a component of glomerulotubular balance[J]. J Am Soc Nephrol, 2015, 26(6): 1269-1278.

[6] Conner MT, Conner AC, Brown JE, et al. Membrane trafficking of aquaporin 1 is mediated by protein kinase C via microtubules and regulated by tonicity[J]. Biochemistry, 2010, 49(5): 821-823.

[7] Conner MT, Conner AC, Bland CE, et al. Rapid aquaporin translocation regulates cellular water flow: mechanism of hypotonicity-induced subcellular localization of aquaporin 1 water channel[J]. J Biol Chem, 2012, 287(14): 11516-11525.

[8] Umenishi F, Schrier RW. Identification and characterization of a novel hypertonicity-responsive element in the human aquaporin-1 gene[J]. Biochem Biophys Res Commun, 2002, 292(3): 771-775.

[9] Leitch V, Agre P, King LS. Altered ubiquitination and stability of aquaporin-1 in hypertonic stress[J]. Proc Natl Acad Sci USA, 2001, 98(5): 2894-2898.

[10] Jenq W, Cooper DR, Bittle P, et al. Aquaporin-1 expression in proximal tubule epithelial cells of human kidney is regulated by hyperosmolarity and contrast agents[J]. Biochem Biophys Res Commun, 1999, 256(1): 240-248.

[11] Umenishi F, Schrier RW. Hypertonicity-induced aquaporin-1(AQP1)expression is mediated by the activation of MAPK pathways and hypertonicity-responsive element in the AQP1 gene[J]. J Biol Chem, 2003, 278(18): 15765-15770.

[12] Yang M, Gao F, Liu H, et al. Immunolocalization of aquaporins in rat brain[J]. Anat Histol Embryol, 2011, 40(4): 299-306.

[13] Yamamoto N, Yoneda K, Asai K, et al. Alterations in the expression of the AQP family in cultured rat astrocytes during hypoxia and reoxygenation[J]. Brain Res Mol Brain Res, 2001, 90(1): 26-38.

[14] Li X, Lei T, Xia T, et al. Molecular characterization, chromosomal and expression patterns of three aquaglyceroporins(AQP3, 7, 9)from pig[J]. Comp Biochem Physiol B Biochem Mol Biol, 2008, 149(3): 468-476.

[15] Yang M, Gao F, Liu H, et al. Temporal changes in expression of aquaporin-3, -4, -5 and 8 in rat brains after permanent focal cerebral ischemia[J]. Brain Res, 2009, 1290: 121-132.

[16] Hua Y, Ding S, Zhang W, et al. Expression of AQP3 protein in hAECs is regulated by Camp-PKA-CREB signalling pathway[J]. Front Biosci(Landmark Ed), 2015, 20(7): 1047-1055.

[17] Jourdain P, Becq F, Lengacher S, et al. The human CFTR protein expressed in CHO cells activates aquaporin-3 in a cAMP-dependent pathway: study by digital holographic microscopy[J]. J Cell Sci, 2014, 127(Pt 3): 546-556.

[18] Marlar S, Arnspang EC, Koffman JS, et al. Elevated cAMP increases aquaporin-3 plasma membrane diffusion[J]. Am J Physiol Cell Physiol, 2014, 306(6): C598-C606.

[19] Koyama Y, Tanaka K. Decreases in rat brain aquaporin-4 expression following intracerebroventricular administration of an endothelin ET B receptor agonist[J]. Neurosci Lett, 2010, 469(3): 343-347.

[20] Eide PK, Eidsvaag VA, Nagelhus EA, et al. Cortical astrogliosis and increased perivascular aquaporin-4 in idiopathic intracranial hypertension[J]. Brain Res, 2016, 1644: 161-175.

[21] Tang Y, Wu P, Su J, et al. Effects of Aquaporin-4 on edema formation following intracerebral hemorrhage[J]. Exp Neurol, 2010, 223(2): 485-495.

[22] Fukuda AM, Adami A, Pop V, et al. Posttraumatic reduction of edema with aquaporin-4 RNA interference improves acute and chronic functional recovery[J]. J Cereb Blood Flow Metab, 2013, 33(10): 1621-1632.

[23] Zelenina M. Regulation of brain aquaporins[J]. Neurochem Int, 2010, 57(4): 468-488.

[24] Yu LS, Fan YY, Ye G, et al. Curcumin alleviates brain edema by lowering AQP4 expression levels in a rat model of hypoxia-hypercapnia-induced brain damage[J]. Exp Ther Med, 2016, 11(3): 709-716.

[25] Yukutake Y, Tsuji S, Hirano Y, et al. Mercury chloride decreases the water permeability of aquaporin-4-reconstituted proteoliposomes[J]. Biol Cell, 2008, 100(6): 355-363.

[26] Migliati E, Meurice N, DuBois P, et al. Inhibition of aquaporin-1 and aquaporin-4 water permeability by a derivative of the loop diuretic bumetanide acting at an internal pore-occluding binding site[J]. Mol Pharmacol, 2009, 76(1): 105-112.

[27] Cao S, Zhu P, Yu X, et al. Hydrogen sulfide attenuates brain edema in early brain injury after subarachnoid hemorrhage in rats: possible involvement of MMP-9 induced blood-brain barrier disruption and AQP4 expression[J]. Neurosci Lett, 2016, 621: 88-97.

[28] Amiry-Moghaddam M, Williamson A, Palomba M, et al. Delayed K^+ clearance associated with aquaporin-4 mislocalization: phenotypic defects in brains of alpha-syntrophin-null mice[J]. Proc Natl Acad Sci USA, 2003, 100(23): 13615-13620.

[29] Warth A, Mittelbronn M, Wolburg H. Redistribution of the water channel protein aquaporin-4 and the K^+ channel protein Kir4.1 differs in low-and high-grade human brain tumors[J]. Acta Neuropathol, 2005, 109(4): 418-426.

[30] Wolburg H, Noell S, Wolburg-Buchholz K, et al. Agrin, aquaporin-4, and astrocyte polarity as an important feature of the blood-brain barrier[J]. Neuroscientist, 2009, 15(2): 180-193.

[31] Li J, Verkman AS. Impaired hearing in mice lacking aquaporin-4 water channels[J]. J Biol Chem, 2001, 276(33): 31233-31237.

[32] Mhatre AN, Stern RE, Li J, et al. Aquaporin 4 expression in the mammalian inner ear and its role in hearing[J]. Biochem Biophys Res Commun, 2002, 297(4): 987-996.

[33] Li J, Patil RV, Verkman AS. Mildly abnormal retinal function in transgenic mice without Müller cell aquaporin-4 water channels[J]. Invest Ophthalmol Vis Sci, 2002, 43(2): 573-579.

[34] Offringa R, Huang F. Phosphorylation-dependent trafficking of plasma membrane proteins in animal and plant cells[J]. J Integr Plant Biol, 2013, 55(9): 789-808.

[35] Cohen P. The regulation of protein function by multisite phosphorylation: a 25 year update[J]. Trends Biochem Sci, 2000, 25(12): 596-601.

[36] Han Z, Wax MB, Patil RV. Regulation of aquaporin-4 water channels by phorbol ester-dependent protein phosphorylation[J]. J Biol Chem, 1998, 273(11): 6001-6004.

[37] Kleindienst A, Fazzina G, Amorini AM, et al. Modulation of AQP4 expression by the protein kinase C activator, phorbol myristate acetate, decreases ischemia-induced brain edema[J]. Acta Neurochir Suppl, 2006, 96: 393-397.

[38] Zelenina M, Zelenin S, Bondar AA, et al. Water permeability of aquaporin-4 is decreased by protein kinase C and dopamine[J]. Am J Physiol Renal Physiol, 2002, 283(2): F309-F318.

[39] Nakahama K, Nagano M, Fujioka A, et al. Effect of TPA on aquaporin 4 mRNA expression in cultured rat astrocytes[J]. Glia, 1999, 25(3): 240-246.

[40] Gunnarson E, Zelenina M, Aperia A. Regulation of brain aquaporins[J]. Neuroscience, 2004, 129(4): 947-955.

[41] Gunnarson E, Zelenina M, Axehult G, et al. Identification of a molecular target for glutamate regulation of astrocyte water permeability[J]. Glia, 2008, 56(6): 587-596.

[42] Gunnarson E, Axehult G, Baturina G, et al. Lead induces increased water permeability in astrocytes expressing aquaporin 4[J]. Neuroscience, 2005, 136(1): 105-114.

[43] Kitchen P, Salman MM, Halsey AM, et al. Targeting aquaporin-4 subcellular localization to treat central nervous system edema[J]. Cell, 2020, 181(4): 784-799, e19.

[44] Moeller HB, Fenton RA, Zeuthen T, et al. Vasopressin-dependent short-term regulation of aquaporin 4 expressed in *Xenopus* oocytes[J]. Neuroscience, 2009, 164(4): 1674-1684.

[45] McCoy ES, Haas BR, Sontheimer H. Water permeability through aquaporin-4 is regulated by protein kinase C and becomes rate-limiting for glioma invasion[J]. Neuroscience, 2010, 168(4): 971-981.

[46] Pannicke T, Iandiev I, Uckermann O, et al. A potassium channel-linked mechanism of glial cell swelling in the postischemic retina[J]. Mol Cell Neurosci, 2004, 26(4): 493-502.

[47] Hamabata T, Liu C, Takeda Y. Positive and negative regulation of water channel aquaporins in human small intestine by cholera toxin[J]. Microb Pathog, 2002, 32(6): 273-277.

[48] Madrid R, Le Maout S, Barrault MB, et al. Polarized trafficking and surface expression of the AQP4 water channel are coordinated by serial and regulated interactions with different clathrin-adaptor complexes[J]. EMBO J, 2001, 20(24): 7008-7021.

[49] Carmosino M, Procino G, Tamma G, et al. Trafficking and phosphorylation dynamics of AQP4 in histamine-treated human gastric cells[J]. Biol Cell, 2007, 99(1): 25-36.

[50] Jung JS, Bhat RV, Preston GM, et al. Molecular characterization of an aquaporin cDNA from brain: candidate osmoreceptor and regulator of water balance[J]. Proc Natl Acad Sci USA, 1994, 91(26): 13052-13056.

[51] Jung JS, Preston GM, Smith BL, et al. Molecular structure of the water channel through aquaporin CHIP. The hourglass model[J]. J Biol Chem, 1994, 269(20): 14648-14654.

[52] Németh-Cahalan KL, Kalman K, Froger A, et al. Zinc modulation of water permeability reveals that aquaporin 0 functions as a cooperative tetramer[J]. J Gen Physiol, 2007, 130(5): 457-464.

[53] Yukutake Y, Hirano Y, Suematsu M, et al. Rapid and reversible inhibition of aquaporin-4 by zinc[J]. Biochemistry, 2009, 48(51): 12059-12061.

[54] Kruczek C, Görg B, Keitel V, et al. Hypoosmotic swelling affects zinc homeostasis in cultured rat astrocytes[J]. Glia, 2009, 57(1): 79-92.

[55] Connors NC, Adams ME, Froehner SC, et al. The potassium channel Kir4.1 associates with the dystrophin-glycoprotein complex via alpha-syntrophin in glia[J]. J Biol Chem, 2004, 279(27): 28387-28392.

[56] Saadoun S, Papadopoulos MC, Krishna S. Water transport becomes uncoupled from K^+ siphoning in brain contusion, bacterial meningitis, and brain tumours: immunohistochemical case review[J]. J Clin Pathol, 2003, 56(12): 972-975.

[57] Amiry-Moghaddam M, Frydenlund DS, Ottersen OP. Anchoring of aquaporin-4 in brain: molecular mechanisms and implications for the physiology and pathophysiology of water transport[J]. Neuroscience, 2004, 129(4): 999-1010.

[58] Adams ME, Mueller HA, Froehner SC. In vivo requirement of the alpha-syntrophin PDZ domain for the sarcolemmal localization of nNOS and aquaporin-4[J]. J Cell Biol, 2001, 155(1): 113-122.

[59] Vajda Z, Pedersen M, Füchtbauer EM, et al. Delayed onset of brain edema and mislocalization of aquaporin-4 in dystrophin-null transgenic mice[J]. Proc Natl Acad Sci USA, 2002, 99（20）: 13131-13136.

[60] Warth A, Kröger S, Wolburg H. Redistribution of aquaporin-4 in human glioblastoma correlates with loss of agrin immunoreactivity from brain capillary basal laminae[J]. Acta Neuropathol, 2004, 107（4）: 311-318.

[61] Eid T, Lee TS, Thomas MJ, et al. Loss of perivascular aquaporin 4 may underlie deficient water and K^+ homeostasis in the human epileptogenic hippocampus[J]. Proc Natl Acad Sci USA, 2005, 102（4）: 1193-1198.

[62] Noell S, Fallier-Becker P, Beyer C, et al. Effects of agrin on the expression and distribution of the water channel protein aquaporin-4 and volume regulation in cultured astrocytes[J]. Eur J Neurosci, 2007, 26（8）: 2109-2118.

[63] Noell S, Fallier-Becker P, Deutsch U, et al. Agrin defines polarized distribution of orthogonal arrays of particles in astrocytes[J]. Cell Tissue Res, 2009, 337（2）: 185-195.

[64] Dalloz C, Sarig R, Fort P, et al. Targeted inactivation of dystrophin gene product Dp71: phenotypic impact in mouse retina[J]. Hum Mol Genet, 2003, 12（13）: 1543-1554.

[65] Rurak J, Noel G, Lui L, et al. Distribution of potassium ion and water permeable channels at perivascular glia in brain and retina of the Large（myd）mouse[J]. J Neurochem, 2007, 103（5）: 1940-1953.

[66] Nagelhus EA, Horio Y, Inanobe A, et al. Immunogold evidence suggests that coupling of K^+ siphoning and water transport in rat retinal Müller cells is mediated by a coenrichment of Kir4.1 and AQP4 in specific membrane domains[J]. Glia, 1999, 26（1）: 47-54.

[67] Nagelhus EA, Mathiisen TM, Ottersen OP. Aquaporin-4 in the central nervous system: cellular and subcellular distribution and coexpression with KIR4.1[J]. Neuroscience, 2004, 129（4）: 905-913.

[68] Ruiz-Ederra J, Zhang H, Verkman AS. Evidence against functional interaction between aquaporin-4 water channels and Kir4.1 potassium channels in retinal Müller cells[J]. J Biol Chem, 2007, 282（30）: 21866-21872.

[69] Zhang H, Verkman AS. Evidence against involvement of aquaporin-4 in cell-cell adhesion[J]. J Mol Biol, 2008, 382（5）: 1136-1143.

[70] Illarionova NB, Gunnarson E, Li Y, et al. Functional and molecular interactions between aquaporins and Na, K-ATPase[J]. Neuroscience, 2010, 168（4）: 915-925.

[71] Nicchia GP, Srinivas M, Li W, et al. New possible roles for aquaporin-4 in astrocytes: cell cytoskeleton and functional relationship with connexin43[J]. FASEB J, 2005, 19（12）: 1674-1676.

[72] Lichter-Konecki U, Mangin JM, Gordish-Dressman H, et al. Gene expression profiling of astrocytes from hyperammonemic mice reveals altered pathways for water and potassium homeostasis in vivo[J]. Glia, 2008, 56（4）: 365-377.

[73] Saadoun S, Papadopoulos MC, Davies DC, et al. Aquaporin-4 expression is increased in oedematous human brain tumours[J]. J Neurol Neurosurg Psychiatry, 2002, 72（2）: 262-265.

[74] Fatemi SH, Folsom TD, Reutiman TJ, et al. Expression of astrocytic markers aquaporin 4 and connexin 43 is altered in brains of subjects with autism[J]. Synapse, 2008, 62（7）: 501-507.

[75] Rash JE, Davidson KG, Kamasawa N, et al. Ultrastructural localization of connexins（Cx36, Cx43, Cx45）, glutamate receptors and aquaporin-4 in rodent olfactory mucosa, olfactory nerve and olfactory bulb[J]. J Neurocytol, 2005, 34（3-5）: 307-341.

[76] Verkman AS, Binder DK, Bloch O, et al. Three distinct roles of aquaporin-4 in brain function revealed by knockout mice[J]. Biochim Biophys Acta, 2006, 1758（8）: 1085-1093.

[77] Crane JM, Van Hoek AN, Skach WR, et al. Aquaporin-4 dynamics in orthogonal arrays in live cells visualized by quantum dot single particle tracking[J]. Mol Biol Cell, 2008, 19（8）: 3369-3378.

[78] Zhao M, Valamanesh F, Celerier I, et al. The neuroretina is a novel mineralocorticoid target: aldosterone up-regulates ion and water channels in Müller glial cells[J]. FASEB J, 2010, 24(9): 3405-3415.

[79] Zhao M, Bousquet E, Valamanesh F, et al. Differential regulations of AQP4 and Kir4.1 by triamcinolone acetonide and dexamethasone in the healthy and inflamed retina[J]. Invest Ophthalmol Vis Sci, 2011, 52(9): 6340-6347.

[80] Dibas A, Yang MH, He S, et al. Changes in ocular aquaporin-4 (AQP4) expression following retinal injury[J]. Mol Vis, 2008, 14: 1770-1783.

[81] Iandiev I, Pannicke T, Reichenbach A, et al. Diabetes alters the localization of glial aquaporins in rat retina[J]. Neurosci Lett, 2007, 421(2): 132-136.

[82] Li SY, Yang D, Yeung CM, et al. Lycium barbarum polysaccharides reduce neuronal damage, blood-retinal barrier disruption and oxidative stress in retinal ischemia/reperfusion injury[J]. PLoS One, 2011, 6(1): e16380.

[83] Kim JE, Park H, Lee JE, et al. Blockade of 67-kDa laminin receptor facilitates AQP4 down-regulation and BBB disruption via ERK1/2-and p38 MAPK-mediated PI3K/AKT activations[J]. Cells, 2020, 9(7): 1670.

[84] Zhang G, Ma P, Wan S, et al. Dystroglycan is involved in the activation of ERK pathway inducing the change of AQP4 expression in scratch-injured astrocytes[J]. Brain Res, 2019, 1721: 146347.

[85] Chu H, Yang X, Huang C, et al. Apelin-13 protects against ischemic blood-brain barrier damage through the effects of aquaporin-4[J]. Cerebrovasc Dis, 2017, 44(1-2): 10-25.

[86] Mecchia A, Palumbo C, De Luca A, et al. High glucose induces an early and transient cytoprotective autophagy in retinal Müller cells[J]. Endocrine, 2022, 77(2): 221-230.

[87] Chen Y, Gao F, Jiang R, et al. Down-regulation of AQP4 expression via p38 MAPK signaling in temozolomide-induced glioma cells growth inhibition and invasion impairment[J]. J Cell Biochem, 2017, 118(12): 4905-4913.

[88] Riemma G, Laganà AS, Schiattarella A, et al. Ion channels in the pathogenesis of endometriosis: a cutting-edge point of view[J]. Int J Mol Sci, 2020, 21(3): 1114.

[89] Nagase H, Agren J, Saito A, et al. Molecular cloning and characterization of mouse aquaporin 6[J]. Biochem Biophys Res Commun, 2007, 352(1): 12-16.

[90] Sakai H, Sato K, Kai Y, et al. Distribution of aquaporin genes and selection of individual reference genes for quantitative real-time RT-PCR analysis in multiple tissues of the mouse[J]. Can J Physiol Pharmacol, 2014, 92(9): 789-796.

[91] Jeremic A, Cho WJ, Jena BP. Involvement of water channels in synaptic vesicle swelling[J]. Exp Biol Med(Maywood), 2005, 230(9): 674-680.

[92] Ishibashi K, Kuwahara M, Gu Y, et al. Cloning and functional expression of a new water channel abundantly expressed in the testis permeable to water, glycerol, and urea[J]. J Biol Chem, 1997, 272(33): 20782-20786.

[93] Shin I, Kim HJ, Lee JE, et al. Aquaporin 7 expression during perinatal development of mouse brain[J]. Neurosci Lett, 2006, 409(2): 106-111.

[94] Zhu SJ, Wang KJ, Gan SW, et al. Expression of aquaporin 8 in human astrocytomas: correlation with pathologic grade[J]. Biochem Biophys Res Commun, 2013, 440(1): 168-172.

[95] Manley GT, Fujimura M, Ma T, et al. Aquaporin-4 deletion in mice reduces brain edema after acute water intoxication and ischemic stroke[J]. Nat Med, 2000, 6(2): 159-163.

[96] Nagelhus EA, Veruki ML, Torp R, et al. Aquaporin-4 water channel protein in the rat retina and optic nerve: polarized expression in Müller cells and fibrous astrocytes[J]. J Neurosci, 1998, 18(7): 2506-2519.

[97] Oshio K, Binder DK, Yang B, et al. Expression of aquaporin water channels in mouse spinal cord[J].

Neuroscience, 2004, 127(3): 685-693.

[98] Gradilone SA, Carreras FI, Lehmann GL, et al. Phosphoinositide 3-kinase is involved in the glucagon-induced translocation of aquaporin-8 to hepatocyte plasma membrane[J]. Biol Cell, 2005, 97(11): 831-836.

[99] Soria LR, Gradilone SA, Larocca MC, et al. Glucagon induces the gene expression of aquaporin-8 but not that of aquaporin-9 water channels in the rat hepatocyte[J]. Am J Physiol Regul Integr Comp Physiol, 2009, 296(4): R1274-R1281.

[100] Arciénega II, Brunet JF, Bloch J, et al. Cell locations for AQP1, AQP4 and 9 in the non-human primate brain[J]. Neuroscience, 2010, 167(4): 1103-1114.

[101] Elkjaer M, Vajda Z, Nejsum LN, et al. Immunolocalization of AQP9 in liver, epididymis, testis, spleen, and brain[J]. Biochem Biophys Res Commun, 2000, 276(3): 1118-1128.

[102] Badaut J, Hirt L, Granziera C, et al. Astrocyte-specific expression of aquaporin-9 in mouse brain is increased after transient focal cerebral ischemia[J]. J Cereb Blood Flow Metab, 2001, 21(5): 477-482.

[103] Badaut J, Petit JM, Brunet JF, et al. Distribution of aquaporin 9 in the adult rat brain: preferential expression in catecholaminergic neurons and in glial cells[J]. Neuroscience, 2004, 128(1): 27-38.

[104] Jelen S, Parm Ulhøi B, Larsen A, et al. AQP9 expression in glioblastoma multiforme tumors is limited to a small population of astrocytic cells and CD15+/CalB(+)leukocytes[J]. PLoS One, 2013, 8(9): e75764.

[105] Rojek AM, Skowronski MT, Füchtbauer EM, et al. Defective glycerol metabolism in aquaporin 9 (AQP9)knockout mice[J]. Proc Natl Acad Sci USA, 2007, 104(9): 3609-3614.

[106] Badaut J, Brunet JF, Guérin C, et al. Alteration of glucose metabolism in cultured astrocytes after AQP9-small interference RNA application[J]. Brain Res, 2012, 1473: 19-24.

[107] Yamamoto N, Sobue K, Fujita M, et al. Differential regulation of aquaporin-5 and-9 expression in astrocytes by protein kinase A[J]. Brain Res Mol Brain Res, 2002, 104(1): 96-102.

[108] Yamamoto N, Sobue K, Miyachi T, et al. Differential regulation of aquaporin expression in astrocytes by protein kinase C[J]. Brain Res Mol Brain Res, 2001, 95(1-2): 110-116.

[109] Arima H, Yamamoto N, Sobue K, et al. Hyperosmolar mannitol simulates expression of aquaporins 4 and 9 through a p38 mitogen-activated protein kinase-dependent pathway in rat astrocytes[J]. J Biol Chem, 2003, 278(45): 44525-44534.

[110] Yang M, Gao F, Liu H, et al. Hyperosmotic induction of aquaporin expression in rat astrocytes through a different MAPK pathway[J]. J Cell Biochem, 2013, 114(1): 111-119.

[111] Wei X, Ren X, Jiang R, et al. Phosphorylation of p38 MAPK mediates aquaporin 9 expression in rat brains during permanent focal cerebral ischaemia[J]. J Mol Histol, 2015, 46(3): 273-281.

[112] Jeyaseelan K, Sepramaniam S, Armugam A, et al. Aquaporins: a promising target for drug development[J]. Expert Opin Ther Targets, 2006, 10(6): 889-909.

[113] Takata K. Aquaporins: water channel proteins of the cell membrane[J]. Prog Histochem Cytochem, 2004, 39(1): 1-83.

[114] Laforenza U, Scaffino MF, Gastaldi G. Aquaporin-10 represents an alternative pathway for glycerol efflux from human adipocytes[J]. PLoS One, 2013, 8(1): e54474.

[115] Flach CF, Qadri F, Bhuiyan TR, et al. Differential expression of intestinal membrane transporters in cholera patients[J]. FEBS Lett, 2007, 581(17): 3183-3188.

[116] Morishita Y, Matsuzaki T, Hara-chikuma M, et al. Disruption of aquaporin-11 produces polycystic kidneys following vacuolization of the proximal tubule[J]. Mol Cell Biol, 2005, 25(17): 7770-7779.

[117] Gorelick DA, Praetorius J, Tsunenari T, et al. Aquaporin-11: a channel protein lacking apparent transport function expressed in brain[J]. BMC Biochem, 2006, 7: 14.

[118] Koike S, Tanaka Y, Matsuzaki T, et al. Aquaporin-11(AQP11) expression in the mouse brain[J]. Int J Mol Sci, 2016, 17(6): 861.

[119] Ishibashi K, Tanaka Y, Morishita Y. The role of mammalian superaquaporins inside the cell[J]. Biochim Biophys Acta, 2014, 1840(5): 1507-1512.

[120] Yang B, Zador Z, Verkman AS. Glial cell aquaporin-4 overexpression in transgenic mice accelerates cytotoxic brain swelling[J]. J Biol Chem, 2008, 283(22): 15280-15286.

[121] Rützler M, Rojek A, Damgaard MV, et al. Temporal deletion of Aqp11 in mice is linked to the severity of cyst-like disease[J]. Am J Physiol Renal Physiol, 2017, 312(2): F343-F351.

[122] Inoue Y, Sohara E, Kobayashi K, et al. Aberrant glycosylation and localization of polycystin-1 cause polycystic kidney in an AQP11 knockout model[J]. J Am Soc Nephro, 2014, 25(12): 2789-2799.

[123] Isokpehi RD, Rajnarayanan RV, Jeffries CD, et al. Integrative sequence and tissue expression profiling of chicken and mammalian aquaporins. BMC Genomics, 2009, 10 Suppl 2(Suppl 2): S7.

[124] Itoh T, Rai T, Kuwahara M, et al. Identification of a novel aquaporin, AQP12, expressed in pancreatic acinar cells[J]. Biochem Biophys Res Commun, 2005, 330(3): 832-838.

第二章　水通道蛋白与中枢神经系统

第一节　水通道蛋白在中枢神经系统的定位分布

迄今，在哺乳动物体内发现的水通道蛋白家族成员共有13个，即AQP0～AQP12。其中，分布在中枢神经系统的有7个，即AQP1、AQP3、AQP4、AQP5、AQP8、AQP9和AQP11。

一、水通道蛋白在脑组织中的表达定位

1. AQP1在脑组织中的表达　AQP1主要表达于哺乳动物脑组织脉络丛上皮细胞顶膜的微绒毛表面（图2-1）[1]，与脑脊液的产生和吸收有关。研究发现，*AQP1*敲除小鼠的颅内压降低、脑脊液生成减少，这为AQP1参与脑脊液动力学提供了直接的功能证据[2]。有研究报道，AQP1还可能在衰老过程中的脉络丛进行性功能下降中发挥重要作用，该过程

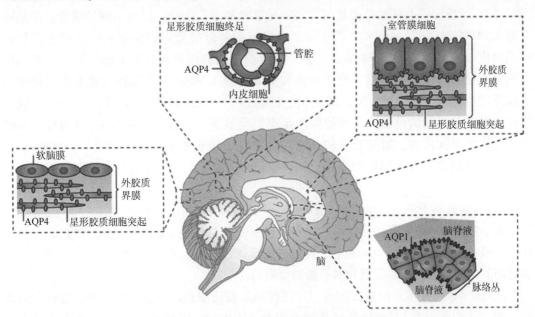

图2-1　AQP1和AQP4在脑组织中的表达定位

AQP1主要表达于脑组织脉络丛上皮细胞顶膜；AQP4主要表达于胶质界膜的星形胶质细胞膜、血管周围星形胶质细胞终足膜、室管膜细胞的基底外侧膜和室管膜下星形胶质细胞

与阿尔茨海默病风险的增加密切相关[3]。AQP1还表达于鼠神经鞘的胶质细胞及人的胶质细胞[4]。此外，有研究发现，AQP1在小鼠三叉神经节神经元中有表达[5, 6]。综上，AQP1在不同物种脑组织中的表达呈现多样性。

AQP1在脑肿瘤组织中也有表达。研究证实，AQP1在脑肿瘤细胞种植鼠的微血管内皮细胞中高表达，而在大鼠正常脑组织血管内皮细胞中则不表达[7, 8]，在脑肿瘤毛细血管中的表达可能与血脑屏障受损相关；而在正常脑组织中不表达可能是由于星形胶质细胞发出信号，抑制内皮细胞表达AQP1所致。AQP1不仅在脑肿瘤微血管内皮细胞有表达，在人和鼠脑肿瘤细胞如星形胶质细胞瘤细胞中也有表达[2]。在星形胶质细胞瘤中，AQP1在肿瘤组织的血管内皮细胞膜和星形胶质细胞膜的表达上调，其表达随其恶性程度的增加而增高[9]。在脉络丛肿瘤中，AQP1的表达上调，脑脊液的产生增加，进一步说明AQP1在脑脊液的产生中发挥了重要作用[3]。

2. AQP3在脑组织中的表达　　AQP3在脑组织中弱表达于脉络丛，散在分布于梨状皮质、海马和背侧丘脑的星形胶质细胞。此外，AQP3还分布于背侧丘脑和皮质下神经元[10]。在大鼠脑缺血模型早期，AQP3随着缺血时间的延长表达上调，但缺血时间超过6小时，随着神经元坏死，AQP3的表达将下调。这提示AQP3的异常表达与缺血性脑水肿的形成密切相关。

3. AQP4在脑组织中的表达　　AQP4是大脑中表达最多且最重要的水通道蛋白之一。AQP4最初在肺组织中发现，研究表明AQP4在兴奋性细胞中不表达，仅在支持细胞中表达[11]。在脑组织，AQP4在脑实质与脑室之间大量表达，主要表达于胶质界膜的星形胶质细胞膜（脑-蛛网膜下腔的脑脊液）、血管周围胶质细胞的终足膜（血-脑）、室管膜细胞的基底外侧膜和室管膜下星形胶质细胞的突起上（脑-脑室脑脊液）[10]（图2-1）。下丘脑视上核和室旁核的分泌加压素的神经元细胞，以及小脑的浦肯野细胞中也有AQP4的表达[12]。AQP4主要分布在内皮细胞邻近的星形胶质细胞终足膜上，如果缺乏内皮细胞的支持，AQP4则会在整个星形胶质细胞膜重新分布（如培养的星形胶质细胞和恶性星形胶质细胞瘤、胶质界膜和室周器官），提示内皮细胞可能发出信号至星形胶质细胞进而调控AQP4在该细胞膜上的表达[13]。AQP4在毛细血管内皮细胞、软脑膜和脑室室管膜侧的星形胶质细胞膜或足突上呈显著极性分布，提示AQP4的分布与脑内水转运具有同向性，对脑脊液的分泌和重吸收也起着非常重要的作用[14]。AQP4在缺血性脑水肿的星形胶质细胞中表达上调，提示AQP4与脑水肿情况下水平衡的调控直接相关。

AQP4还分布于海马齿状回与小脑神经细胞，主要参与调节细胞间隙和钾离子的浓度，与钾离子通道Kir4.1一起协同调节相关神经元的兴奋性（图2-2）[15]。有研究报道，AQP4缺失小鼠的脑细胞在受到电刺激后，钾离子的清除延迟[16]，提示AQP4参与水、钠、钾平衡的调节，具有渗透压感受器和水平衡调节器的功能[17]。

此外，AQP4在星形胶质细胞瘤、反应性星形胶质细胞以及反应性小胶质细胞中的表达上调，其变化与这3种细胞的迁移能力相关（图2-2）。研究证实，AQP4缺失小鼠皮质损伤后的胶质瘢痕延迟，进一步说明AQP4与细胞的迁移相关[18]。

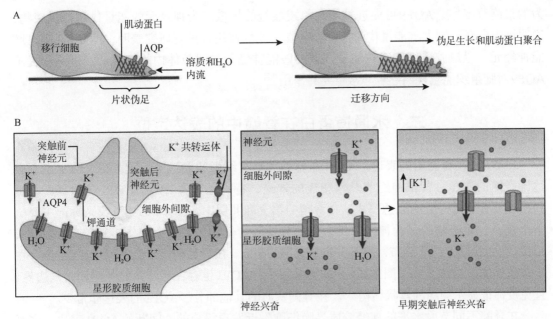

图 2-2 AQP4 在星形胶质细胞的表达参与其迁移和神经兴奋

A：AQP4 可能在星形胶质细胞迁移中发挥作用。左图是肌动蛋白（actin）解聚和活性溶质内流形成的渗透梯度，水主要通过伪足中的 AQP4 进入细胞质，促进了细胞迁移方向上的伪足延伸（右图）。B：AQP4 在突触间隙外的神经兴奋中发挥作用（左图），神经元释放 K^+，星形胶质细胞摄取 K^+ 和水（中间图），AQP4 则转运水进入星形胶质细胞，导致细胞外间隙（ECS）的体积减小和 K^+ 浓度增加，进一步促进星形胶质细胞对 K^+ 的摄取（右图）

4. AQP5、AQP8 在脑组织中的表达　大鼠脑组织中，AQP5 与 AQP8 的分布相似。研究表明，AQP5 与 AQP8 分布于梨状皮质、背侧丘脑的神经元和星形胶质细胞，弱表达于脉络丛[19]。在大鼠脑缺血早期，与 AQP3 一样，AQP5、AQP8 随着缺血时间的延长表达上调，但随着缺血时间的进一步延长，AQP5、AQP8 的表达会降低[20, 21]。该项研究提示 AQP3、AQP5、AQP8 协同参与缺血损伤后脑水肿的形成。

5. AQP9 在脑组织中的表达　AQP9 最初于 1997 年在人类脂肪组织基因序列系统分析中被发现[21]。AQP9 在脑组织中主要分布于脑室周围室管膜细胞、蛛网膜下腔和脑室周围的星形胶质细胞、白质及海马区的星形胶质细胞，以及下丘脑视上核、室旁核和视交叉上核等部位的神经元[11, 22, 23]。AQP9 作为水甘油通道蛋白，不仅对水具有通透性，对尿素、甘油、甘露醇、乳酸等小分子物质也具有通透性[24]。AQP9 在大鼠脑出血模型早期的表达下调，以减少细胞间液进入细胞内从而延缓细胞水肿，而后期 AQP9 的表达则迅速增高，促进脑水肿的形成，提示 AQP9 可能是渗透压感受器，对脑组织水的转运和脑脊液循环起着重要作用，参与水、电解质的调节。

AQP9 在下丘脑的分布提示其与神经内分泌密切相关。此外，AQP9 还在儿茶酚胺能神经元的胞体和突起上表达，提示其可能在脑组织的能量代谢中具有重要作用[25]。AQP9 绝大部分分布于儿茶酚胺能神经元，在大鼠和小鼠中这些神经元以表达酪氨酸羟化酶为特征，而儿茶酚胺能神经元与神经肽 Y 等其他类型神经元连接，参与动物体内能量代谢的调节。儿茶酚胺能神经元的电位活动在甘油和乳酸浓度上升时发生改变，造成包括摄食行为的最终改变[26]。AQP9可促进甘油和单羧酸盐的扩散，为神经元的能量代谢提供底物，说明该蛋白对控制大脑能量代谢发挥着非常重要的作用。另有研究表明，AQP9在多巴胺能神经元、星形胶质细胞线粒体及儿茶酚胺能神经元的线粒体内均有表达，进一步提示AQP9对脑组织能量代谢起着非常重要的作用[27]。

二、水通道蛋白在脊髓中的表达定位

1. AQP1在脊髓中的表达 AQP1在脊髓和脑组织中的表达不一致，在脊髓中主要表达于脊髓的背角、背根神经节及其相连的神经，特别是坐骨神经。AQP1在颈髓、胸髓和腰髓具有非常一致的表达模式。在背角的第Ⅰ和Ⅱ层中，AQP1的免疫荧光标记主要位于小直径、无髓鞘的感觉神经纤维，少部分位于含髓鞘的神经元（图2-3A，D）。此外，AQP1在突触膜上广泛分布，但不分布于突触间隙[4]。在第Ⅲ和第Ⅳ层中，AQP1主要存在于背角外侧、内侧边界（图2-3C）。在第Ⅴ层中，AQP1也存在于背角的外侧和内侧边界，且免疫标记显示缠结的丝状结构[28]。在脊髓灰质的其他部分，AQP1的表达非常少。

在脊髓不同节段水平的神经胶质界膜细胞附近发现稀疏的、间断的AQP1表达，这可能与白质和灰质中动脉血管突出的小动脉的分布有关（图2-3D～G），此种现象与其在人和大鼠的脑实质中的散发表达一致[29-31]。另外，有研究发现AQP1还表达于大鼠脊髓的星形胶质细胞[32]。

2. AQP4在脊髓中的表达 AQP4在脊髓中广泛表达，但与其在脑组织中的表达位置也不一致。AQP4表达在整段脊髓的白质和灰质的星形胶质细胞上（图2-4A）[33]。在脊髓白质纵向切片中发现，AQP4在星形胶质细胞突起包裹有髓神经纤维处大量表达[34]（图2-4E）。在脊髓灰质中，AQP4在背角的第Ⅰ和第Ⅱ层表达（图2-4A），在脊髓中央管、软脊膜以及神经元的细胞突起中呈极性表达，且在血管周围检测到更明显的免疫标记，通常以围绕血管的完整环的形式分布（图2-4B，C）[33-35]。AQP4在脊髓中的分布特点充分反映出其与脊髓水转运密切相关。此外，在脊髓腹侧的树突、神经纤维网和运动神经元周围也观察到丝状和斑点状AQP4的表达（图2-4C，E）[36]。

3. AQP8在脊髓中的表达 AQP8主要表达于脊髓的室管膜细胞，星形胶质细胞和少突胶质细胞中也有少量AQP8的表达[37]。

4. AQP9在脊髓中的表达 在脊髓中，AQP9主要表达于胶质界膜的星形胶质细胞突起和白质。免疫荧光标记显示，AQP9阳性信号分布可从脊髓白质的星形胶质细胞突起，至胶质细胞界膜、中央管和脊髓背角处[38]。然而，目前对于AQP9在脊髓中的表达定位的数据尚不充分，还需进一步深入研究[39]。

迄今，关于中枢神经系统中AQP定位表达的研究主要集中于AQP1、AQP4和AQP9。AQP1在脑脉络丛上皮细胞顶膜高表达，与脑脊液的分泌有关。AQP4是中枢神经系统内表达最多的水通道蛋白，尤其在室管膜下星形胶质细胞和室管膜细胞基底外侧膜表达，提示AQP4在促进水分子进出大脑和脊髓中起着重要作用。AQP9在中枢神经系统中也有表达，特别是中脑儿茶酚胺能神经元、多巴胺能神经元及星形胶质细胞，提示AQP9在中枢神经系统能量代谢中发挥着重要作用。这三种水通道蛋白的分布归纳见表2-1[33]。

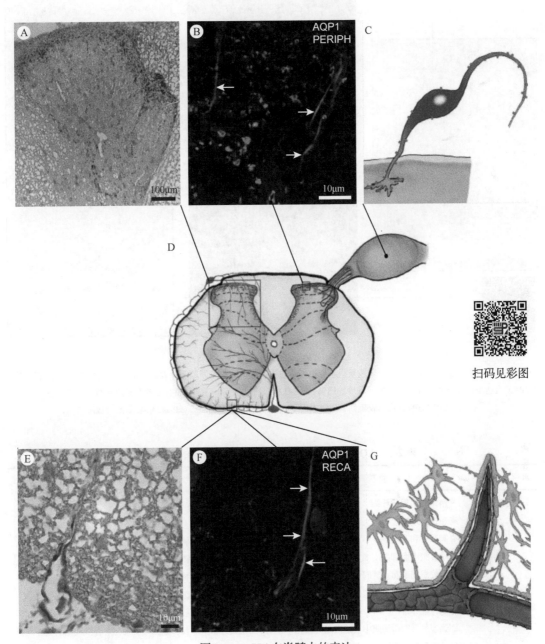

图2-3　AQP1在脊髓中的表达

AQP1信号在脊髓背角的第Ⅰ、Ⅱ层高表达，而在背角内侧边缘至第Ⅵ层其表达信号逐渐减弱（A，D）；少部分AQP1在无髓鞘神经元细胞（B）以及从背根神经节向背角前表层发出的轴突形成的周围感觉神经纤维表达（C）；AQP1在脊髓冠状切面的表达示意图（D）；AQP1在脊髓白质中表达很少，主要分布在靠近胶质界膜处（E）的动脉血管分支小动脉的位置（F，G）。PERIPH，外周蛋白；RECA，大鼠内皮细胞抗原

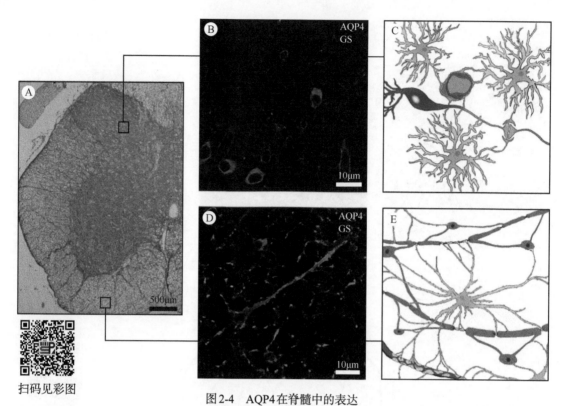

图 2-4　AQP4 在脊髓中的表达

AQP4 在整段脊髓中大量表达，并在背角浅层、腹角边缘、胶质界膜区和血管周围明显表达（A）；灰质原浆型星形胶质细胞 AQP4 的表达呈极化模式，其信号在星形胶质细胞足突围绕血管处强度最高（B）；而在三者形成的突触部位表达较弱（C）；白质中 AQP4 在星形胶质细胞膜表达（D，E）。GS（glutamate synthase），谷氨酸合成酶

表 2-1　AQP1、AQP4、AQP9 在脑和脊髓中的表达比较

水通道蛋白	脊髓	脑
AQP1	背角无髓鞘感觉纤维	脉络膜丛上皮细胞
	有髓神经纤维（稀疏）	室周器官内皮细胞
	脊髓 V 层和 X 层	脑实质内血管内皮细胞（稀疏）
	胶质界膜小血管内皮细胞	星形胶质细胞，动眼神经周围的施万细胞和三叉神经纤维
	星形胶质细胞和室管膜细胞	软脑膜血管表面的神经元
AQP4	星形胶质细胞端足突包围毛细血管	软脑膜下星形胶质细胞形成胶质界膜
	有髓神经纤维	皮质内血管周星形胶质细胞足突
	原浆型星形胶质细胞中表达向足突极化	高极化表达于星形胶质细胞终足膜，室管膜细胞基底外侧膜
	纤维型星形胶质细胞中分布均匀	
AQP9	白质和胶质界膜的星形胶质细胞突起	神经元、室管膜细胞、下丘脑胶质细胞
		蛛网膜下腔和脑室附近的星形胶质细胞

（陈玉琴　杨　美）

参 考 文 献

[1] Papadopoulos MC, Verkman AS. Aquaporin water channels in the nervous system[J]. Nat Rev Neurosci, 2013, 14(4): 265-277.

[2] Oshio K, Binder DK, Liang Y, et al. Expression of the aquaporin-1 water channel in human glial tumors[J]. Neurosurgery, 2005, 56(2): 375-381; discussion 375-381.

[3] Longatti P, Basaldella L, Orvieto E, et al. Aquaporin(s) expression in choroid plexus tumours[J]. Pediatr Neurosurg, 2006, 42(4): 228-233.

[4] Shields SD, Moore KD, Phelps PE, et al. Olfactory ensheathing glia express aquaporin 1[J]. J Comp Neurol, 2010, 518(21): 4329-4341.

[5] Shields SD, Mazario J, Skinner K, et al. Anatomical and functional analysis of aquaporin 1, a water channel in primary afferent neurons[J]. Pain, 2007, 131(1-2): 8-20.

[6] Nandasena BG, Suzuki A, Aita M, et al. Immunolocalization of aquaporin-1 in the mechanoreceptive Ruffini endings in the periodontal ligament[J]. Brain Res, 2007, 1157: 32-40.

[7] Endo M, Jain RK, Witwer B, et al. Water channel(aquaporin 1)expression and distribution in mammary carcinomas and glioblastomas[J]. Microvasc Res, 1999, 58(2): 89-98.

[8] Kobayashi H, Minami S, Itoh S, et al. Aquaporin subtypes in rat cerebral microvessels[J]. Neurosci Lett, 2001, 297(3): 163-166.

[9] Dolman D, Drndarski S, Abbott NJ, et al. Induction of aquaporin 1 but not aquaporin 4 messenger RNA in rat primary brain microvessel endothelial cells in culture[J]. J Neurochem, 2005, 93(4): 825-833.

[10] Benga O, Huber VJ. Brain water channel proteins in health and disease[J]. Mol Aspects Med, 2012, 33(5-6): 562-578.

[11] Xu M, Xiao M, Li S, et al. Aquaporins in nervous system[J]. Adv Exp Med Biol, 2017, 969: 81-103.

[12] Tait MJ, Saadoun S, Bell BA, et al. Water movements in the brain: role of aquaporins[J]. Trends Neurosci, 2008, 31(1): 37-43.

[13] Nagelhus EA, Ottersen OP. Physiological roles of aquaporin-4 in brain[J]. Physiol Rev, 2013, 93(4): 1543-1562.

[14] Buffoli B. Aquaporin biology and nervous system[J]. Curr Neuropharmacol, 2010, 8(2): 97-104.

[15] Benfenati V, Ferroni S. Water transport between CNS compartments: functional and molecular interactions between aquaporins and ion channels[J]. Neuroscience, 2010, 168(4): 926-940.

[16] Binder DK, Oshio K, Ma T, et al. Increased seizure threshold in mice lacking aquaporin-4 water channels[J]. Neuroreport, 2004, 15(2): 259-262.

[17] Assentoft M, Larsen BR, MacAulay N, 2015. Regulation and function of AQP4 in the central nervous system[J]. Neurochem Res, 40(12): 2615-2627.

[18] Saadoun S, Papadopoulos MC, Watanabe H, et al. Involvement of aquaporin-4 in astroglial cell migration and glial scar formation[J]. J Cell Sci, 2005, 118(Pt 24): 5691-5698.

[19] Yang M, Gao F, Liu H, et al. Immunolocalization of aquaporins in rat brain[J]. Anat Histol Embryol, 2011, 40(4): 299-306.

[20] Yang M, Gao F, Liu H, et al. Hyperosmotic induction of aquaporin expression in rat astrocytes through a different MAPK pathway[J]. J Cell Biochem, 2013, 114(1): 111-119.

[21] Yang M, Gao F, Liu H, et al. Temporal changes in expression of aquaporin-3, -4, -5 and-8 in rat brains after permanent focal cerebral ischemia[J]. Brain Res, 2009, 1290: 121-132.

[22] Dasdelen D, Mogulkoc R, Baltaci AK. Aquaporins and roles in brain health and brain injury[J]. Mini Rev Med Chem, 2020, 20(6): 498-512.

[23] Badaut J, Lasbennes F, Magistretti PJ, et al. Aquaporins in brain: distribution, physiology, and pathophysiology[J]. J Cereb Blood Flow Metab, 2002, 22(4): 367-378.

[24] Potokar M, Jorgačevski J, Zorec R. Astrocyte aquaporin dynamics in health and disease[J]. Int J Mol Sci, 2016, 17(7): 1121.

[25] Badaut J, Petit JM, Brunet JF, et al. Distribution of Aquaporin 9 in the adult rat brain: preferential expression in catecholaminergic neurons and in glial cells[J]. Neuroscience, 2004, 128(1): 27-38.

[26] Badaut J, Regli L. Distribution and possible roles of aquaporin 9 in the brain[J]. Neuroscience, 2004, 129(4): 971-981.

[27] Méndez-Giménez L, Rodríguez A, Balaguer I, et al. Role of aquaglyceroporins and caveolins in energy and metabolic homeostasis[J]. Mol Cell Endocrinol, 2014, 397(1-2): 78-92.

[28] Oklinski MK, Lim JS, Choi HJ, et al. Immunolocalization of water channel proteins AQP1 and AQP4 in rat spinal cord[J]. J Histochem Cytochem, 2014, 62(8): 598-611.

[29] Nielsen S, Smith BL, Christensen EI, et al. Distribution of the aquaporin CHIP in secretory and resorptive epithelia and capillary endothelia[J]. Proc Natl Acad Sci USA, 1993, 90(15): 7275-7279.

[30] Oshio K, Watanabe H, Song Y, et al. Reduced cerebrospinal fluid production and intracranial pressure in mice lacking choroid plexus water channel Aquaporin-1[J]. FASEB J, 2005, 19(1): 76-78.

[31] Verkman AS. Aquaporins in endothelia[J]. Kidney Int, 2006, 69(7): 1120-1123.

[32] Nesic O, Lee J, Unabia GC, et al. Aquaporin 1—a novel player in spinal cord injury[J]. J Neurochem, 2008, 105(3): 628-640.

[33] Oklinski MK, Skowronski MT, Skowronska A, et al. Aquaporins in the spinal cord[J]. Int J Mol Sci, 2016, 17(12): 2050.

[34] Smith AJ, Verkman AS. Superresolution imaging of aquaporin-4 cluster size in antibody-stained paraffin brain sections[J]. Biophys J, 2015, 109(12): 2511-2522.

[35] Vitellaro-Zuccarello L, Mazzetti S, Bosisio P, et al. Distribution of aquaporin 4 in rodent spinal cord: relationship with astrocyte markers and chondroitin sulfate proteoglycans[J]. Glia, 2005, 51(2): 148-159.

[36] Yeo SI, Ryu HJ, Kim JE, et al. The effects of electrical shock on the expressions of aquaporin subunits in the rat spinal cords[J]. Anat Cell Biol, 2011, 44(1): 50-59.

[37] Oshio K, Binder DK, Yang B, et al. Expression of aquaporin water channels in mouse spinal cord[J]. Neuroscience, 2004, 127(3): 685-693.

[38] Hu AM, Li JJ, Sun W, et al. Myelotomy reduces spinal cord edema and inhibits aquaporin-4 and aquaporin-9 expression in rats with spinal cord injury[J]. Spinal Cord, 2015, 53(2): 98-102.

[39] Arciénega II, Brunet JF, Bloch J, et al. Cell locations for AQP1, AQP4 and 9 in the non-human primate brain[J]. Neuroscience, 2010, 167(4): 1103-1114.

第二节　水通道蛋白在中枢神经系统的锚定

一、AQP4的极性表达模式

AQP4作为脑内最主要、含量最丰富的水通道蛋白，主要表达于星形胶质细胞胞膜。星形胶质细胞胞体上具有非常丰富的突起，部分突起的末端膨大呈扁平状，称为终足。终足在结构、功能，特别是膜分子组成方面，均与星形胶质细胞胞膜的其他部位存在很大差异[1]。最具代表性的特征是在软脑膜和室管膜下，以及血管周围的终足膜上，AQP4极度密集地表达[2]。仅占星形胶质细胞膜表面积约10%的终足，密集表达了大量的AQP4，而其余90%的细胞膜区域，则表达密度较低的AQP4，这称为AQP4的极性分布[1, 3]。这种极性分布很早以前就被发现，但直到最近才明确，终足上密集表达的AQP4并非以蛋白单体的模式表达，而是以粒子正交阵列（orthogonal arrays of particles，OAP）的形式排列。早在1995年人们就发现星形胶质细胞上有OAP形式的粒子排列，Wolburg[4]当时还对OAP的研究进行了系统的综述，即使那时AQP已经被发现[5, 6]，但在很长的一段时间里科学家也未将二者联系到一起。直到十多年后，Rash等[7, 8]才证实，OAP是由AQP4组成的。AQP4通常是以4个单体组成的四聚体在细胞膜上表达，每个单体包含一个水通道，OAP是以AQP4四聚体为单位基础而构建的，每个OAP可以包含4～100多个四聚体[4, 9]。AQP4有两个单体，较长的M1和稍短的M23。与M1相比，M23表现出更强大的输水能力，它通常形成稳定的大型筏状OAP晶格，而M1不形成或形成较小的阵列。所以，M1与M23的比例决定OAP数目的大小，M23容易形成大数目的、结构较稳定的OAP，而M1则形成较小数目的OAP[10]。

二、AQP4极性分布的锚定机制

AQP4的这种极性分布对脑内水转运起着至关重要的作用。密集表达AQP4的星形胶质细胞终足在软脑膜和室管膜下，以及血管周围形成了一层胶质界膜，这些界膜均与脑的液体腔隙相邻，水则通过这些胶质界膜上的水通道进入或流出脑实质：通过血管周围的胶质界膜进/出血管腔；通过脑表面的胶质界膜进/出蛛网膜下腔；通过脑室周围的胶质界膜进/出脑室[11]。AQP4极性的消失，必定会引起脑内水转运的紊乱。Amiry-Moghaddam等在研究中发现AQP4细胞内的锚定蛋白α-syntrophin基因被敲除后，AQP4的极性表达消失，并出现严重的脑水肿，因此他们提出AQP4在星形胶质细胞细胞膜上的极性分布与其锚定机制有关[1, 12]。AQP4是通过以前被称为"抗肌萎缩相关复合物"（dystroglycan-associated protein complex，DAPC）[1, 12, 13]（图2-5）而现在更多地被称为"抗肌萎缩蛋白（肌营养不良蛋白）-抗肌萎缩蛋白聚糖复合物"（dystrophin-dystroglycan complex，DDC）[1, 3, 14-19]的结构锚定在星形胶质细胞胞膜上的。

DDC包括位于细胞外的α-抗肌萎缩蛋白聚糖（α-dystroglycan，α-DG）、跨膜的β-抗肌萎缩蛋白聚糖（β-dystroglycan，β-DG）和胞内的抗肌萎缩蛋白（dystrophin）。此复合物在细胞内通过细胞骨架蛋白中抗肌萎缩蛋白上的α-肌营养蛋白或其他syn与AQP4结合，将AQP4锚定在细胞膜上；在细胞外其通过层粘连蛋白和蛋白聚糖与基底膜相连接[20, 21]。

在细胞内，α-syntrophin是最早被关注的AQP4的锚定蛋白。它位于细胞骨架蛋白中抗肌萎缩蛋白上，通过其PDZ结构域与AQP4的M23或M1亚基相锚定。α-syntrophin和M1亚基的结合目前还未被证实，而研究已证明α-syntrophin的PDZ结构域与AQP4的M23亚基通过直接或间接结合相锚定[3]。且对α-syntrophin敲除小鼠的研究表明，α-syntrophin和AQP4的极性表达、重新分布，以及脑水肿的形成密切相关[1, 3, 13, 22]。

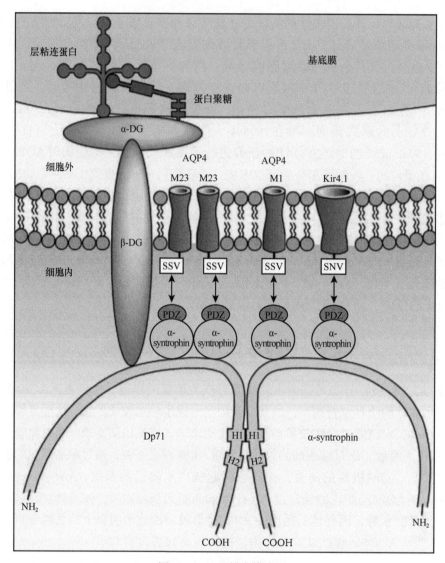

图 2-5　AQP4 锚定模式图

AQP4. 水通道蛋白 4；Kir4.1. 一种钾离子通道；M23/M1. 水通道蛋白 4 的两个亚基；SSV. AQP4 蛋白结构中的 SSV 结构域；SNV. Kir4.1 蛋白结构中的 SNV 结构域；PDZ. syn 蛋白结构中的 PDZ 结构域

在细胞外，DDC 通过 α-DG 与层粘连蛋白和蛋白聚糖相结合被锚定在基底膜上。蛋白聚糖是位于基底膜内的一种硫酸乙酰肝素蛋白聚糖，其 N 端含有一个与 K 型蛋白酶抑制剂同源的序列，该序列介导了蛋白聚糖与基底膜层粘连蛋白的结合。蛋白聚糖的 C 端含有 3 个与黏附分子同源的 G 结构域，介导了蛋白聚糖与 α-DG 以钙依赖的方式结合[4, 23]。蛋白聚糖最早是因为对神经-肌肉接头处乙酰胆碱受体的聚集起作用而被发现的；最近发现其在中枢神经系统的毛细血管内皮细胞、脑膜细胞及星形胶质细胞的基底膜上均有表达[15]，对突触联系的调节[14, 24, 25]、血脑屏障完整性的维持均有重要的作用[14, 15]。虽然 AQP4 的锚定机制在 2003 年就已经被提出[1]，但在接下来的很长一段时间里蛋白聚糖与 AQP4 极性表达的关系并未引起很大关注。即使在 2004 年，Warth 等曾发现在人脑胶质瘤中，毛细血管基底膜上蛋白聚糖表达缺失，同时 AQP4 失去极性并重新分布[26]，但这也仅仅揭示了两种同时存在的现象而已，真正揭开其二者关系的研究是最近的一些实验。Noell 等通过基因敲除、基因沉默等方法，发现蛋白聚糖对 AQP4 在星形胶质细胞终足上形成 OAP 结构是必要的[14, 15, 27]，并可促进 OAP 的形成，还能促进 AQP4 在星形胶质细胞膜上的插入；蛋白聚糖消失所造成的 OAP 结构消失及 AQP4 在细胞膜上的重分布，是导致人脑胶质瘤水肿形成的原因[4, 17]。

层粘连蛋白是细胞外基质中的大分子多链糖蛋白，主要存在于基底膜的透明层。它是由 α、β、γ 三条不同的多肽链组成的异构三聚体，在其长臂的末端有 G 结构域，可通过此结构域同 α-DG 结合。在中枢神经系统，与蛋白聚糖类似，层粘连蛋白也表达于血管内皮及星形胶质细胞的基底膜内[4]，已有研究表明，层粘连蛋白参与钾通道 Kir4.1 和 AQP4 在星形胶质细胞终足膜上的正确定位[28, 29]。笔者所在课题组在加压培养的 Müller 细胞中，也发现层粘连蛋白与 AQP4 的异常分布关系密切。

无论是蛋白聚糖还是层粘连蛋白，它们均不与 AQP4 直接连接，而是与 α-DG 直接相连。在表达低糖基化 α-DG 的 Largemyd 小鼠中，发现 α-DG 和层粘连蛋白或蛋白聚糖的相互作用严重受损，且此部位的 AQP4 表达量明显减少（约75%），并伴有相应的 α-肌营养蛋白缺失[30]。在人脑胶质瘤组织中，发现蛋白聚糖和 α-DG 均消失，而 AQP4 的表达与 OAP 的表达变化不一致[17]；在 α-DG 基因敲除后，脑表面区域的 AQP4 不能形成 OAP 结构，而脑实质深部的 AQP4 及 OAP 结构无明显变化，血管周围细胞终足膜上 AQP4 虽然形成部分 OAP 结构，但其表达还是受到抑制[18]。笔者所在课题组发现在脑出血[31]、脑外伤[32-34]后 AQP4 极性表达模式的改变均与 α-DG 的表达变化密切相关，并与脑水肿的形成、发展、转归有密切关系。

综上所述，AQP4 的锚定相关蛋白，即 α-syntrophin、蛋白聚糖、层粘连蛋白和 α-DG，对 AQP4 在细胞膜上的正确定位及水转运功能的正常发挥起关键作用。如果 AQP4 的锚定机制发生改变，必将影响到 AQP4 对水的正常转运而导致脑水肿的形成。目前科学家都在寻求各种直接调控 AQP4 表达的方法，以调控脑水肿的发展。但至今尚未找到可用的、有效的 AQP4 的抑制剂或激动剂来调节 AQP4 的表达；而 Verkman 等[35]想通过 *AQP4* 基因敲除达到治疗脑水肿的目的，但发现 *AQP4* 基因敲除后，细胞毒性脑水肿减轻，而血管源性脑水肿却加重，还会引起细胞迁移和神经信号传导的改变。在对水平衡的维持中，AQP4 能导致脑内水的聚集，同时也能增强水的排出。*AQP4* 基因敲除后，虽然会减轻脑内水的聚集，但同时也会导致脑内水的清除障碍[36]。由此可见，从基因水平直接干预 AQP4 的表达也很难达到治疗脑水肿的目的。另外，Romeiro 等[37]对102名外伤性脑水肿患者的 *AQP4* 基因中与水转运密切相关的外显子4进行检测，结果并未发现异常，提示脑外伤后脑水肿的形成不是由 *AQP4* 基因改变造成其蛋白结构变化所致。那么脑水肿的形成可能与 AQP4 的极性表达模式有关。所以，从 AQP4 的锚定机制入手，调节 AQP4 极性表达的模式，从而调控颅内水的转运途径、速度及模式，可能是各种病理状态下预防及治疗脑水肿的一个可行的新思路。

（刘　辉　易耀兴）

参 考 文 献

[1] Amiry-Moghaddam M, Ottersen OP. The molecular basis of water transport in the brain[J]. Nat Rev Neurosc, 2003, 4(12): 991-1001.

[2] Yang J, Lunde LK, Nuntagij P, et al. Loss of astrocyte polarization in the Tg-ArcSwe mouse model of Alzheimer's disease[J]. J Alzheimers Dis, 2011, 27(4): 711-722.

[3] Amiry-Moghaddam M, Frydenlund DS, Ottersen OP. Anchoring of aquaporin-4 in brain: molecular mechanisms and implications for the physiology and pathophysiology of water transport[J]. Neuroscience, 2004, 129(4): 999-1010.

[4] Wolburg H, Noell S, Fallier-Becker P, et al. The disturbed blood-brain barrier in human glioblastoma[J]. Mol Aspects Med, 2012, 33(5-6): 579-589.

[5] Denker BM, Smith BL, Kuhajda FP, et al. Identification, purification, and partial characterization of a novel Mr 28,000 integral membrane protein from erythrocytes and renal tubules[J]. J Biol Chem, 1988, 263(30): 15634-15642.

[6] Preston GM, Carroll TP, Guggino WB, et al. Appearance of water channels in *Xenopus* oocytes expressing red cell CHIP28 protein[J]. Science, 1992, 256(5055): 385-387.

[7] Rash JE, Davidson KG, Yasumura T, et al. Freeze-fracture and immunogold analysis of aquaporin-4 (AQP4) square arrays, with models of AQP4 lattice assembly[J]. Neuroscience, 2004, 129(4): 915-934.

[8] Rash J. Molecular disruptions of the panglial syncytium block potassium siphoning and axonal saltatory conduction: pertinence to neuromyelitis optica and other demyelinating diseases of the central nervous system[J]. Neuroscience, 2010, 168(4): 982-1008.

[9] Wolburg H, Wolburg-Buchholz K, Fallier-Becker P, et al. Structure and functions of aquaporin-4-based orthogonal arrays of particles[J]. Int Rev Cell Mol Biol, 2011, 287: 1-41.

[10] Berezowski V, Fukuda AM, Cecchelli R, et al. Endothelial cells and astrocytes: a concerto en duo in ischemic pathophysiology[J]. Int J Cell Biol, 2012, 2012: 176287.

[11] Papadopoulos MC, Verkman AS. Aquaporin-4 and brain edema[J]. Pediatr Nephrol, 2007, 22(6): 778-784.

[12] Saadoun S, Papadopoulos MC. Aquaporin-4 in brain and spinal cord oedema[J]. Neuroscience, 2010, 168(4): 1036-1046.

[13] Bragg AD, Amiry-Moghaddam M, Ottersen OP, et al. Assembly of a perivascular astrocyte protein scaffold at the mammalian blood-brain barrier is dependent on alpha-syntrophin[J]. Glia, 2006, 53(8): 879-890.

[14] Wolburg H, Noell S, Wolburg-Buchholz K, et al. Agrin, aquaporin-4, and astrocyte polarity as an important feature of the blood-brain barrier[J]. Neuroscientist, 2009, 15(2): 180-193.

[15] Noell S, Fallier-Becker P, Deutsch U, et al. Agrin defines polarized distribution of orthogonal arrays of particles in astrocytes[J]. Cell Tissue Res, 2009, 337(2): 185-195.

[16] Fallier-Becker P, Sperveslage J, Wolburg H, et al. The impact of agrin on the formation of orthogonal arrays of particles in cultured astrocytes from wild-type and agrin-null mice[J]. Brain Res, 2011, 1367: 2-12.

[17] Noell S, Wolburg-Buchholz K, Mack AF, et al. Dynamics of expression patterns of AQP4, dystroglycan, agrin and matrix metalloproteinases in human glioblastoma[J]. Cell Tissue Res, 2012, 347(2): 429-441.

[18] Noell S, Wolburg-Buchholz K, Mack AF, et al. Evidence for a role of dystroglycan regulating the membrane architecture of astroglial endfeet[J]. Eur J Neurosci, 2011, 33(12): 2179-2186.

[19] Amiry-Moghaddam M, Otsuka T, Hurn PD, et al. An alpha-syntrophin-dependent pool of AQP4 in astroglial end-feet confers bidirectional water flow between blood and brain[J]. Proc Natl Acad Sci USA, 2003, 100(4): 2106-2111.

[20] Amiry-Moghaddam M, Williamson A, Palomba M, et al. Delayed K^+ clearance associated with aquaporin-4 mislocalization: phenotypic defects in brains of alpha-syntrophin-null mice[J]. Proc Natl Acad Sci USA, 2003, 100(23): 13615-13620.

[21] Nico B, Tamma R, Annese T, et al. Glial dystrophin-associated proteins, laminin and agrin, are downregulated in the brain of mdx mouse[J]. Lab Invest, 2010, 90(11): 1645-1660.

[22] Amiry-Moghaddam M, Xue R, Haug FM, et al. Alpha-syntrophin deletion removes the perivascular but not endothelial pool of aquaporin-4 at the blood-brain barrier and delays the development of brain edema in an experimental model of acute hyponatremia[J]. FASEB J, 2004, 18(3): 542-544.

[23] McMahan UJ, Horton SE, Werle MJ, et al. Agrin isoforms and their role in synaptogenesis[J]. Curr Opin Cell Biol, 1992, 4(5): 869-874.

[24] Kröger S, Schröder JE. Agrin in the developing CNS: new roles for a synapse organizer[J]. News Physiol Sci, 2002, 17: 207-212.

[25] Ksiazek I, Burkhardt C, Lin S, et al. Synapse loss in cortex of agrin-deficient mice after genetic rescue of perinatal death[J]. J Neurosci, 2007, 27(27): 7183-7195.

[26] Warth A, Kröger S, Wolburg H. Redistribution of aquaporin-4 in human glioblastoma correlates with loss of agrin immunoreactivity from brain capillary basal laminae[J]. Acta Neuropathol, 2004, 107(4): 311-318.

[27] Noell S, Fallier-Becker P, Beyer C, et al. Effects of agrin on the expression and distribution of the water channel protein aquaporin-4 and volume regulation in cultured astrocytes[J]. Eur J Neurosci, 2007, 26(8): 2109-2118.

[28] Guadagno E, Moukhles H. Laminin-induced aggregation of the inwardly rectifying potassium channel, Kir4.1, and the water-permeable channel, AQP4, via a dystroglycan-containing complex in astrocytes[J]. Glia, 2004, 47(2): 138-149.

[29] Noël G, Belda M, Guadagno E, et al. Dystroglycan and Kir4.1 coclustering in retinal Müller glia is regulated by laminin-1 and requires the PDZ-ligand domain of Kir4.1[J]. J Neurochem, 2005, 94(3): 691-702.

[30] Rurak J, Noel G, Lui L, et al. Distribution of potassium ion and water permeable channels at perivascular glia in brain and retina of the Large(myd)mouse[J]. J Neurochem, 2007, 103(5): 1940-1953.

[31] Qiu GP, Xu J, Zhuo F, et al. Loss of AQP4 polarized localization with loss of β-dystroglycan immunoreactivity may induce brain edema following intracerebral hemorrhage[J]. Neurosci Lett, 2015, 588: 42-48.

[32] Liu H, Qiu Gp, Zhuo F, et al. Lost polarization of Aquaporin4 and dystroglycan in the core lesion after traumatic brain injury suggests functional divergence in evolution[J]. Biomed Res Int, 2015, 2015: 471631.

[33] Zhang G, Ma P, Wan S, et al. Dystroglycan is involved in the activation of ERK pathway inducing the change of AQP4 expression in scratch-injured astrocytes[J]. Brain Res, 2019, 1721: 146347.

[34] 万珊珊, 马佩莹, 徐进, 等. 脑外伤后ERK信号通路调节AQP4及其锚定蛋白DG的表达[J]. 中国生物化学与分子生物学报, 2019, 35(8): 870-879.

[35] Verkman AS, Binder DK, Bloch O, et al. Three distinct roles of aquaporin-4 in brain function revealed by knockout mice[J]. Biochim Biophys Acta, 2006, 1758(8): 1085-1093.

[36] Verkman A. More than just water channels: unexpected cellular roles of aquaporins[J]. J Cell Sci, 2005, 118(Pt 15): 3225-3232.

[37] Romeiro RR, Romano-Silva MA, De Marco L, et al. Can variation in aquaporin 4 gene be associated with different outcomes in traumatic brain edema?[J]. Neurosci Lett, 2007, 426(2): 133-134.

第三节　水通道蛋白与脑损伤

一、水通道蛋白在缺血性脑水肿发生发展中的表达变化及其调控机制

缺血性脑卒中是中枢神经系统的常见病，其特征为急性或渐进性脑血管阻塞导致脑血流量减少，从而引发脑功能障碍和脑组织损伤[1]。缺血性脑卒中的病理生理变化涵盖多个

方面，包括血流动力学改变、能量代谢障碍、水和电解质转运平衡被破坏、血脑屏障损伤、炎症因子释放增加、神经元凋亡及坏死等[1-4]。深入研究和阐明这些病理生理变化的发生机制，将有助于为缺血性脑卒中的诊断和治疗提供理论依据。

脑水肿是缺血性脑卒中的主要并发症之一，可能引起脑内神经细胞损伤、颅内压增高及脑疝等严重后果。AQP在脑内细胞膜内外的水转运中发挥重要作用，AQP蛋白家族的发现极大地加深了研究者对脑内水转运平衡和脑水肿形成的认识。目前，AQP已成为治疗脑水肿的重要靶点之一[5]。本部分将围绕AQP在缺血性脑水肿发生发展过程中的表达变化及其调控机制进行探讨。

(一) 脑缺血后AQP的作用

脑内水分分布广泛，包括脑脊液、血液、细胞外间隙和细胞内等。水分在不同部位之间的转运依赖渗透压和静水压梯度的作用，正常的水转运对维持脑内生理功能具有重要意义[6]。然而，在脑缺血后，由于能量代谢障碍和血脑屏障损伤等因素，脑内水转运异常，进而诱发脑水肿的形成。脑水肿在缺血性脑卒中发生后第一周内出现，并在24～72小时达到高峰[7]。缺血引发的脑水肿可根据病理生理机制的不同，分为早期的细胞毒性脑水肿（发生于缺血后数分钟）和晚期的血管源性脑水肿（出现于缺血后4～6小时）[8]。

在正常生理状态下，当细胞内外的渗透压发生改变，导致水分向细胞内流动并引发细胞肿胀时，星形胶质细胞能通过启动容积调控机制（容积激活的钾离子和氯离子通道）对细胞内外的水转运迅速作出响应，从而使细胞容积恢复至正常水平[9]。然而，在缺血急性期，由于能量代谢障碍、ATP合成不足导致Na^+-K^+-ATP酶（钠钾泵）功能丧失，细胞内离子浓度增高，水分通过在星形胶质细胞膜上大量表达的AQP顺着渗透压梯度从细胞外进入细胞内，此时容积调控机制失效，星形胶质细胞在缺血后30分钟内出现肿胀[10,11]。

星形胶质细胞肿胀将进一步引发后续一系列的细胞毒性脑损伤。血管周围的星形胶质细胞终足膜的肿胀会压迫血管，限制血流通过，导致微血管堵塞，进而出现再灌注时的"无复流"现象[12]；此外，星形胶质细胞肿胀可引起谷氨酸通道和其他兴奋性氨基酸通道开放，释放谷氨酸等物质，导致和加重细胞的兴奋性毒性死亡；星形胶质细胞的显著肿胀会严重减小细胞外间隙，从而导致细胞外谷氨酸和K^+的浓度升高，使细胞外的谷氨酸浓度增至兴奋性毒性水平[13]。

在脑缺血4～6小时后，随着脑组织坏死、血管基底膜降解以及内皮细胞紧密连接紊乱，血脑屏障的完整性遭受破坏。白蛋白和其他血清蛋白开始从血液渗漏至脑组织细胞间隙[14]，导致脑组织含水量增加一倍以上，进而引发缺血后期的血管源性脑水肿。Verkman等[15,16]学者的研究指出，在脑肿瘤和脑外伤等血管源性脑水肿的发生过程中，作为具有双向转运功能的水通道蛋白，位于胶质界膜和室管膜等部位的AQP4可介导水分从细胞外间隙向血液和脑脊液中转运，有利于组织间隙中水肿液的清除。

鉴于AQP4对不同类型脑水肿发生与发展具有不同影响，研究者应根据脑缺血后不同时相出现的脑水肿类型，针对AQP4采取相应的调控策略：在缺血初期，当出现细胞毒性脑水肿时，可通过抑制AQP4的表达或水转运功能，减少细胞内水分的摄入，从而减轻细胞水肿；而对于后期出现的血管源性脑水肿，则可通过上调AQP4的表达水平或增强其水转运功能，以加强组织间隙中水肿液的清除。目前，关于脑缺血后对其他AQP采取何种调控机制目前尚不清楚，有待进一步深入研究。

（二）脑缺血后AQP的表达变化

在哺乳动物体内发现的水通道蛋白家族成员共计13种（AQP0～AQP12）[17]。在啮齿动物脑内，主要有AQP1、AQP4和AQP9 3种类型。其中，AQP4在中枢神经系统中的表达最为丰富[18]，也是研究者在对脑缺血后AQP表达变化和调控的研究中最为关注的。

1. 脑缺血后脑内AQP4的表达 在不同的脑缺血模型中，缺血后脑组织中AQP4的表达变化规律存在差异[19]。研究发现，在小鼠短暂性脑缺血模型中[20]，位于血管周围终足膜上的AQP4表达迅速上调，于缺血后1小时达到高峰，并且缺血核心区和缺血周边区域的早期脑水肿与AQP4的表达增高密切相关。AQP4表达上调的另一个高峰出现在缺血后48小时，此时AQP4的表达增高与缺血周边区域的脑水肿相关。这些结果提示，AQP4可能是短暂性脑缺血后参与水转运的关键通道。然而，在大脑中动脉栓塞（middle cerebral artery occlusion，MCAO）的脑卒中模型中，未观察到AQP4的早期表达变化，这表明在严重的缺血早期，脑内AQP4蛋白合成可能受阻[19]。

笔者所在课题组为探究脑缺血后AQP4的表达变化规律，采用大鼠MCAO模型，通过PCR、蛋白质印迹（Western blot）以及免疫荧光等方法对脑内AQP4 mRNA和蛋白的表达水平进行了检测，结果表明，在缺血后24小时以内，AQP4的免疫荧光强度、mRNA和蛋白表达水平随着时间变化而逐步上调，同时，脑水肿的严重程度与AQP4的表达增高呈正相关[21]。在新生大鼠缺血缺氧性脑损伤模型中，AQP4 mRNA表达水平在24小时达到峰值后逐步下降，这一变化规律与成年动物模型中AQP4的表达变化规律不同[22]。Badaut等在缺血再灌注动物模型中发现，不同缺血脑区AQP4的表达模式亦存在差异[23]：在纹状体缺血区，AQP4在缺血再灌注24小时后表达显著降低；而在皮质缺血区，AQP4的表达水平则先下降，再灌注72小时后又逐步上升，这提示AQP4可能参与缺血再灌注损伤后脑血管和脑实质之间水分的重新分布[24]。以上实验表明，脑缺血后AQP4的表达改变受多因素影响，包括脑缺血模型的类型、动物种类、年龄和损伤严重程度等。

在中枢神经系统内，AQP4分布于毛细血管周围星形胶质细胞终足膜和神经胶质界膜上[11, 25]，此即AQP4的"极性表达"。这些极性表达部位均位于脑实质与脑血管或脑脊液接触的界面上，表明AQP4的极性表达形式在脑实质与脑血管/脑脊液的水、电解质代谢平衡中发挥重要作用。笔者所在课题组在对缺血后脑内AQP4的表达水平完成检测后，近年来围绕脑缺血病理过程中AQP4是否存在极性表达变化进行了系统研究[26]。结果发现，在MCAO脑缺血动物模型中AQP4在细胞膜上由"极性表达"向"非极性表达"转变：部分原分布于血管周围的星形胶质细胞终足膜上的AQP4发生内化转移，分布至胞体和其他突起部位，AQP4进入细胞内的早期和（或）晚期内涵体，并有部分分选至溶酶体降解。

内化是一个重要的细胞生物学现象，具有调节膜蛋白和细胞表面受体的表达量，进而调节下游信号转导和细胞功能等作用[27, 28]。由于AQP4在细胞毒性脑水肿和血管源性脑水肿中发挥的作用不同，AQP4从"极性表达"向"非极性表达"变化，其内化在缺血后脑水肿的不同发展阶段具有不同的病理生理意义：在缺血早期的细胞毒性脑水肿阶段，AQP4内化并被分选至溶酶体降解，使得细胞膜上AQP4的表达减少，导致进入细胞内的水分减少，因此，AQP4内化对细胞而言可能是一种自我保护性调节；然而，在缺血后期和脑出血中的血管源性脑水肿阶段，AQP4内化将导致细胞膜上AQP4表达减少，AQP4对细胞间隙中多余水分的清除能力下降，从而促进脑水肿的发生和发展，在此背景下，AQP4内化在血管源性脑水肿的发生和发展过程中扮演着不利因素的角色。

鉴于AQP4极性表达改变（发生内化）可以调节细胞膜上AQP4的表达水平，在缺血后不同时期内对脑水肿的发生和发展产生不同的作用，因此，通过调控AQP4的极性表达和内化以减轻脑水肿，可能成为治疗缺血性脑水肿的有效策略，并为基础研究向临床转化提供实验依据。然而，尽管在前期研究中，笔者所在课题组已检测到AQP4极性表达变化和内化现象，但其在缺血后脑水肿中的发生机制的研究处在进一步探讨阶段。

2. 脑缺血后AQP1和AQP9的表达 在小鼠短暂性脑缺血模型中，Marlise等对AQP1、AQP4和AQP9表达的时相变化进行了系统研究[20]。在此模型中，缺血后1小时和48小时出现两个脑水肿高峰期，7天后水肿明显消退。AQP4的表达变化与脑水肿的形成具有相关性，具体情况如上所述。值得注意的是，AQP1在脑内的分布不仅局限于脉络丛上皮细胞，研究人员观察到AQP1还在部分皮质神经元中表达，主要位于对照组和脑缺血小鼠的额叶、顶叶皮质锥体神经元的突起以及与软脑膜表面相接触的部位。另有文献报道，AQP1在肿瘤组织和蛛网膜下腔出血（subarachnoid hemorrhage，SAH）模型中表达水平发生改变[29-32]，但在脑缺血模型研究中并未观察到AQP1表达量的变化，推测AQP1可能与肿瘤和SAH模型中脑水肿形成过程中对水的转运更为密切。

在大鼠和小鼠脑内，AQP9分布于穹窿下器、白质的星形胶质细胞、儿茶酚胺能神经元以及室旁核等部位。在脑缺血后，穹窿下器中星形胶质细胞和神经元中的AQP9表达量无明显改变，然而，从缺血后24小时至缺血后7天，缺血周边的星形胶质细胞中AQP9的表达水平明显增高，而缺血核心区星形胶质细胞中AQP9的表达水平则在缺血7天后才开始显著增高。在脑内，AQP9不仅具有水转运功能，还参与乳酸和甘油等能量代谢物质的转运。脑缺血后AQP9的表达改变与脑水肿的形成无明显关联，提示AQP9的表达变化可能与缺血后脑内的能量代谢改变关系更为密切。

3. 脑缺血后AQP3、AQP5和AQP8的表达 在脑缺血动物模型中，对AQP3、AQP5和AQP8的研究较少，为探索和阐明它们在缺血性脑水肿发生和发展过程中的作用，笔者所在课题组在大鼠MCAO脑缺血模型中对这些AQP成员的表达情况进行了检测[21]。这3种AQP在大鼠脑内的表达部位相似，均在神经元和星形胶质细胞中表达，主要分布于梨状皮质、海马、背侧丘脑、苍白球和脉络丛等部位。脑缺血后AQP3、AQP5和AQP8不但在缺血侧大脑半球表达增高，在缺血对侧的大脑半球表达水平也有上调；脑缺血3小时后，它们的表达区域延伸到齿状回、室管膜、室旁核、尾壳核和下丘脑外侧核等部位。在缺血周边区域，随着缺血时间的延长，AQP3、AQP5和AQP8的表达逐步增加，而在缺血核心区，它们的表达则从缺血后1小时迅速增加，至缺血6小时到达高峰，随后逐步下降。在缺血早期，缺血周边和核心区与脑水肿的形成具有相关性。这些结果提示 AQP3、AQP5 和 AQP8 的表达改变可能参与了缺血早期脑水肿的发生。另外，由于 AQP3、AQP5 和 AQP8 的表达变化与缺血后神经元开始出现肿胀的时间相吻合，提示这 3 种 AQP 可能在脑缺血后神经元的水肿发生过程中起关键作用。

（三）脑缺血后对 AQP 的调控

在众多 AQP 中，AQP4 在缺血性脑水肿的发生和发展过程中具有关键作用，因此，本部分将重点探讨脑缺血后对 AQP4 的调控机制。

1. 药物对 AQP4 功能的调控　近年来，研究发现有多种化学合成药物对 AQP4 的表达和水转运功能具有较好的抑制效果，包括乙酰唑胺（acetazolamide，AZA）、托吡酯（topiramate，TPM）、唑尼沙胺（zonisamide，ZNS）、2-烟酰胺-1,3,4-噻二唑（2-nicotinamide-1,3,4-thiadiazole，TGN-020）等[33-35]。乙酰唑胺被证实为最有效的 AQP4 抑制剂之一，可以使通过 AQP4 转运的水分减少 80%[33]，我国学者段升强等的研究表明，乙酰唑胺可通过降低 AQP4 的表达而减轻缺血后脑水肿的程度[36]。托吡酯和唑尼沙胺具有和乙酰唑胺一致的理化性质，不仅具有抗癫痫作用，还能阻断 AQP4 的表达[34]。另有研究证实，新型 AQP4 抑制剂 TGN-020 可显著减轻小鼠脑缺血模型中脑水肿程度并减小梗死面积[35]。

除了 AQP4 抑制剂外，神经保护剂促红细胞生成素（erythropoietin，EPO）、抗凝血药 tPA 或 EPO 与 tPA 联合应用，可通过抑制 AQP4 表达来减轻大鼠脑缺血/再灌注后脑水肿的形成[37]。此外，越来越多的研究显示，缺血后低氧诱导因子 1α（hypoxia inducible factor-1α，HIF-1α）、AQP4 和基质金属蛋白酶 9（matrix metalloproteinase-9，MMP-9）与缺血后脑水肿的发生和血脑屏障破坏密切相关[38-42]。

脑缺血可引起 HIF-1α 的表达增高，进而上调 AQP4 和 MMP-9 的表达。增高的 AQP4 促进缺血后细胞毒性脑水肿的形成；MMP-9 的表达增高可导致 ZO-1 和 occludin 等紧密连接蛋白降解，引起血脑屏障损伤和血管源性脑水肿形成。HIF-1α 的抑制剂 2-甲氧基雌二醇（2-methoxyestradiol，2ME2）和 MMP-9 的抑制剂米诺环素（minocycline）可以下调 AQP4 的表达水平，减轻血脑屏障的损伤程度。HIF-1α/MMP-9/AQP4 信号通路可能为缺血性脑水肿的有效药物靶向治疗开辟新途径。

有研究表明炎症是一种潜在的 AQP4 诱导剂[43,44]，因此，任何抑制炎症反应的药物都有可能成为 AQP4 抑制剂。吡罗昔康作为一种非甾体抗炎药物，可通过抑制环氧合酶对花生四烯酸的代谢作用，降低组织局部前列腺素的合成，抑制白细胞的趋化性以及溶酶体酶、细菌产物、中性粒细胞阳离子蛋白、细胞因子等炎症介质的释放。有研究显示[45]，在 MCAO 大鼠脑缺血/再灌注模型中，吡罗昔康具有减小梗死面积和减轻脑水肿等神经保护功能，其保护机制是通过抑制细胞内钙水平升高和下调 AQP4 表达来实现的。

此外，也可阻断下游炎症级联反应，以保护 BBB 的完整性[46,47]。这些药物对脑水肿的有益作用可能是通过减少炎症因子的释放，进而阻断 AQP4 上调来实现的。

2. 干细胞对 AQP4 功能的调控　除药物外，干细胞疗法在调节 AQP4 功能方面亦表现出潜力。研究表明，将间充质干细胞（mesenchymal stem cell，MSC）移植至脑缺血小鼠脑内后，可通过p38途径下调AQP4的表达，从而维护血脑屏障的完整性，并最终改善脑缺血小鼠神经功能评分[48]。临床试验结果表明，骨髓干细胞移植对脑卒中患者的预后具有积极意义，且未见明显的不良反应[49]。

3. 微RNA（microRNA，miRNA）对AQP4功能的调控　miRNA是由约22个核苷酸组成的非编码单链RNA，其通过与mRNA的3′端非翻译区结合，调控基因表达[50]。随着对miRNA研究的深入，研究者发现miRNA可能涉及体内多种病理生理过程的调节，其中脑卒中患者miRNA-29的表达减少与预后不良相关。研究表明，脑缺血小鼠脑内的miR-29b减少与AQP4表达增高相关联[51]。Gan等的研究显示，脑缺血患者血液中miR-145的表达水平较对照人群显著增高[52]；进一步研究表明，miR-145通过转录后调节AQP4表达，通过下调AQP4表达从而保护星形胶质细胞免受缺血性损伤[53]。Sepramaniam等[54]发现，miR-320a在体外和体内可直接抑制AQP4和AQP1的表达；而抑制miR-320a则可以上调AQP4和AQP1的表达，并减小脑缺血时的梗死体积。miR-130a可以选择性抑制AQP4 M1启动子活性，Zheng等[55]的研究显示，在小鼠MCAO脑组织和OGD处理的原代星形胶质细胞中，上调miR-130b的表达水平可保护星形胶质细胞免受缺血性损伤，并抑制AQP4的表达。

4. 蛋白激酶的磷酸化对AQP4的调节　蛋白激酶（protein kinase）作为一类催化蛋白质磷酸化反应的酶，能够催化从ATP转移出磷酸基团并共价结合到特定蛋白质分子中某些丝氨酸（Ser）、苏氨酸（Thr）或酪氨酸（Tyr）残基的羟基上，从而调节蛋白质的功能。在AQP4的结构中，存在多个可被蛋白激酶磷酸化的位点：Thr6，Ser111，Ser180，Ser276，Ser285（图2-6）[56]。已有的研究表明，上述磷酸化位点可以被下列几种蛋白激酶

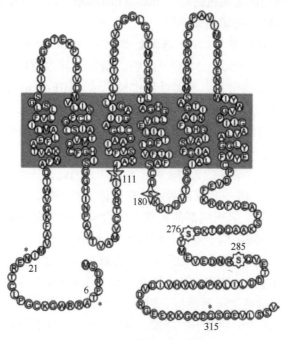

图2-6　AQP4蛋白结构中的磷酸化位点

磷酸化：蛋白激酶A（PKA）、蛋白激酶C（PKC）、酪蛋白激酶2（CK2）和钙调蛋白依赖的蛋白激酶Ⅱ（CaMKⅡ）[56]。例如，在转染AQP4的非洲爪蟾卵母细胞（Xenopus oocytes）中，PKC可以作用于AQP4蛋白结构中的Ser180，从而影响AQP4的水通透性[57]；在MDCK细胞系中，CK2能使AQP4蛋白结构中的Ser276磷酸化，增强AQP4与AP3的相互作用[58]。

在脑缺血动物模型中，众多研究显示了蛋白激酶磷酸化可参与脑内AQP4表达的调节。褪黑素、硫化氢等可通过激活PKC以抑制AQP4，从而在大鼠MCAO脑缺血模型中发挥神经保护作用[59,60]。Apelin-13可从形态学和功能上保护脑缺血后血脑屏障免受破坏，可能是通过激活ERK和PI3K/Akt等蛋白激酶磷酸化途径调节AQP4来实现的[61]。Nito等研究观察了丝裂原活化蛋白激酶途径是否与AQP4的表达相关，发现p38 MAPK参与了调节缺血性损伤后大鼠脑皮质中星形胶质细胞AQP4的表达[62]。

二、水通道蛋白与缺血再灌注损伤及其调控机制

在缺血基础上恢复血流后，组织损伤反而加重，甚至发生不可逆损伤的现象称为缺血再灌注损伤，其发生机制尚未彻底阐明，目前认为自由基的作用、细胞内钙超载和白细胞的激活是缺血再灌注损伤的重要发病学环节[63]。1966年，Jennings[64]首次提出心肌再灌注损伤的概念，证实再灌注会引起心肌超微结构不可逆坏死，包括暴发性水肿、组织结构崩解等；1968年，Ames[12]率先报道脑缺血再灌注损伤，以后陆续有其他器官缺血再灌注损伤的报道，说明再灌注损伤几乎可在每一种组织、器官发生，而心、脑等器官对氧需求量高，则更容易发生损伤，其缺血再灌注后明显的组织学变化是组织水肿及细胞的坏死。水通道蛋白为一种重要的、分布广泛的、具有能量代谢和物质转运作用的选择性水转运蛋白[65]，研究和阐明其在缺血再灌注损伤后的组织水肿产生和神经元凋亡过程中的发生机制和作用，有助于为减轻缺血再灌注损伤后的水肿及坏死反应提供理论基础。

（一）AQP在脑缺血再灌注损伤后脑水肿及能量代谢障碍中的作用

脑血管疾病是一种严重危害人类身体健康的疾病，具有高致残率、高死亡率和难以预见的特点[66]。其中，缺血性脑损伤疾病占脑血管疾病的绝大部分，缺血后在有限的时间窗内及时恢复血流再灌注对于恢复缺血区脑组织血氧供应、维持受损脑组织的正常形态与功能具有重要意义。但当脑组织缺血时间较长时，再给予恢复血流再灌注的处理会进一步加重脑组织的损伤，此即为脑缺血再灌注损伤，它的发病机制是一个快速的级联反应，这个级联反应包括许多环节[67]。主要环节包括细胞内钙稳态失调、脑组织中氨基酸含量失稳态、自由基生成、炎症反应、凋亡基因激活及能量障碍等。这些机制彼此重叠，相互联系，形成恶性循环，最终引起细胞凋亡或坏死，导致缺血区脑组织水肿及不可逆的损伤[68]。动物实验表明缺血再灌注后6小时，脑组织含水量增加，随着再灌注时间延长，含水量逐渐增加，再灌注48~72小时达到高峰[69]，由此所引起的颅内压增高、脑组织灌注量下降形成缺血缺氧，再次加重神经细胞损伤，临床上甚至可能导致患者出现

脑疝等严重后果，影响预后。水通道蛋白是一组与水及甘油、尿素、乳酸等小分子物质的通透性有关的膜转运蛋白。目前在哺乳动物体内发现的水通道蛋白家族成员共有13个（AQP0～AQP12），AQP1～AQP5、AQP8、AQP9在啮齿动物脑组织中均有表达，其中AQP1、AQP4、AQP9为主要类型，它们分布于脑组织中的星形胶质细胞、血管内皮细胞和儿茶酚胺类神经元的细胞膜上。研究表明，水通道蛋白家族成员与缺血再灌注后脑组织中水的转运、水和电解质的平衡调节、渗透压的改变及能量代谢有关[70, 71]。在缺血再灌注的不同阶段，水通道蛋白有不同的表达变化，对水肿的影响不同，使得水通道蛋白可能成为治疗缺血再灌注性脑损伤的调控点之一。

1. AQP4在脑缺血再灌注损伤脑水肿形成中的作用　　AQP4是由4个独立的具有活性的亚单位组成的异聚体结构，广泛分布于脑星形胶质细胞，脑表面的软脑膜，脑室系统的室管膜、脉络丛，下丘脑的视上核和室旁核等，在中枢神经系统中表达最为丰富，可控制水进出脑组织的通道，在脑组织水平衡、星形细胞迁移、神经兴奋及炎症等方面均起着重要作用[72]。大量研究表明，AQP4与脑水肿关系密切，参与了多种疾病所引起的脑水肿的病理过程，因此也成为脑组织缺血再灌注后表达变化和调控的最受到关注的水通道蛋白家族成员。

缺血再灌注可导致脑组织血脑屏障通透性增加，AQP4作为胶质细胞与血液及脑脊液之间水调节和运输的重要通道，其表达水平与血脑屏障的完整性密切相关。研究表明[23, 30, 73, 74]，AQP4可减缓脑缺血再灌注损伤初期脑水肿的形成，其可能机制是再灌注后，细胞间隙中K^+、H^+和谷氨酸等兴奋性氨基酸浓度上升，激活蛋白激酶C，促进了AQP4蛋白磷酸化，在一定程度上减少了水分子经由已破坏的血脑屏障进入脑组织的细胞间隙，从而减轻了脑水肿。将小鼠*AQP4*基因敲除后，短暂局灶性脑缺血后*AQP4*缺陷小鼠脑水肿和梗死的体积均减小。

2. AQP4与Kir4.1的再分布在缺血再灌注后脑水肿形成中的作用　　由于生理状态下，AQP4介导的水的运输和Kir4.1参与的钾离子的转运相互协同（偶联），可调节细胞内外水分子和K^+的浓度，两者共同参与维持中枢神经系统内水、电解质平衡，以及内环境的稳定。在病理状态下两者的表达发生了变化[63]。在高度恶性脑肿瘤组织中Kir4.1的表达明显增强，而AQP4不表达或低表达，这一现象称为"再分布"（redistribution）[75]；而在脑外伤、细菌性脑膜炎和脑肿瘤患者脑组织中，AQP4与Kir4.1的表达部位及表达量也发生了改变[75]；视网膜缺血后，组织中AQP4与Kir4.1的表达降低，但两者降低幅度并不一致[76]。上述变化导致脑水肿发生，提示AQP4和Kir4.1的再分布可能是病理状态下水肿发生的重要原因。因此，笔者所在课题组通过大鼠缺血再灌注模型对AQP4与钾离子通道在缺血再灌注损伤后的表达与分布进行了研究。研究结果表明，在缺血再灌注12小时内，血脑屏障通透性增加，脑含水量逐渐增加，AQP4在软脑膜、室管膜等与水平衡代谢密切相关的界面上表达增加。研究提示后期随着再灌注时间的延长，血脑屏障通透性增加，大量血管内水分子，以及其他小分子物质透过血脑屏障进入组织间隙，形成血管源性脑水肿。AQP4除在软脑膜、室管膜表达增加以外，在大脑皮质等脑实质的血管周围也出现显著表达，胶质细胞明显水肿。在缺血再灌注后期，水肿液在通过脑脊液回流途径清除的同时，细胞外间隙水分子被大量转运进细胞，导致星形胶质细胞水肿加剧，这种现象被称为病理性适应性调节（adaptive malregulation）[77, 78]。PCR显示，AQP4的增加量明显大于Kir4.1，提示星形胶质细胞对水分子和K^+的转运能力严重失衡。细胞外水分子与钾离子不成比例地进入细胞内，导致细胞外间隙K^+浓度相对增加。而细胞外相对高K^+可促进Na^+、Cl^-大量内流，后者更进一步带动细胞外水分子进入星形胶质细胞，加重细胞毒性脑水肿。因此，再灌注后期脑水肿为血管源性与细胞毒性共同存在的混合性脑水肿。通过药物等途径抑制AQP4的表达，可能有助于缺血再灌注后期脑水肿的控制。

3. AQP4在脑缺血再灌注损伤神经元凋亡中的作用 研究表明，AQP4表达量与细胞色素C释放量呈负相关，提示AQP4的促凋亡作用可能诱导细胞色素C释放引起线粒体肿胀，诱导凋亡刺激物产生[78]。笔者所在课题组研究尚未涉及该机制，因此在缺血再灌注损伤的神经元中，AQP4是否发挥相关的作用，也有待进一步研究证实。

4. AQP9在大鼠脑局灶性缺血再灌注后脑组织中的表达变化及其意义 AQP9在水通道蛋白家族中属于水甘油通道蛋白亚族，其主要分布在脑组织星形胶质细胞、血管内皮细胞和儿茶酚胺类神经元的细胞膜上，含量仅低于AQP4，除了具有水转运功能，它还参与乳酸和甘油等能量代谢物质的转运[79-81]。Warth 等[82]发现，与 $AQP9$ 基因敲除大鼠（$AQP9^{-/-}$）比较，野生型大鼠（$AQP9^{+/+}$）血浆内的甘油、甘油三酯含量明显偏高，结合AQP9在儿茶酚胺能神经元胞体和突起上的表达，推测AQP9在脑组织能量代谢中具有重要的作用，即AQP9可促进乳酸等能量物质从星形胶质细胞向神经元传递。笔者所在课题组的研究发现，AQP9不仅在包括穹窿下器和脉络丛等脑室周围器官（circumventricular organs，CVOs）、脑室周围实质的星形胶质细胞中有表达，而且在下丘脑的某些神经核团（如视上核、室旁核）、海马齿状回锥体细胞层和大脑皮质中表达强烈，这与国外研究报道的结果基本一致[83]。于再灌注早期（12小时内），AQP9在皮质、海马、穹窿下器、脉络丛等脑室周围器官以及渗透压感受区视上核、室旁核强烈表达；AQP9 mRNA的表达量随时间延长逐渐升高，12小时达到峰值。HE染色显示，细胞间隙增宽，组织水肿。在脑组织缺血期，炎症反应、白细胞渗出、血小板聚集等阻塞毛细血管，可导致微循环障碍和再灌注早期的"无复流现象"（no-reflow phenomenon）[84]发生；同时再灌注后各种继发性损伤因子的释放等也加重了缺血缺氧，而缺血缺氧导致了更多的损伤因子释放，形成恶性循环，使得再灌注后脑内缺血缺氧的状态持续存在，此时能量代谢障碍不仅未能得到迅速改善，还随着缺血再灌注时间的延长持续性加重；细胞膜Na^+-K^+-ATP酶等活性降低，细胞内外水、电解质分布失衡进一步加剧，细胞内外渗透压改变加剧，伴随着脑血管内皮细胞Ca^{2+}超载以及炎症反应等所造成的血脑屏障的损伤，细胞间隙水分大量聚集。此时 AQP9 在与水接触界面，以及渗透压感受区的表达上调，介导细胞间隙内的水分子大量进入细胞，在减少组织间隙过量水分的同时，却导致了细胞水肿。因此，AQP9在缺血再灌注早期的表达上调，实际上是一种适应性的病理性调节（adaptive malregulation）[85]。由于缺血再灌注后缺血缺氧状态的加剧，能量代谢障碍的进一步加重，大量乳酸生成并堆积，致使组织内环境酸化。生理条件下，AQP9可以介导水分子和乳酸的转运；在酸性环境下，AQP9对乳酸的通透性可增加4倍；AQP9还可以通过乳酸穿梭机制，将乳酸从星形胶质细胞传递给神经元，将乳酸作为神经元能量代谢的供能物质。因此，在脑缺血再灌注早期，AQP9在胶质细胞和神经元上的表达上调，有利于将乳酸从胶质细胞向神经元转移，以维持神经元的能量代谢。而随着再灌注时间延长（12小时后），免疫组化结果显示AQP9表达部位无明显变化，RT-PCR结果显示AQP9 mRNA表达量逐渐下降，推测可能与再灌注时间延长、缺血部分脑组织血液供应及脑内能量代谢逐渐恢复有关，其机制尚有待于进一步研究。

5. AQP3、AQP5和AQP8脑缺血再灌注后的表达　在脑缺血再灌注动物模型中，对AQP3、AQP5和AQP8的研究较少，但笔者所在课题组为探索和阐明其在缺血性脑水肿发生发展过程中的作用[21]，利用大鼠局灶性脑缺血动物模型（如MCAO），对上述AQP成员的表达变化进行了检测，相关结果见前一节。研究结果提示，这3种AQP可能在脑缺血后神经元的水肿发生中起重要作用。那么，这3种AQP是否在缺血再灌注损伤中发挥作用，有待进一步深入研究。

（二）脑缺血再灌注后对AQP的调控

1. 丝裂原活化蛋白激酶信号通路对AQP4水平的调节　丝裂原活化蛋白激酶（mitogen-activated protein kinase，MAPK）信号通路是生物体内重要的信号转导系统，参与介导细胞生长、发育、分裂和分化等多种病理生理过程。作为MAPK家族的一员，p38可在多种（缺血再灌注损伤、渗透压变化和生理应激等）病理生理条件下激活，是脑组织损伤等病理改变的重要信号通路。研究发现，大鼠脑缺血再灌注后缺血周边区脑组织中p38的水平明显增加，AQP4表达亦增高，脑水肿程度加重。与脑缺血再灌注大鼠相比，在造模前预先给予p38特异性抑制剂（SB203580）的大鼠缺血周边区脑组织中p38的磷酸化受到明显抑制，AQP4表达水平亦降低，可下调AQP4的高表达，减轻脑水肿的程度[62, 86]。

2. 药物对缺血再灌注后AQP4水平的调控　β-七叶皂苷钠具有稳定细胞膜、恢复毛细血管通透性、清除自由基的作用，其可能通过抑制脑组织核因子κB（nuclear factor kappa-B，NF-κB）活性，下调脑组织肿瘤坏死因子α（tumor necrosis factor α，TNF-α）蛋白表达等多种机制，降低胶质细胞足突上AQP4的表达，减少经胶质细胞膜进入细胞内的水含量，抑制细胞毒性脑水肿；β-七叶皂苷钠对星形胶质细胞及血脑屏障的保护作用，可减少大分子和水分子物质经由破坏的血脑屏障进入脑组织细胞间隙，从而抑制血管源性脑水肿。因此推测，β-七叶皂苷钠可能通过抑制AQP4的表达，对细胞毒性脑水肿及血管源性脑水肿均有抑制作用[87]。

在中药治疗方面，有研究[88]在血脑屏障通透性最大时给予丹参川芎嗪注射液干预，结果显示，在局灶性脑缺血再灌注损伤的情况下，丹参川芎嗪表现出明显的降低血脑屏障通透性及抑制AQP4表达的作用。笔者推测，丹参川芎嗪可能通过清除自由基、抑制脂质过氧化等抑制脑组织NF-κB活性，下调脑组织TNF-α表达等多种机制，降低胶质细胞足突上AQP4的表达，降低胶质细胞对水的通透性，抑制细胞毒性脑水肿；该药对星形胶质细胞和血脑屏障的保护作用，可减轻缺血再灌注后血管源性脑水肿。

以上表明，无论是β-七叶皂苷钠，还是丹参川芎嗪，均兼有抑制细胞毒性脑水肿和血管源性脑水肿的功效，其作用机制值得深入研究。

三、水通道蛋白在出血性脑水肿发生发展中的作用及其调控机制

脑出血（intracerebral hemorrhage，ICH）是一种临床上的常见病、多发病，通常由高血压、脑动脉瘤、颅内肿瘤和脑血管畸形等疾病的脑血管意外引发，其死亡率和致残率均较高，尤其是在亚洲国家，其发病率已高达20%～30%[89-91]。脑水肿是ICH最重要的并发症之一，其发生发展是导致ICH病情恶化甚至死亡的关键因素。严重的脑水肿可引起颅内压增加、脑血流量减少，甚至引起致死性脑疝等一系列继发性反应[5, 92]。一般来说，只要患者能安全度过脑水肿期，其死亡率即可明显降低，因此对脑水肿采取有效控制措施是降低ICH致残率、致死率的关键措施之一。

研究证实，脑水肿发生的关键因素是脑细胞内外水、电解质分布失衡，从而破坏了脑细胞内环境的稳定和结构的完整性[93]。然而，目前有关脑水肿形成的众多学说，如血脑屏障障碍学说、脑细胞膜磷脂代谢障碍学说、脑循环障碍学说、乳酸中毒、兴奋性氨基酸神经毒性作用、自由基损害、脑机能代谢障碍学说、细胞膜Na^+-K^+-ATP酶活性降低、紧密黏附素（内膜素）的作用、细胞内Ca^{2+}超载等都不能全面解释脑组织内水、电解质失衡的机制。传统观点认为，水分子是以简单扩散的方式通过细胞膜的脂质双分子层，但这不能解释水分子高速通过细胞生物膜的能力。近年来研究表明，水分子的转运存在着一种由水通道蛋白（aquaporin，AQP）家族介导的主动转运和快速调节过程[94, 95]。AQP的发现为进一步阐明脑内水、电解质运输平衡及其异常的分子机制提供了新契机[30, 96, 97]。

AQP是一类水通道蛋白，能调节生理和病理状态下的水稳态，可以作为缓解出血性脑水肿新的治疗靶点。AQP是一组高度保守的蛋白家族，存在于从细菌到哺乳动物不同的生物体内，其在生物体内的广泛表达在一定程度上表明了其功能的重要性。迄今为止，研究者已经在哺乳动物体内发现了13种AQP（AQP0～AQP12），在哺乳动物中枢神经系统内有7种AQP表达，分别是AQP1、AQP3、AQP4、AQP5、AQP8、AQP9、AQP11[8, 98]。

（一）中枢神经系统内表达的AQP4

AQP4是目前发现的脑组织中含量最丰富、分布最广泛、功能最重要的一种AQP亚型[11, 25, 99, 100]，其参与了水和电解质转运、细胞迁移、神经兴奋性传导等重要的病理生理过程，尤其在维持脑内动态水平衡中发挥了重要作用。

生理情况下，AQP4在脑实质与脑脊液、血液交界处的星形胶质细胞终足膜上密集表达，尤其是软脑膜下外胶质界膜、脑室室管膜细胞基底部及其下的内胶质界膜，以及脑实质毛细血管周围星形胶质细胞终足膜上，参与构成胶质界膜[11, 15, 25, 75, 96, 99, 100]，从而介导水在细胞、组织间隙、血管和脑室间顺渗透压和流体静力压梯度的转运，以维持脑内动态水平衡（图2-7）[115]。定量分析显示，AQP4在终足膜上的表达量远高于非终足膜，研究者将这种现象称为AQP4的"极性表达"（polarized expression）[101]。终足膜上极性表达的AQP4并非以单个蛋白的模式表达，而是以晶体样的粒子正交阵列（orthogonal arrays of particles，OAP）的形式存在[102, 103]。OAP是以AQP4四聚体为单位基础而构建的，每个OAP可以包含4至100多个AQP4四聚体，OAP是星形胶质细胞终足膜上的AQP4行使其高速水转运功

能的结构基础。生理情况下,AQP4在星形胶质细胞终足膜上极性表达,对维持血脑/血脑脊液间水平衡具有至关重要的意义[99]。

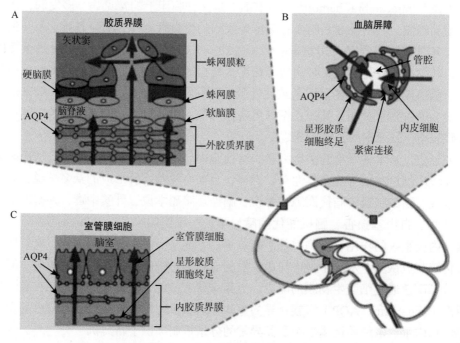

图 2-7　AQP4在脑内的表达部位

A:软脑膜下星形胶质细胞终足膜构成的外胶质界膜。B:脑实质内毛细血管周围星形胶质细胞终足膜。C:室管膜细胞基底部及其下星形胶质细胞终足膜构成的内胶质界膜

　　研究发现,脑出血后脑组织中存在水平衡紊乱,出血侧脑组织含水量明显增高,毛细血管周围间隙明显增宽,并伴随着内皮细胞肿胀,这些结果提示,早期的出血性脑水肿主要为血管源性脑水肿[104]。血肿周围组织中,AQP4蛋白和mRNA的表达水平均明显上调,AQP4不再极性表达于星形胶质细胞终足膜上,而是成团地聚集到外胶质界膜,弥散性地表达到整个室管膜细胞,以及表达于血肿周围的星形胶质细胞胞体和突起上[105],以上结果提示,脑出血脑组织中AQP4极性表达缺失。而在AQP4的非极性表达部位,如穹窿下器、海马、齿状回等,AQP4的表达方式没有明显变化[106],提示ICH后病理改变仅影响极性表达的AQP4的分布,而非极性表达的AQP4则不受影响。换言之,ICH可造成AQP4的极性表达缺失。

　　综上所述,血肿周围组织中AQP4表达明显上调,但其表达于星形胶质细胞胞体和突起上,而非星形胶质细胞终足膜上,即AQP4极性表达缺失。在血肿周围组织中,由于血凝块回缩血浆渗出,毛细血管通透性增加,以及血管内容物渗出等原因,大量水分积聚到细胞外间隙,与此同时血管周围星形胶质细胞终足膜上极性表达的AQP4明显降低,故而不能将细胞外间隙的水分快速转运进血管,从而延缓了细胞外水分的清除。因此,AQP4极性表达缺失可能在一定程度上促进了出血性脑水肿的发展。

　　AQP4的极性表达与DDC密切相关[103, 107-109]。DDC是连接细胞骨架与细胞外基质

（extracellular matrix，ECM）的一组相互作用蛋白，其能将多种膜蛋白分子定位到相应细胞膜域。DDC的核心成员β-DG（β-dystroglycan）[103]是一种跨膜分子，可同时连接ECM分子与细胞骨架，并通过与DDC中别的成员的连接，将AQP4极性锚定到星形胶质细胞终足膜上。研究发现，在生理情况下，AQP4与β-DG严格共定位于AQP4的极性表达部位，而在脑出血后脑组织中，伴随着AQP4的极性表达缺失，β-DG表达下调，提示β-DG表达下调与AQP4极性表达缺失密切相关[101, 105]。推测β-DG表达下调后，AQP4不能通过β-DG与细胞外基质连接，从而不能极性定位到星形胶质细胞终足膜上；AQP4与β-DG仍共表达于穹窿下器、海马、齿状回等非极性表达部位，提示AQP4的非极性表达可能不受脑出血后病理生理改变的影响，或者有别的分子负责AQP4的非极性表达定位；另一方面，血肿周围组织中，β-DG蛋白含量明显降低，而β-DG mRNA水平无明显变化，提示β-DG的表达下调属于转录后调控。

综上所述，脑出血后脑组织中，β-DG表达下调可能促进了AQP4极性表达缺失，进而促进了出血性脑水肿的发展。

此外，内化作为一种重要的生物学现象，其在AQP4蛋白极性表达的调控中可能发挥了重要作用，因而引起了研究者的广泛关注。实验表明一些化合物可以促使AQP4蛋白从质膜转运至细胞内囊泡而发生内化。在转染AQP4的人胃上皮细胞系（HGT-1）[110]和犬肾细胞系（MDCK）[58]的体外培养实验中，AQP4可以通过内化至晚期内涵体而出现表达部位的改变，造成细胞膜水渗透性发生变化，从而导致细胞内外水分布的变化。在转染了AQP4的非洲爪蟾卵母细胞的体外实验中，同样观察到在PKC激动剂孵育下AQP4出现了内化现象[157]。综上，内化是AQP4的重要调节机制，通过内化可使得AQP4的亚细胞定位发生改变，从而影响着AQP4在水平衡、水转运中的调节作用。

研究发现，脑出血后6小时、12小时、24小时、48小时，在脉络丛、血肿周围、大脑皮质都可以观察到AQP4和甘露醇-6-磷酸受体（晚期内涵体标志物，MPR）的共表达，以及AQP4与溶酶体相关膜蛋白-1（溶酶体标志物，LAMP1）的共表达。通过统计分析发现，AQP4和LAMP1的共表达量与脑出血后的时间呈现依赖关系，这说明在脑出血后早期（6~48小时），部分AQP4蛋白从质膜内化进入到了晚期内涵体，之后进一步分选至溶酶体而降解。

目前已知AQP4缺失可以对细胞毒性脑水肿起到保护作用，然而在血管源性脑水肿发生时，AQP4的缺失却不利于水肿的清除。胶原酶所致的脑出血后脑水肿在早期（出血后1~3天）主要是血管源性的，所以脑出血后早期AQP4的内化并不利于血管源性脑水肿的清除。综上所述，阐明AQP4内化以及溶酶体分选的发生发展机制，阻止AQP4的内化甚至分选至溶酶体降解，可能成为脑水肿防治的有效策略。

（二）中枢神经系统内表达的其他AQP

AQP9又称为水甘油通道蛋白，意味着其不仅转运水，还允许如甘油、尿素、乳酸、单羧酸盐等其他小分子通过[111]。AQP9主要表达于脑实质星形胶质细胞、脑室室管膜细胞、脑干儿茶酚胺能神经元、中脑多巴胺能神经元以及下丘脑神经元等部位[112, 113]。研究发现，ICH早期AQP9表达降低，并于ICH后6小时降至最低[114]；随后AQP9表达上调，并于ICH后48小时达到峰值。免疫组化和免疫荧光双标显示，ICH后48小时，AQP9强表达于脑室周围器官和渗透压感受器。以上结果提示，ICH早期AQP9表达降低可延迟细胞内水肿的形成，而ICH后期AQP9表达上调，可能与水、电解质平衡调节和能量代谢调节有关。

AQP1既可以转运水分子，也可以转运像CO_2这样的挥发性物质。在哺乳动物中枢神经系统中，AQP1主要表达于脉络丛上皮细胞和室管膜细胞，与脑脊液的产生有关[111]。此外，在非人灵长类动物的大脑组织中，AQP1还在支配软膜下血管活动的神经元中表达，提示AQP1还有调节脑血流量的潜在作用[19]。

AQP5表达于大脑皮质的软脑膜、脉络丛、血管周围、海马锥体细胞层、齿状回颗粒细胞层、视上核、视交叉上核和大脑纵裂两侧皮质深部，与AQP4的分布范围相似，但表达量明显低于AQP4[115]，提示AQP5在大鼠脑组织中分布广泛，可能协同AQP4在脑组织水运输平衡、脑脊液产生与回流及渗透压的调节过程中发挥重要作用[116]。

还有学者在大鼠脑实质星形胶质细胞和神经元中均检测到AQP3、AQP5和AQP8的蛋白和mRNA的表达[117]。AQP11是超级水通道蛋白亚家族的成员之一，在神经元、脉络丛上皮和脑毛细血管内皮细胞中表达[118, 119]。但目前尚未见这些AQP与出血性脑水肿关系的报道，尚需要进一步的研究来证实这些AQP是否在出血性脑水肿发生发展中发挥了作用。

四、AQP4与烧伤后继发性脑水肿及其调控机制

烧伤是一种常见的损伤，严重烧伤可引起多器官的损害，脑水肿作为严重烧伤后主要并发症之一，对于防治烧伤并发症非常重要。临床上，烧伤后脑水肿症状常常被其他并发症所掩盖，缺乏特异性的症状和体征，因此常被医生忽视，多数情况是在死亡后尸检时发现，烧伤后脑水肿的生前诊断率仅为24.59%。烧伤后大量体液自创面外渗造成低血容量性休克，由此引发的缺血缺氧性损害及失控性炎症反应是严重烧伤后脏器的基本损害因素[120, 121]。烧伤后机体产生大量的自由基、兴奋性氨基酸、白细胞介素、血小板活性因子等对血管内皮细胞造成损伤，导致脑毛细血管通透性增加，血脑屏障破坏，血管内液体渗入细胞外间隙引起血管源性脑水肿[122]；缺血缺氧所致能量代谢障碍，Na^+-K^+-ATP酶活性下降，使Na^+内流，形成细胞内的高渗状态，大量游离水进入细胞内，细胞肿胀、增大，致细胞源性脑水肿。因此，烧伤后脑水肿兼有血管源性脑水肿和细胞源性脑水肿的病理学特点。无论是血管源性脑水肿还是细胞源性脑水肿，神经胶质细胞水肿是最突出的表现。

脑水肿的关键问题是脑细胞膜内外水、电解质的分布失衡，破坏了脑细胞赖以生存及发挥正常生理功能的内环境。水分子在跨膜转运过程中，需要通过水通道结合膜蛋白的调节。水通道蛋白是一组与水通透有关的细胞膜转运蛋白[117]，广泛分布于动物、植物及微生物与水代谢密切相关的部位。在生理情况下，水通道基本处于激活状态，水分子经水通道向高渗方向转运，不需要门控来调节[123]。哺乳动物的水通道蛋白家族成员有13

个（AQP0～AQP12）[17]，分布于脑组织的AQP有7种，即AQP1、AQP3，AQP4，AQP5、AQP8、AQP9和AQP11[8, 98]，其中AQP4在中枢神经系统的含量最丰富[18]。AQP4在脑组织水代谢疾病中的作用日益受到人们关注，AQP4与严重烧伤后脑水肿的发生发展是否存在相关性，迄今鲜有报道。

（一）AQP4与烧伤后脑水肿

AQP4广泛分布于中枢神经系统、肾脏、胃肠的腺上皮、肺、眼等部位。在中枢神经系统，其主要分布于毛细血管周围星形胶质细胞终足膜和神经胶质界膜[11, 25]，介导水在细胞、组织间隙、血管和脑室间顺渗透压和流体静力压梯度的转运[124]，此即AQP4的"极性表达"[101]。极性表达均位于脑实质与脑血管或脑脊液接触的界面上，含有AQP4的星形胶质细胞作为脑水分子代谢的中介体，对维持神经元生存微环境稳定具有重要作用。Nielsen等[11]运用电镜对视上核和室旁核等脑组织渗透压感受区进行观察，发现AQP4在此区的胶质细胞表达明显增强。Venero等[125]观察到AQP4 mRNA在下丘脑前核、视上核、杏仁核、海马的锥体细胞层阳性表达。视上核和室旁核主要由神经内分泌细胞组成，AQP4在该区域的表达提示其参与调节神经内分泌活动。视上核和室旁核的大细胞神经元可以合成和分泌精氨酸加压素（AVP），且对渗透压很敏感，微小的渗透压变化即可通过视上核、室旁核合成和释放AVP，调节肾脏水排泄，达到体内水代谢的平衡[23]。笔者在实验中也观察到烧伤后AQP4及其mRNA在视上核、室旁核、视交叉上核、海马齿状回的颗粒细胞和锥体细胞层以及脉络丛、穹窿下器等处表达呈强阳性，据此推测AQP4可能参与调节血浆渗透压平衡和垂体AVP的分泌，可能与血脑屏障的功能有关，并兼有细胞外渗透压感受器的功能。

细胞实验中，低渗条件下AQP4及其mRNA的表达均随着渗透压的下降和作用时间的延长而增强。在高渗条件下，AQP4及其mRNA的表达增强，在高渗液作用12小时后最显著[126, 127]。上述研究进一步证明了AQP4作为渗透压感受器，在渗透压发生改变时其表达上调，调节水的转运，从而维持渗透压处于平衡状态。

研究证实，脑缺血和脑出血导致脑水肿，脑组织AQP4的表达增强[21, 128]，基因敲除或使用AQP4抑制剂可以减轻脑水肿的形成[35, 129]。通过磁共振成像（MRI）观察烫伤后家兔脑损伤并动态检测表观弥散系数（apparent diffusion coefficient，ADC），发现烫伤后AQP4蛋白及其mRNA表达均明显高于正常对照组，与相应ADC值变化呈显著负相关[130]。提示严重烫伤后早期脑水肿（烫伤后6小时以内）以细胞毒性脑水肿为主，AQP4在其形成过程中发挥重要作用。笔者的实验发现：严重烧伤后脑含水量显著增加，烧伤后6小时达到高峰，此后脑含水量逐渐减少，48小时内仍高于正常水平。免疫组化、原位杂交结果显示：正常大鼠AQP4表达呈弱阳性，严重烧伤后AQP4及其mRNA的表达显著增强，其高峰出现时间与脑含水量的高峰时间一致，AQP4的表达与脑含水量的变化亦呈正相关[131]。上述实验结果表明，AQP4参与了严重烧伤后脑水肿的发生发展过程，随着AQP4表达增强，脑含水量也增加，脑水肿程度加重。考虑其发生机制是严重烧伤后机体缺血缺氧、代谢性或呼吸性酸中毒等因素导致内环境破坏，体内渗透压发生改变，激活了渗透压感受器（或AQP4），通过信号转导促使 *AQP4* 基因表达增强，AQP4合成增加，水的通透性增加，脑含水量增加，出现脑水肿。由于在神经系统中广泛表达及其功能特点，AQP4已成为神经疾病的潜在药物靶点[132]。

（二）AQP9与烧伤后脑水肿

1997年Kuriyama等[80]在进行人类脂肪组织基因序列分析时，发现了一种在脂肪组织特异表达的新基因。其分子结构和水的通透性与其他水通道蛋白相似，该基因被命名为 *AQP9*。AQP9不仅转运水，还转运甘油、尿素、乳酸、单羧酸盐等其他小分子[111, 124]，也称为水甘油通道蛋白。AQP9表达于星形胶质细胞、室管膜细胞、脑干儿茶酚胺能神经元、中脑多巴胺能及下丘脑神经元等部位[113]。AQP9与AQP4在中枢神经系统的分布非常相似，不同的是，它不仅分布在微血管周围，还分布在与血管不相接触的星形胶质细胞突起和细胞体上。Ko等[133]提出：整个脑组织的胶质细胞中均有AQP9表达，AQP9参与了中枢神经系统细胞内外环境和系统循环间的信号传递，参与了血浆渗透压感受器的调节功能，维持细胞内外水环境的稳定以及参与脑脊液的生成和重吸收。

在脑缺血和脑外伤模型中均观察到AQP9的表达上调，尤其海马和齿状回星形胶质细胞AQP9表达增加显著[134, 135]。Badaut等[136]在脑缺血模型中观察到AQP9在梗死灶的周围，主要是皮质、苍白球、杏仁体核的星形胶质细胞中AQP9免疫标记上调。在脑缺血酸中毒的病理条件下，AQP9对乳酸盐和水的通透性增加，当体内pH降到5.5时，AQP9对乳酸盐通透性增高4倍，星形胶质细胞过量吸收乳酸盐的同时伴有快速的水内流，从而加剧了脑水肿的发生，AQP9在脑缺血后的过度表达证明其参与了脑水肿的形成。AQP9缺失的脑组织微血管内皮细胞增殖受损，海马神经元损伤严重，提示AQP9可能在脑血管生成中发挥重要作用。AQP9极可能在脑损伤中发挥代偿作用，促进脑血管生成，减少神经元死亡，从而防止神经功能的损害[137]。Yamamoto等[117]在细胞实验中发现AQP4、AQP9在星形胶质细胞中的表达水平最高。在缺氧条件下，AQP4、AQP9的表达明显下降，再给氧6小时后，二者逐渐恢复到原来水平，再给氧超过9小时后，其表达超正常水平（可超出正常水平2倍以上）。笔者的研究发现，在大脑皮质、脉络丛、视上核、室旁核、穹窿下器、海马等处均有AQP9标志物，且AQP9及其mRNA在烧伤后的表达规律与AQP4的表达规律一致，统计学分析均有显著意义。二者在严重烧伤后的表达部位和表达规律均表现出一致，因此认为它们在脑水肿的形成过程中起协同作用，但AQP4、AQP9的重复分布究竟有何意义，尚需进一步研究证实。

关于AQP9的调节机制尚不明确，已知一些关键信号分子参与了AQP9表达的调控。NF-κB抑制剂姜黄素通过抑制NF-κB通路进而抑制AQP4和AQP9的表达，可有效减轻脑缺血小鼠的脑水肿[138]。在高渗条件下，p38 MAPK也可以调节星形胶质细胞AQP9的表达[139]。脑缺血模型鼠侧脑室注射p38 MAPK抑制剂SB203580可减弱AQP9表达，表明通过p38 MAPK信号转导途径介导的AQP9表达的动态变化可能参与了脑缺血后脑水肿的发生[140]。

（三）血浆精氨酸加压素与烧伤后脑水肿的关系

AVP是由下丘脑视上核和室旁核的神经元分泌的一种九肽激素，也称血管升压素或抗利尿激素。AVP的分泌主要受血浆渗透压、血容量、血压及脑灌注压等因素的调节。AVP在应激条件下释放增加，并且与脑水肿的形成关系密切[141, 142]。AVP主要由下丘脑视上核和室旁核的大细胞神经元合成与分泌，经下丘脑-垂体束轴浆运送到神经垂体储存，然后释放入血液循环，最终到达靶器官发挥生理效应[143]。后来发现AVP也存在于室旁核和视交叉上核的小神经细胞内，其纤维投射到垂体门脉系统、终板血管器、前脑、中脑和脊髓中间外侧柱等。在靶器官，AVP与细胞膜上的受体结合，通过第二信使最终发挥效应。AVP是与血管活性、血压及体液平衡调节有关的神经多肽，参与多种生理活动，在体内起着重要的神经递质和调质的作用。国内外临床试验和动物实验均表明，AVP有明显的中枢加压效应，其主要通过以下途径发挥作用：中枢性兴奋交感神经，使交感传出冲动增强，抑制迷走神经；AVP引起周围血管收缩，导致血压升高，使颅内压进一步增高，加重颅脑损伤[144]。AVP增加脑毛细血管通透性及降低脑细胞膜Na^+-K^+-ATP酶的活性，导致细胞内水钠潴留而引起脑水肿，尤其在损伤区通过作用于损伤区毛细血管，使其对水的渗透性进一步增加而加重脑水肿。

AVP受体有3个亚型，即V1a、V1b与V2。AVP分泌后，其受体被激活，每种受体类型都具有特定功能，在渗透压与心血管稳态平衡的生理性调节中发挥重要作用[145]。V1a受体激活后引发血管收缩，V1b受体负责调节情绪和行为，V2受体诱导肾脏的水分重吸收[146]。AVP和V2受体在肾小管远端和集合管中的抗利尿作用已经得到证实，近年来对V1a和V1b受体生理作用的认识越来越多。V1a受体最初发现于血管平滑肌，V1b受体发现于垂体前叶[147]。脑卒中常并发AVP分泌不受控制和低钠血症，对脑产生极大危害[148]。AVP通过V1a和V2受体触发低钠血症、血管痉挛和血小板聚集，从而加重脑水肿。使用AVP-V1a受体拮抗剂可减轻脑水肿和血脑屏障的破坏[149]。

在创伤性和缺血性脑损伤动物的研究中，早期的细胞毒性水肿由V1受体介导，极大可能是通过AQP4上调发挥作用。使用V1受体拮抗剂可减少AQP4上调，抑制脑水肿形成[150, 151]。脑缺血后使用V1受体拮抗剂可减少脑缺血引起的梗死面积和脑水肿，并能显著提高AQP4的表达[152]。在外伤所致蛛网膜下腔出血患者中，检测到血浆和脑脊液AVP水平在伤后24小时内显著升高，且与脑损伤严重程度密切相关[153]。笔者所在课题组采用放射免疫法检测严重烧伤后大鼠血浆AVP水平和脑组织AQP4、AQP9的表达变化，结果显示：烧伤后大鼠血浆AVP显著上升[154]，且随着时间的延长递增；脑组织含水量变化，AQP4、AQP9及其mRNA的表达变化与AVP水平的变化规律一致，它们均在烧伤后6小时出现高峰，各指标呈显著正相关。据此我们推测，烧伤后血浆渗透压和血容量的改变以及强烈的应激反应，导致下丘脑视上核和室旁核合成与分泌AVP增加，经下丘脑-垂体轴输送至脑脊液和血液循环的AVP增加，中枢AVP与血浆AVP同步上调，血浆AVP亦能反映中枢AVP变化情况。在脑组织内AVP与相应受体结合，通过信号转导机制使*AQP4*、*AQP9*基因表达增强，AQP4和AQP9合成加速，从而增加细胞膜对水的通透性，脑组织含水量增加，引起脑水肿。AVP有促进和加重病理性脑水肿的作用，其通过调节脑组织水通道蛋白的表达，使水的通透性增加，促进脑水肿的形成。在临床实践中，可以通过监测血浆AVP含量来评估脑水肿的严重程度，同时为应用AVP受体拮抗剂治疗脑损伤提供理论依据。

五、水通道蛋白与创伤性脑损伤

（一）APQ4 与创伤性脑损伤

创伤性脑损伤（traumatic brain injury，TBI）是指在外界冲击力的直接或间接作用下，脑组织结构和功能的一种病理性损伤[155]。TBI 包括原发性损伤和继发性损伤。原发性损伤是指外界冲击力对脑组织的直接损伤，包括轴突断裂、突触丢失甚至神经细胞死亡。继发性损伤是指在原发性脑损伤的基础上，由钙超载、氧化应激、炎症反应等一系列反应引起的线粒体损伤、细胞凋亡坏死和血脑屏障破坏等病理生理过程[156-158]。其中，脑水肿是最为严重的一种继发性损伤，可造成高颅压、脑疝甚至死亡，TBI 后脑水肿可大致分为血管源性脑水肿与细胞毒性脑水肿。

血管源性脑水肿主要表现为组织间隙液体的聚集，主要由血脑屏障破坏和炎症反应引起。首先，TBI 的冲击力可以对血管造成直接损伤，使血脑屏障通透性增加，从而促使血浆内的白蛋白等渗透活性分子进入组织间隙，引起局部渗透压升高，继而引起组织间隙水潴留[159]。随后，炎症细胞因子和趋化因子被释放，招募并激活免疫细胞，进一步加重炎症反应和血脑屏障破坏[160]。细胞毒性脑水肿主要表现为神经元和胶质细胞胞内的液体聚集。缺血、兴奋毒性、氧化应激和线粒体功能障碍等因素会损伤细胞膜并刺激兴奋性神经递质释放，激活离子通道，引起细胞代谢能力紊乱，从而造成细胞外钠离子和钙离子大量内流，引起细胞内液体聚集，造成细胞肿胀和功能障碍[161, 162]。

AQP 是一类位于生物细胞膜上的通道型蛋白，可以介导水、气体（CO_2 和 NO）和小分子溶质（如乳酸、甘油和尿素等）在细胞内外的转运。在哺乳动物中，AQP 家族成员有 13 个，其中 AQP1、AQP4、AQP5、AQP8 和 AQP9 在啮齿类和灵长类动物的神经细胞中表达，而目前针对 AQP4 的研究最为深入[21, 23, 163]。AQP4 是脑内表达最为丰富的一种 AQP，主要分布在室管膜上皮细胞与脑脊液的接触面、与软脑膜直接接触的星形胶质细胞细胞膜以及血管周围的星形胶质细胞终足膜上，而在终足膜上的表达最为丰富。AQP4 通过与肌萎缩蛋白（dystrophin）复合体中的 α-syntrophin 相互作用，锚定到星形胶质细胞终足膜和室管膜上皮细胞的细胞膜上，从而发挥正常的生理功能[107]。AQP4 有两种主要的亚基 M1 和 M23，还有一种少见的亚基 Mz。M1、M23 和 Mz 均通过 *AQP4* 基因的可变剪切产生。M1 亚基有完整长度，包含 6 个跨膜结构，主要分布在星形胶质细胞的终足膜上；M23 亚基相对较短，缺失前两个跨膜结构，主要分布在星形胶质细胞的胞膜上；Mz 的长度更长，它包含一个特殊的 N 端序列。M1 亚基和 M23 亚基均在脑内不同区域表达，但 M23 亚基的表达最为丰富，至少是 M1 亚基的 3 倍。这两种表型对水的通透性是相似的，功能也一致，均介导了水在细胞内外的转运。Mz 在下丘脑和脉络丛等脑的一些特定部位表达，其功能目前还不清楚[164]。AQP4 可以与内向整流钾通道 Kir4.1 偶联，从而介导中枢神经系统中水和钾离子的转运，维持组织细胞内外水和电解质的平衡[165-167]。在生理状态下，AQP4 可以介导水在细胞内外的双向转运，维持动态平衡[18]。在蛛网膜下腔出血、中枢神经系统炎症、脑肿瘤及脑创伤等病理状态下，AQP4 和 Kir4.1 失去偶联，主要增加了水向细胞内的转运，导致脑内水和电解质平衡出现紊乱[76, 83, 168, 169]。

AQP4在多种急性脑损伤后脑水肿的形成过程中发挥关键作用[19, 170]。AQP4主要在细胞毒性脑水肿中发挥作用，通过介导星形胶质细胞肿胀，促进脑水肿的进展[5]。目前AQP4已成为治疗细胞毒性脑水肿的潜在靶点之一，但由于AQP4在伤后的表达复杂多变，以AQP4为靶点的治疗尚需考虑不同的受伤机制及时间窗[6]。本小节将总结AQP4在TBI后的表达变化、调控机制以及相应转化研究的进展。

1. TBI后AQP4的表达变化　在TBI后细胞毒性脑水肿的发生发展过程中，相较于AQP家族其他成员，AQP4具有更加重要的作用[171]。因此，有学者利用不同的TBI动物模型对AQP4的表达变化进行研究，并探讨AQP4表达变化与细胞毒性脑水肿的关系。

在大鼠自由落体撞击所致TBI模型中，与假手术组相比，伤后24小时AQP4的mRNA水平在伤灶处明显增加，但在伤灶周围明显减少，而在伤灶远隔部位却保持不变。伴随伤灶处AQP4表达增加，伤灶处脑水含量也显著增加[172]。在爆炸诱导的TBI大鼠模型中，AQP4的蛋白水平在伤侧皮质，海马CA1、CA2和CA3区均升高[173]。在大鼠控制性皮质损伤（controlled cortical impact，CCI）模型中，伤侧大脑半球AQP4的蛋白水平逐渐升高，在伤后第12小时和第72小时分别达到两个高峰[174, 175]。但另有学者利用大鼠CCI模型研究发现，与假手术组相比，伤后24小时伤灶核心区域AQP4的蛋白表达明显降低[176]。不仅如此，双侧大脑半球的AQP4蛋白在伤后均逐渐降低，持续到伤后48小时[177]。

在小鼠CCI模型中，伤后第7天，伤灶周围组织、伤灶侧大脑半球纹状体，以及对侧大脑半球海马CA1区的血管周围，AQP4的表达降低，而星形胶质细胞胞体和较粗突起上AQP4的表达则升高；在伤后4周，受损周围组织中血管周围的AQP4表达继续降低，而在星形胶质细胞胞体和较粗的突起上AQP4的表达继续升高，但大脑的其他区域的AQP4表达则恢复正常[178]。

以上研究表明，不同动物模型中AQP4的表达变化有差异，这种差异呈时间依赖性和区域特异性[175]。因此，TBI后细胞毒性脑水肿的研究应同时考虑AQP4表达的时间和空间因素。

有学者用实验进一步探索了TBI后AQP4表达变化的意义。在小鼠CCI模型中，应用小干扰RNA技术降低AQP4的表达水平，可以减少脑水肿的形成，并且改善伤后远期的认知功能[170]。还有实验发现，在大鼠闭合性头部撞击损伤模型后30小时，将抗AQP4抗体经尾静脉注射到大鼠体内，脑水肿的程度明显减轻[38]。由此表明，AQP4的表达增加与TBI后脑水肿的形成直接相关。

除了对脑水肿的影响外，AQP4还能促进星形胶质细胞迁移，导致胶质瘢痕形成[179]。TBI后胶质瘢痕形成，可在早期限制炎症细胞进入受损区域，但在后期会阻止轴突再生[180]。有研究提取*AQP4*敲除小鼠或野生型小鼠的星形胶质细胞，注射到小鼠刀刺脑损伤模型的受损部位，发现与野生型小鼠相比，*AQP4*敲除小鼠的星形胶质细胞向伤处迁移明显减少，胶质瘢痕明显减少[181]。

2. TBI后AQP4的表达调控

（1）HMGB1/TLR4/IL-6/AQP4信号通路：高迁移率族蛋白B1（high-mobility group box 1，HMGB1）是一种炎症介导物。在TBI患者的脑脊液中，HMGB1的表达在前3天均升高,第4天开始降低。在小鼠CCI模型中,学者发现HMGB1在伤后立即从细胞核释放到细胞外,通过结合小胶质细胞的Toll样受体4(Toll-like receptor 4,TLR4),促进白介素6(interleukin-6,IL-6)的释放,从而增加星形胶质细胞中AQP4的表达[182]。

(2)FoxO3a信号通路:FoxO3a是叉头转录因子大家族成员之一。在小鼠CCI模型中,FoxO3a的核转位在伤后12小时和24小时增加,从而在转录水平上调AQP4的表达,继而促进脑水肿的发生,加重神经元和神经功能损害[183]。

(3)HIF-1α/VEGF信号通路:缺氧诱导因子1α(hypoxia-inducible factor 1α,HIF-1α)作为一种重要的转录因子,在大鼠CCI模型中表达升高,从伤后6小时开始升高,1天达到高峰,随后逐渐降低[184]。HIF-1α可在脑损伤后激活血管内皮生长因子(vascular endothelial growth factor,VEGF)[185],VEGF与AQP4在脑挫伤患者脑水肿组织的星形胶质细胞中共表达,协同调控脑水肿的形成和消退[186]。通过人尸检样本行免疫组化发现,TBI后AQP4和HIF-1α的表达均逐渐升高,并持续到伤后30天,且二者的表达呈正相关性[187]。

(4)Sirtuin 2/K310/NF-κB信号通路:Sirtuin 2(Sirt2)是一种去乙酰化酶,是Sirtuin家族成员之一,在神经炎症和脑损伤中发挥作用。在小鼠CCI模型中,Sirt2的表达在伤后6小时增加,3天达到高峰,并持续到第7天。抑制Sirt2可以增加K310乙酰化,并增强NF-κB p65核转位,引起NF-κB活化,后者将引起AQP4的过表达,从而加重TBI后神经炎症的程度和血脑屏障的破坏[188]。

(5)神经生长因子(nerve growth factor,NGF):神经生长因子在脑损伤中发挥保护作用,在TBI后不仅能抑制氧自由基形成[189],还能减轻乙酰胆碱功能缺失,减少细胞死亡,通过减少AQP4的表达减轻脑水肿[190-192]。

综上,TBI后AQP4的表达变化受多种信号通路调控,深入探讨AQP4表达变化的调控机制对寻找治疗TBI的靶点具有重要意义。

3. TBI后针对AQP4的转化研究 鉴于AQP4的表达增加可促进脑水肿的发生发展,学界尝试在TBI后抑制AQP4的表达,以此减缓细胞毒性脑水肿的进展,从而减轻脑水肿引起的神经损害。

(1)AQP4的抑制剂:AQP4的抑制剂可以大致分为四类[193]。①基于半胱氨酸-反应性重金属的抑制剂;②小分子抑制剂;③以AQP4为靶点的治疗视神经脊髓炎(neuromyelitis optica,NMO)的小分子抑制剂;④AQP4新型抑制剂。

重金属如银或金的复合物可以抑制AQP包括AQP4的表达,这些抑制剂可能通过与半胱氨酸残基相互作用来发挥功能[194-195]。钾通道阻滞剂四乙基铵(tetraethyla-mmonium,TEA$^+$)是一种小分子抑制剂,可以可逆性地抑制AQP包括AQP4的表达[196]。除此之外,另外几种小分子抑制剂袢利尿剂、乙酰唑胺、磺胺类抗生素和几种抗癫痫药包括唑尼沙胺、拉莫三嗪、苯妥英钠和托吡酯等对AQP4也有抑制作用[33, 34, 197, 198]。NMO是一种由AQP4特异性抗体引起的自身免疫性疾病,而非致病性的AQP4-IgG重组单克隆抗体Aquaporumab可以竞争性抑制NMO-IgG结合到AQP4,从而治疗NMO[199],因此Aquaporumab具有治疗TBI后脑水肿的潜力。最后,AQP4的一种新型抑制剂TGN-020在缺血性脑卒中的动物模型中发挥对抗脑水肿的作用已被多项研究证实[35, 200, 201],在小鼠CCI模型中TGN-020也能发挥作用减轻脑水肿[202]。

（2）其他药物对AQP4表达的调控：有实验表明，孕激素对AQP4的表达具有调控作用。在大鼠CCI模型中，伤后1小时、6小时和48小时反复经腹腔注射孕激素，结果发现侧脑室旁和伤灶周围组织中AQP4的表达均明显降低，同时脑水含量明显减轻，提示孕激素减轻了AQP4的致脑水肿作用[175]。丙泊酚作为一种全身麻醉药对AQP4的表达也有调控作用。在大鼠CCI模型中，伤后10分钟经腹腔注射丙泊酚，能明显抑制AQP4的mRNA和蛋白水平，且降低了脑水含量[203]。高张盐水对AQP4的表达也有调控作用，在大鼠CCI模型中，伤后经尾静脉注射高张盐水，不仅能降低AQP4的蛋白水平，减轻脑水肿，还能减轻促炎性细胞因子IL-1β和TNF-α的表达，抑制细胞凋亡[204, 205]。胃饥饿素（ghrelin）是胃黏膜分泌的一种神经内分泌激素，在小鼠自由落体撞击所致TBI模型中，伤前及伤后1小时立即腹腔注射胃饥饿素，发现胃饥饿素能够明显降低脑血管周围AQP4的蛋白水平，同时抑制脑水肿[206]。另外，米诺环素在小鼠CCI模型中也发挥保护作用，包括对抗炎症反应，抗细胞凋亡，减轻脑水肿，保护血管和神经功能等，这些作用也可能与其能减少AQP4的表达有关[202]。

（3）转录因子对AQP4表达的调控：转录因子通过调控AQP4表达成为治疗TBI的重要靶点。在大鼠自由落体撞击所致TBI模型中，HIF-1α的表达增加，从而上调了AQP4的表达水平[207]。在大鼠CCI模型中，抑制HIF-1α的活性能下调AQP4的表达，从而减轻TBI后脑水肿的程度，改善预后[208]。在脑静脉窦血栓模型中，NF-κB的抑制可以减少AQP4的表达，从而减轻脑水肿[209]。以NF-κB为靶点的治疗方式也能减轻大鼠TBI后脑水肿的程度，推测与其降低AQP4的表达有关[205]。FoxO3a是调控AQP4表达的另一种转录因子，有研究人员用腺病毒下调小鼠FoxO3a的表达，在CCI模型中AQP4的mRNA和蛋白水平均降低，脑水肿减轻，记忆功能也得到改善[183]。

（4）非编码RNA对AQP4的调控：Malat1是一种长链非编码RNA，在中枢神经系统中发挥重要的调节作用。在TBI大鼠模型中，过表达Malat1可以通过抑制NF-κB的表达下调AQP4水平，从而减轻脑水肿[210]。

（5）手术对AQP4表达的影响：有研究表明，在大鼠TBI后，进行去骨瓣减压可以减少AQP4的表达，从而降低皮质水含量，但具体机制目前还不清楚[211]。

综上所述，TBI后脑水肿的研究已开展多年，但其形成和消退的分子机制目前仍不清楚，针对AQP的药物靶点有望成为对抗脑水肿的新型疗法[212]。

（二）AQP9与创伤性脑水肿

1. TBI后AQP9在大脑的表达变化 脑外伤后，大脑组织将发生一系列病理生理变化。其中能量代谢的变化以及水、电解质平衡的紊乱始终贯穿于脑外伤的原发性损伤期和继发性损伤期。有研究表明，水通道蛋白（AQP9）与大脑能量代谢、水和电解质平衡密切相关[112]。

笔者所在课题组研究发现[213]，轻度脑外伤不会造成大面积脑缺血灶。损伤后1～6小时，脑缺血灶不明显，24小时后脑缺血灶仅局限在脑组织撞击处，至72小时缺血灶最大，但仍较局限。

与缺血情况相对应，脑外伤组双侧皮质和对侧海马AQP9 mRNA含量于1～3小时开始明显增加,其中对侧皮质于1小时,伤侧皮质及对侧海马于3小时出现显著增加（$P<0.05$，$P<0.01$），至6小时基本又降至正常水平,其中伤侧海马AQP9 mRNA明显低于正常水平（$P<0.05$）。24小时及以后各时相点均明显高于正常水平（$P<0.05$），其中伤侧皮质AQP9 mRNA于48小时达峰值,对侧皮质及双侧海马均于72小时达峰值。

免疫组化显示,AQP9在术后1小时呈低水平表达,至3小时AQP9广泛表达于双侧大脑皮质、海马、脑室和导水管系统的室管膜、软脑膜与脉络丛的星形胶质细胞。6小时皮质、海马AQP9表达减弱,脉络丛仍有较强的表达。24小时及其以后各时相点全脑表达广泛上调,48～72小时表达最强,随后略有减弱,但至1周时仍明显高于正常水平。此时穹窿下器的胶质细胞、下丘脑神经元也有较强的表达。假手术组AQP9呈低水平表达,各时相点没有明显差别（$P>0.05$）。

2. AQP9在TBI后脑水肿形成发展中的作用 水通道蛋白（AQP）家族是一组与水、电解质运输平衡相关的膜通道蛋白。AQP9属于水甘油通道蛋白,且在脑组织中含量丰富,其主要分布在脑组织星形胶质细胞、血管内皮细胞和儿茶酚胺类神经元的细胞膜上[79, 112, 214-216]。除水分子外,AQP9对尿素、甘油、多元醇、乳酸等多种小分子溶质均有通透性,可能与组织中水的转运、水和电解质的平衡调节、渗透压的感知及能量代谢有关[79, 82, 136, 217]。

笔者的实验结果显示,在TBI原发性损伤期,TBI后3小时双侧大脑皮质及对侧海马AQP9 mRNA的含量明显增高,AQP9在皮质、海马、穹窿下器、脉络丛的胶质细胞均有广泛表达。于伤后6小时,AQP9 mRNA表达下降,其蛋白表达也随之减弱。在损伤早期,一方面TBI造成组织损伤和血脑屏障的破坏,使得血管内大量蛋白、血浆及其他物质进入细胞间隙,细胞间液增多,渗透压增高。另一方面,当大脑遭受外力打击时,大鼠脑干和下丘脑受到外力的强烈刺激,可通过脑干、下丘脑与大脑半球的广泛投射联系,使神经细胞电兴奋性瞬间变化,开启膜上电压依赖性Ca^{2+}通道,使Ca^{2+}内流增加,从而激活钙依赖性钾通道[218, 219],细胞内K^+大量外流,到达细胞间隙,造成细胞内外K^+分布失衡,引起细胞内外渗透压及电化学梯度变化。由于AQP9与水和电解质平衡有关,此时AQP9的适时高表达可能参与细胞内外水和电解质平衡的调节。

笔者所在课题组对TBI的相关研究表明[219],在TBI后脑干蓝斑去甲肾上腺素（NA）能系统以及下丘脑-垂体-肾上腺系统功能明显增强,NA合成增加,促肾上腺皮质激素（ACTH）与糖皮质激素（GC）大量释放,使机体处于TBI应激状态。在此应激状态下,细胞能量代谢加快,神经内分泌活动加强。研究表明,AQP9与能量代谢平衡、应激及分泌调节有关[79, 217]。因此,笔者认为AQP9在原发性损伤期的表达上调,可能是TBI应激反应的一部分；AQP9参与了能量代谢调节和神经内分泌的精细调节。TBI 6小时后AQP9表达的下调,可能与应激反应的逐渐减弱有关。

以上研究表明,在TBI的原发性损伤期AQP9及AQP9 mRNA表达的上调,可能与TBI应激时的水和电解质、渗透压平衡调节及能量代谢有关,是TBI应激时的一种保护性反应。

在TBI继发性损伤期（24小时以后）,AQP9 mRNA含量继早期6小时低谷之后再次升高,至24小时及以后各时相点,无论伤侧还是对侧、海马还是皮质,AQP9 mRNA的含量均明显增加,除伤侧皮质于48小时达峰值外,其余区域均于72小时达峰值。在TBI继发性损伤期,由于血管的破坏以及原发性损伤期的病理生理变化,大脑处于持续的缺血缺氧状态,加之各种继发性损伤因子的大量释放,使得缺血缺氧进一步加重;而缺血缺氧的加重又引起更多继发性损伤因子的释放。这种恶性循环造成大脑能量供应不足,加之损伤早期能量的过度消耗,糖的无氧酵解增加,乳酸堆积,促使组织内环境酸化。在生理条件下,AQP9可以介导水分子和乳酸根的转运;在酸性环境下,AQP9对乳酸根的通透性可增加4倍;AQP9还可通过乳酸穿梭机制,将乳酸从星形胶质细胞传递给神经元,使乳酸作为神经元能量代谢的供能物质[79]。因此,在TBI继发性损伤阶段,AQP9及AQP9 mRNA在胶质细胞及神经元的表达上调,可能是将乳酸从胶质细胞传递给神经元,以维持神经元能量代谢平衡。综上所述,TBI后AQP9 mRNA含量及AQP9表达增高,可能与TBI后渗透压及水和电解质平衡、能量代谢调节有关。然而,目前TBI后AQP9表达上调的调控机制及其与AQP其他成员间的关系,还有待进一步研究。

六、水通道蛋白在颅内感染疾病中的表达变化及其调控机制

颅内感染是由细菌、病毒、真菌或寄生虫等病原体侵犯脑实质、脑膜和颅内血管等引起的急慢性炎症或非炎症性疾病,常见的颅内感染包括化脓性脑膜炎、结核性脑膜炎、病毒性脑膜炎和脑囊虫病等[220-223]。颅内感染后病原体成分的释放启动并激活免疫防御系统和炎症反应。小胶质细胞活化和白细胞入侵等免疫炎症反应,一方面对病原体的吞噬和清除等过程至关重要,另一方面也可通过释放自由基、蛋白酶、炎症因子和兴奋性氨基酸等物质,导致神经元损伤、动脉血管炎、静脉血栓形成、继发性缺血、血脑屏障破坏和脑水肿等不良后果[224]。

诸多研究表明,AQP参与了巨噬细胞活化和白细胞迁移等免疫炎症反应过程。例如,在铜绿假单胞菌感染后,信号分子30-C12-HSL能够上调巨噬细胞膜中AQP9的表达,进而调节进入细胞内的水分,引起细胞体积、形态的变化,参与变形运动、伸出伪足等巨噬细胞对病原体的吞噬过程[225]。此外,AQP4与神经系统疾病中的炎症反应以及后续的血管损伤和脑水肿形成等病理过程密切相关。在新生大鼠脑缺血缺氧模型中,利用基因沉默技术抑制AQP4的表达,可显著降低脑水肿程度,并伴有炎症因子TNF-α、IL-1β和IL-10的表达下调[226]。在脂多糖诱导的体外星形胶质细胞炎症反应中,AQP4可通过SPHK1/MAPK/AKT途径诱导炎症因子TNF-α和IL-6的表达[227]。

鉴于AQP在中枢神经系统感染后的炎症反应和脑水肿等病理过程中发挥重要作用,本小节将阐述AQP在颅内感染性疾病中的表达变化及其调控机制。

(一)单纯疱疹病毒感染后脑内AQP的表达变化及调控

单纯疱疹病毒(herpes simplex virus,HSV)是欧美发达国家中导致急性散发病毒性脑炎最常见的病原体[228]。边缘系统和岛叶皮质是HSV脑炎最易侵犯的区域[229]。在自然病程中,HSV脑炎的死亡率高达70%,即使经过有效的抗病毒治疗,该病造成严重神经功能

缺损的发生率依然相当高[230]。

为了明确AQP在HSV脑炎病程中的潜在作用，研究者在小鼠模型脑内分别检测了AQP1和AQP4 mRNA的表达水平。结果显示：在感染急性期（7天），AQP1 mRNA表达水平无明显变化，而AQP4 mRNA表达水平较对照组降低了35%；在感染慢性期（6个月后），AQP1和AQP4 mRNA的表达水平分别升高至对照组的3.76倍和5.5倍。感染急性期AQP4表达下调可能是一种内源性保护机制，旨在减轻胶质细胞的肿胀程度[177]。然而，在HSV脑炎慢性期，AQP4表达上调可能与炎症反应、基质金属蛋白酶（matrix metalloproteinase，MMP）表达上调因素等造成的组织损伤相关[231, 232]。关于AQP1在慢性期表达上调的病理生理学意义，尤其是其与组织病理学的相关性，尚需进一步研究。

有关研究者对HSV感染患者血清和脑脊液进行研究时，发现AQP9的表达有所增高。然而，在大鼠HSV感染脑炎模型中，脑内AQP9 mRNA表达水平相较于对照组偏低[233]。Jennische等分析认为，HSV感染后患者血清中AQP9表达上调的原因可能有两方面：其一是病毒的直接诱导作用；其二是感染后细胞能量需求增加，由于AQP9具有转运甘油和尿素的能力，与细胞能量代谢有关，其表达增高可能与机体启动对神经元保护的防御机制相关[233]。另外，有分析认为，AQP9在白细胞中表达[234, 235]，脑炎患者脑脊液中白细胞数量增加可能是脑脊液中AQP9蛋白水平升高的原因。HSV感染动物脑内*AQP9*基因表达水平较低，可能与脑内存在AQP9 mRNA池（a pool of AQP9 mRNA）相关[233]：AQP9的表达调节发生在翻译水平而不是转录水平，在感染等应激状态下，被感染的细胞利用mRNA池来翻译AQP9蛋白，从而使脑内AQP9 mRNA表达水平下调。

（二）人类免疫缺陷病毒感染后脑内AQP的表达变化及调控

关于人类免疫缺陷病毒（human immunodeficiency virus，HIV）感染后脑内AQP表达情况的研究，不同研究机构得出的结论存在差异。Hillaire等对HIV感染引起痴呆的患者尸体脑中央前回（脑内HIV病毒复制部位）进行取材，采用Western blot方法检测，发现AQP4的表达水平较HIV阴性对照标本显著增高[231]。在进一步离体实验中证实，AQP4的表达上调与HIV感染后p38丝裂原活化蛋白激酶（p38 mitogen-activated protein kinase，p38 MAPK）等蛋白激酶的活化相关。

Xing等利用猿猴人免疫缺陷病毒（simian-human immunodeficiency virus，SHIV）感染猕猴，构建了HIV相关的神经认知功能疾病（HIV-associated neurocognitive disorder，HAND）模型。免疫组化染色显示，模型脑内额叶皮质的神经毡等部位AQP4的表达呈斑片状不均匀减少；免疫组化双标结果显示，兴奋性氨基酸转运体-2（excitatory amino acid transporter-2，EAAT-2）表达减少与AQP4的表达减少相关，但EAAT-2的减少程度不及AQP4[236]。鉴于EAAT-2的降低与caspase-3的激活相关，研究者分析AQP4表达下调可能是引起EAAT-2下调的原因，从而进一步诱导神经元凋亡。另外，研究者还探讨了两种HIV感染后AQP4表达不一致的原因，可能是HIV感染后脑内存在慢性炎症反应，而SHIV感染的猕猴模型脑内未见炎症反应，这种不同导致AQP4的表达变化存在差异。

（三）链球菌感染后脑内AQP的表达变化及调节

Papadopoulos等在小鼠脑脊液中注入肺炎链球菌构建动物脑膜炎模型，探讨颅内感染后AQP4的表达变化。结果显示，细菌注入30小时后，小鼠的脑脊液白细胞和细菌计数相对升高，动脉血气分析结果显示，小鼠出现代谢性酸中毒伴代偿性呼吸性碱中毒、相对低温等全身脓毒症的表现。通过脑切片免疫组化定量分析和蛋白质印迹法（Western blot）分析，发现感染模型小鼠脑内AQP4表达相较于对照组显著增高[237]。

笔者所在课题组发现在B组溶血性链球菌感染后的大鼠和小鼠脑内，AQP4的表达水平增高[238, 239]，与Papadopoulos等的结果吻合。此外，笔者所在课题组还对AQP4的表达部位变化进行了研究。在正常脑内，AQP4主要呈极性分布，位于毛细血管周围的星形胶质细胞终足膜、软脑膜内表面形成的胶质界膜和室管膜细胞的基底部等部位[11, 15]。近年来，发现在B组溶血性链球菌感染后的大鼠和小鼠脑内，不仅AQP4蛋白及mRNA的表达水平发生改变，AQP4的极性分布形式亦发生变化，部分AQP4从细胞膜进入细胞内的早期内涵体和晚期内涵体，即AQP4发生内化；同时，还检测到部分AQP4进入溶酶体并被降解。

资料显示，AQP4的极性分布与其锚定机制密切相关[107, 240]。AQP4通过抗肌萎缩蛋白-抗肌萎缩蛋白聚糖复合物（dystrophin-dystroglycan complex，DDC）将其锚定在星形胶质细胞膜上。DDC是由dystroglycan（DG）、dystrophin/utrophin、α-syntrophin和α- dystr-obrevein（α-DB）等组成的蛋白复合体，其核心成员为DG（包括α-DG和β-DG）。β-DG是一个跨膜分子，经α-DG与细胞外基质（extracellular matrix，ECM，包括laminin、perlecan和agrin等）连接，通过dystrophin/utrophin等蛋白与细胞内骨架蛋白（肌动蛋白，actin）连接，形成一个复合结构；该结构可通过α-syntrophin与AQP4相结合，将AQP4锚定到星形胶质细胞相应的细胞膜区域[107, 240, 241]。

研究表明，AQP4的锚定机制对其亚细胞定位至关重要。例如，在脑内DDC中的各组成蛋白分布于星形胶质细胞的终足膜上[107, 242]，与AQP4的高表达区域相一致；在dystrophin缺乏的mdx小鼠，AQP4在胶质细胞终足膜上的极性分布消失[243, 244]；表达低糖基化型α-DG的Largemyd小鼠，α-DG与ECM分子（laminin或agrin）的相互作用严重受损，同时星形胶质细胞终足膜上AQP4的分布明显减少（约75%）[245]。DDC和细胞外基质成分中的分子表达或相互作用发生改变，会对AQP4的极性分布产生显著影响。

聚合酶δ相互作用蛋白2（Poldip2）最初被发现是一种DNA聚合酶δ的p50亚基的结合蛋白，它参与DNA修复、线粒体融合、细胞骨架重塑以及细胞增殖和迁移等多种细胞功能[246]。笔者所在课题组研究发现，B组溶血性链球菌感染后的小鼠脑内Poldip2表达增高，并且Poldip2可上调MMP的表达水平和活性，而MMP可通过降解β-DG，影响DDC的稳定性，从而影响AQP4的锚定机制，导致AQP4的极性分布发生改变[239]。

（四）葡萄球菌感染后脑内AQP的表达变化及调控

脑脓肿的成因包括颅外感染的局部扩展、感染栓塞的血行扩散、穿透性头部创伤或神经外科并发症[247]。金黄色葡萄球菌是引起细菌性脑脓肿最常见的微生物之一[247, 248]。除了由感染性微生物引起的直接组织破坏和由此产生的免疫反应外，脑脓肿周围还会出现严重的脑水肿。Bloch等在葡萄球菌感染引起的小鼠脑脓肿模型中，观察了AQP4的表达变化和胶质增生的情况[249]，他们对感染葡萄球菌第3天的脑切片进行AQP4免疫组化染色，结果显示与同侧和对侧半球远离脓肿的区域相比，脓肿囊周围的AQP4免疫反应性显著上调。与AQP4表达增加的模式不同，胶质纤维酸性蛋白（glial fibrillary acidic protein, GFAP）免疫染色显示，胶质细胞激活的分布更广泛，在整个葡萄球菌注射的脑半球均观察到反应性星形胶质细胞，从脓肿包膜延伸到脑表面，但很少延伸到对侧半球。进一步观察发现，AQP4的表达仅局限于脓肿周围区域的反应性星形胶质细胞的足突部位。在此研究中还探讨了AQP4对脑脓肿周围脑水肿的意义：AQP4表达升高有利于脓肿周围血管源性脑水肿中细胞间隙内多余水分的清除。但是，研究未讨论脑脓肿周围AQP4表达增高的发生及调节机制。

（五）脑膜炎患者脑内AQP的表达变化及调控

目前有两个研究团队分别对脑膜炎患者的病理组织和脑脊液中AQP的表达变化情况进行了检测。

在人类正常大脑中，AQP4分布于星形胶质细胞足突、胶质细胞界限和室管膜；Kir4.1作为啮齿动物星形胶质细胞向内整流性钾通道，可缓冲细胞外K^+浓度增高，并定位于微血管周围星形胶质细胞突起；在啮合动物中，α-syntrophin可结合AQP4并将其募集至星形胶质细胞足突[242]，同时α-syntrophin还募集Kir4.1至视网膜Müller足突[250]。然而在人类，α-syntrophin与AQP4和Kir4.1的偶联情况尚不清楚。因此，研究者采用免疫组织化学染色的方法检测了正常和脑膜炎患者脑内AQP4及Kir4.1和α-syntrophin的表达情况[83]，发现在脑膜炎患者脑组织中AQP4在星形胶质细胞中较对照组增高，同时观察到脑内与AQP4相偶联的Kir4.1和α-syntrophin在星形胶质细胞中亦较对照组增高。星形胶质细胞中AQP4表达上调，表明AQP4可能参与脑膜炎患者脑水肿的形成或吸收过程。

有研究者对35例脑膜炎患者和27例对照者的脑脊液中AQP1和AQP4的表达情况进行了探讨[251]。这些脑膜炎患者涉及多种细菌感染，如肺炎链球菌、李斯特菌、脑膜炎奈瑟菌、流感嗜血杆菌、化脓性链球菌、粪肠球菌和无乳链球菌等。脑膜炎组脑脊液中AQP1的平均浓度为3.8ng/ml（SD 3.4，中位数2.6、25/75百分位：0.8/6.3）显著高于对照组平均浓度0.8ng/ml（SD 0.5、中位数0.8、25/75百分位：0.3/1.1）。脑膜炎组AQP4的平均浓度为1.8ng/ml（SD 3.1、中位数0、25/75百分位数：0/2.2），与对照组平均浓度0.1ng/ml（SD 0.2、中位数0、25/75百分位数：0/0.1）相比有升高的趋势，但差异并不显著。

脑膜炎患者的常规脑脊液参数与AQP1或AQP4之间无显著相关性；然而，在对照人群中，脑脊液中的AQP1与白蛋白（albumin）之间存在显著相关性。此研究还检测了脑膜炎患者和对照者血清中的AQP1和AQP4浓度，发现脑膜炎患者和对照组血清中AQP1或AQP4的浓度无显著差异。另外，分析细菌性脑膜炎（bacterial meningitis，BM）患者与对照者的脑脊液和血清样本，发现在脑膜炎患者中，脑脊液和血清中AQP1或AQP4之间无显著相关性；而在对照人群中，血清中的AQP1与脑脊液中的AQP1之间呈现显著相关性。

研究者对脑脊液中AQP1表达增高的原因进行了分析：首先，脑膜炎患者脑内包括脉络丛上皮细胞在内的多种细胞死亡总量增加，而AQP1表达于脉络丛上皮细胞，所以脑脊液中AQP1增高的一个可能来源是脉络膜丛上皮细胞。其次，由于AQP1在红细胞的含量丰富，脑脊液中AQP1的另一个可能来源是AQP1从血液中扩散至脑脊液。血清中AQP1和脑脊液中AQP1的相关性以及对照组中AQP1与白蛋白的相关性支持了这一假设。最后，脑脊液样本AQP1表达增高可能是样本采集时的血液污染所致。

（六）小结

目前颅内感染后AQP表达变化情况的研究尚不充足，大部分研究仅局限于观察不同感染模型中AQP表达水平的变化。然而，对于感染后AQP表达部位的变化、AQP表达改变后引起的功能改变、AQP发生改变的调节机制等问题的研究相对较少。未来研究者需要在AQP的亚细胞定位、病理情况下的功能改变及调节机制等方向进行更深入的研究。

七、高眼压状态下水通道蛋白的表达变化及其调控机制

眼球是人体含水量最丰富的器官之一，由眼内容物及眼球壁构成。水和离子在眼房内和体循环中的持续运动，对于维护眼的多项生理功能至关重要。房水由睫状体上皮持续分泌并维持一定的眼压，水分子的运动有助于去除角膜、睫状体、晶状体和视网膜等产生的代谢产物，进而保持眼房的透明度；同时通过角膜上皮及其内皮细胞维持角膜的透明度；在视网膜中，水分子被输送到玻璃体腔并穿过视网膜色素上皮，以调节细胞外环境和视网膜的水合作用。AQP在眼的多个构成部分的水转运中发挥着重要作用：AQP0有助于维持晶状体透明度；AQP1参与房水的分泌和排出；AQP3和AQP5行使角膜和结膜的屏障功能；AQP4在视网膜水稳态中起着重要作用[252-254]；AQP6～AQP12最近才被发现，其在人眼功能中的研究尚不深入。

研究表明，眼的不同部分有不同AQP的表达。在角膜中主要有AQP1[254-256]、AQP3、AQP5[254, 255]、AQP7[257]及AQP11[257]的表达；眼部小梁网表达AQP1[254, 258, 259]；睫状体所表达的AQP包括AQP1、AQP4[254, 260]、AQP7[257]、AQP9[257]及AQP11[257]；晶状体中分布有AQP0[261, 262]、AQP1及AQP7[257]；视网膜中Müller细胞表达AQP4[253]、AQP7[257]及AQP11[257]，视网膜节细胞则表达AQP0[263]、AQP1[264, 265]及AQP9[215, 217, 254, 257]。上述AQP在眼中发挥着各自重要的生理功能。AQP1和AQP4参与眼压的调节；AQP0、AQP1和AQP5参与角膜及晶状体透明度的维持；AQP4与视觉信号的转导有关；AQP3参与了结膜的屏障作用；AQP5与泪液的分泌有关[262, 265-267]。

生理条件下，眼球内容物对眼球壁产生15～21mmHg的压力，称为眼内压。眼内压的相对恒定有利于维持视网膜组织正常的血液灌注和水、电解质的运输平衡，对维持视功能的正常具有重要意义。多种眼部疾病可造成眼内压升高至＞21mmHg，称为高眼压，高

眼压可造成眼内睫状体、小梁网、视网膜和视神经等结构发生病理改变，致视力进行性下降及致盲，高眼压是青光眼发病的主要危险因素之一。

当眼内压迅速升高超过视网膜自身调节能力后，即形成急性高眼压。急性高眼压是临床常见的眼内水运输失衡的急症，可使视网膜组织血液灌注和水、电解质的运输受到影响，从而导致视网膜损伤及视功能异常。高眼压的主要病理变化之一是视网膜水肿（retina edema）。视网膜水肿主要表现为视网膜水、电解质运输失衡，是患者视力下降甚至失明的重要原因，而视网膜胶质细胞的肿胀及其所表达蛋白的异常在视网膜水肿等病理改变中发挥着关键性的作用[77, 268]。但迄今为止，关于高眼压导致视网膜水肿发生的分子机制尚未完全阐明，导致其防治相对滞后。近年来的研究表明，高眼压及青光眼的发生发展与眼内AQP0、AQP1、AQP4、AQP5、AQP6、AQP7、AQP9及AQP11等多种AQP的异常表达有紧密联系[269, 270]。本小节将对高眼压发生过程中，眼中AQP的表达变化及其在病程进展中的作用加以介绍，有助于为临床将AQP作为治疗靶点，进而减轻高眼压条件下眼部的损伤提供理论依据。

1. 高眼压条件下AQP0的表达变化及其调控机制 AQP0主要分布于晶状体纤维细胞，与晶状体透明性的维持有关；此外，AQP0也表达于视网膜的神经上皮层及色素上皮层。研究表明，在氩激光光凝法建立的大鼠视网膜静脉阻塞模型中，可发现视网膜神经上皮层及色素上皮层中AQP0的表达均上升；而在高眼压1小时后再灌注24小时的大鼠，其视网膜神经上皮层未见AQP0的表达变化，但在色素上皮层可见AQP0表达减少[270]。由此可见，AQP0的表达调控在不同损伤模型中具有不同的机制，其确切机制尚待进一步研究。

2. 高眼压条件下AQP1的表达变化及其调控机制 AQP1在生理条件下表达于小梁网、Schlemn管内皮细胞、细胞外基质、虹膜基质和睫状体无色素上皮细胞。AQP1是眼组织中含量最高的水通道蛋白，目前对其研究较多也较深入，认为其主要与房水的生成有关。在长期高眼压的作用下，人的角膜内皮细胞AQP1的表达显著下降；在家兔高眼压模型中发现，其角膜AQP1的表达趋势是先增加后逐渐降低，直至恢复正常水平，并且AQP1的表达变化可能与角膜的修复密切相关[271]，而在原发性开角型青光眼及闭角型青光眼中可见虹膜中AQP1表达的增加，升高的AQP1可能促进房水的产生，使眼内压持续增高[272]。在*AQP1*完全敲除或者敲低的动物模型中，均可见到眼内压降低[273, 274]以及房水生成降低[273]，但未见前房形态出现异常，其临床意义尚不确定。

3. 高眼压条件下AQP3的表达变化及其调控机制 AQP3主要分布于结膜上皮[275]。文献表明，在高眼压条件下眼中AQP3的表达未见明显改变。

4. 高眼压条件下AQP4的表达变化及其调控机制 在眼内组织，AQP4有着极其丰富的表达，主要分布于视网膜的星形胶质细胞、Müller细胞、睫状体无色素上皮细胞、小梁网和Schlemn管中，其分布与青光眼的发病密切相关。研究发现，AQP4及其mRNA在视网膜的外网层/外丛状层到神经纤维层均有表达[276]；Nagelhus等[165]通过定量分析进一步证实，AQP4主要分布于视网膜Müller细胞朝向玻璃体面和血管周围胶质细胞的终足膜；Hamann等应用免疫组织化学方法显示AQP4在睫状体色素上皮细胞、虹膜后色素上皮细胞、晶状体前囊上皮细胞呈阳性表达[254]；冉建华等[277]发现，在大鼠角膜上皮细胞和晶状体纤维细胞上也有AQP4的分布。

在生理情况下，AQP4在视网膜中表达量较低，可维持神经元的正常电兴奋性和神经传导性[277]；在急性高眼压早期，大鼠视网膜内AQP4及其mRNA的表达持续增强，并且呈压力依赖性。AQP4的表达增加使其对水的转运能力明显加强，将使水分子在视网膜细胞内过量积聚，导致细胞水肿。由此可见，AQP4的异常高表达可能与急性高眼压的病理损伤过程以及眼内水、电解质代谢失衡有关[278]；在慢性高眼压条件下，AQP4的表达也将升高，导致视网膜厚度增加，并认为其增加对视网膜胶质细胞活化的发生发展起着重要作用[279]。值得一提的是，在急性高眼压状态，AQP4（及AQP9）的表达上调，可理解为视网膜细胞在病理条件下的一种适应性反应，即病理性适应性调节（maladaptive regulation）[280]，增加的AQP4可促进水分子通过旁路途径排出；同时，由于高眼压导致视网膜缺血缺氧，视网膜胶质细胞内环境紊乱及代谢异常，增加的AQP4也将促使视网膜组织间液中过多的水分进入胶质细胞内而促进胶质细胞水肿的发展。AQP4的表达变化也将影响视网膜的功能及形态结构。*AQP4*基因敲除小鼠的视网膜电流图中b波的振幅明显降低[281]，将伴随着视网膜视功能的下降；在急性高眼压小鼠模型中发现，*AQP4*基因敲除后视网膜水肿程度以及其神经节细胞减少程度均较野生型小鼠明显减轻[267]。AQP4的缺失也将导致眼压和房水生成的减低，但尚未发现其变化导致眼前房形态异常[273]。

研究表明，多种药物可调控视网膜AQP4的表达量或表达部位，并伴随着视网膜功能的改善。褪黑素及其受体激动剂Neu-P11能通过降低AQP4在视网膜的表达，调节眼内水和电解质的平衡，保护视网膜结构和功能的完整性，降低眼内压[282]。利尿剂布美他尼可通过抑制细胞外调节蛋白激酶（extracellular regulated protein kinases，ERK）/基质金属蛋白酶-9（matrix metalloproteinase 9，MMP9）通路，减少对AQP4具有锚定作用的β-肌营养不良蛋白聚糖（β-dystroglycan，β-DG）裂解，从而恢复AQP4的极性表达，发挥神经保护作用[283]。

5. 高眼压条件下AQP5的表达变化及其调控机制　AQP5表达于角膜、泪腺与色素上皮细胞[254, 284, 285]，目前关于AQP5在高眼压中的报道较少，有文献报道其在高眼压条件下的表达无变化[286]。

6. 高眼压条件下AQP6的表达变化及其调控机制　AQP6表达于正常视网膜Müller细胞中[287]，目前尚无有关AQP6在高眼压中的报道。

7. 高眼压条件下AQP7的表达变化及其调控机制　在青光眼患者中发现，睫状体非色素上皮细胞中AQP7的表达减少，但Müller细胞终足部位AQP7的表达却增加[286]。

8. 高眼压条件下AQP9的表达变化及其调控机制　Tran等研究者在大鼠和人的视网膜中检测到AQP9 mRNA的表达[288]。在眼前房灌注介导的高眼压大鼠模型中发现，视网膜AQP9的表达上调，其表达上调的峰值时段与视网膜厚度增加的峰值时段基本一致。上述结果提示，AQP9的异常高表达与急性高眼压的病理损伤过程有关，其在眼内水运输平衡中可能兼有压力感受器和水通道蛋白的双重功能。Yang等的研究发现，青光眼动物模型视网膜中AQP9的表达上调，并且其与AQP4、GFAP的表达部位一致，因此分析高眼压条件下视网膜胶质细胞AQP的表达变化，将有助于寻找受损节细胞恢复的作用靶点[289]。

在青光眼患者中发现，睫状体非色素上皮细胞及视网膜节细胞（retinal ganglion cell, RGC）中AQP9的表达均减少，其表达变化与青光眼中能量代谢的改变有关[286]。AQP9除了表达于视网膜外，在视神经头（optic nerve head）中也有表达，而高眼压条件将减少视神经头中AQP9的表达，其减少可能在青光眼所致的视神经病变中发挥作用[257, 290]。RGC下调AQP9表达与RGC代谢和细胞存活相关[257, 290]。

9. 高眼压条件下AQP11的表达变化及其调控机制 目前关于高眼压条件下眼中AQP11的表达变化，尚无文献报道。

10. 高眼压条件下AQP12的表达变化及其调控机制 在高眼压条件下，视网膜神经上皮层中AQP12上调，但是视网膜色素上皮层中AQP12的表达没有改变[270]。

综上所述，AQP在正常眼中发挥着重要的生理作用，同时也参与了高眼压条件下眼中神经节细胞及胶质细胞的病理损伤，但目前关于其确切的调控机制仍然不清楚，还需要进行更深入的探索。

八、水通道蛋白与脑积水

脑脊液（cerebrospinal fluid，CSF）包围脑和脊髓，能吸收中枢神经系统的震荡，还能起到类似淋巴液的免疫作用[291]。正常CSF为无色透明的液体，由细胞成分和非细胞成分组成：细胞成分包括淋巴细胞、单核细胞及红细胞等，而非细胞成分中99%为水分，其余溶质包含钠、钾、氯、糖、乳酸及蛋白质等[292, 293]。CSF循环是人体内三大循环（还包括血液和淋巴循环）之一[294]。传统观点认为：人类的CSF由侧脑室脉络丛上皮细胞分泌产生，经室间孔到达第三脑室，再通过中脑导水管进入第四脑室，最后经第四脑室正中孔和外侧孔进入蛛网膜下腔；CSF进入蛛网膜下腔循环以后，逐步回流到大脑的上矢状窦附近，并被上矢状窦旁的蛛网膜颗粒重新吸收进入血液循环[295]。在生理状态下，CSF每天的分泌量为500～600ml，其产生和回流保持着动态平衡，这种平衡对维持大脑的正常形态和功能具有十分重要的意义。

临床上将多种颅内疾病引起CSF分泌过多和（或）循环吸收障碍，导致脑室系统和蛛网膜下腔中的CSF异常蓄积，并伴有脑室和蛛网膜下腔持续性扩张的一种状态，称为脑积水（hydrocephalus）[296, 297]。脑积水是神经科的一种常见疾病，其发病率和死残率均较高[298-300]。根据发展速度的不同可分为急性脑积水和慢性脑积水：急性脑积水容易引起颅内压快速升高，造成严重的神经功能损伤，甚至死亡；慢性脑积水则容易导致患者长期认知功能障碍，影响生活自理能力，明显降低生活质量。而依据发病原因的差异，脑积水又可以分为以下两种：第一种为梗阻性脑积水，是指位于脑室系统内或脑实质内的占位性病变压迫脑室系统，导致CSF循环受阻，其梗阻部位主要位于第四脑室出口以上，常继发于颅内肿瘤、蛛网膜囊肿、颅咽管瘤、导水管闭锁或狭窄等，其特点是脑室系统不能充分地与蛛网膜下腔相通，出现梗阻部位以上脑室系统扩大；第二种为交通性脑积水，其发生的原因主要是CSF吸收障碍或循环通路受阻部位在第四脑室出口以下（脑室系统以外），常继发于蛛网膜下腔出血、脑膜炎、静脉栓塞或者颅脑外伤等，其特点是脑室系统普遍扩大,且与蛛网膜下腔相通[301,302]。

大脑中的水分,主要分布于脑实质(组织间隙、细胞内外)、脑室系统及蛛网膜下腔。资料显示,脉络丛每分钟产生0.3~0.5ml CSF,其中的水分,实际上是脑实质水分的一种"输出"。最新的研究表明,由脉络丛产生的CSF不仅经蛛网膜颗粒进入体循环系统,还可通过脉络丛上皮细胞、室管膜细胞及穹窿下器,重新回到脑实质[303-305]。不难理解,脑组织细胞内外水、电解质平衡,直接与CSF产生和回流之间的平衡相关。在临床上,当梗阻性脑积水的脑室系统阻塞被解除后,脑积水并不能完全改善,提示脑组织水、电解质平衡出现了障碍,但是其具体机制目前不十分清楚。

20世纪90年代水通道蛋白的发现,为研究中枢神经系统水、电解质平衡提供了可能。迄今为止,在哺乳动物体内分离出的水通道蛋白,共计13种(AQP0~AQP12),而在中枢神经系统表达的水通道蛋白有9种(AQP1、AQP3、AQP4、AQP5、AQP6、AQP7、AQP8、AQP9和AQP11)[212]。大量的研究表明,AQP4和AQP1对于维持脑组织细胞内外水、电解质平衡有着至关重要的作用。

（一）AQP1

1. AQP1在中枢神经系统中的定位分布及功能 脉络丛上皮细胞膜分为顶膜、侧膜和基膜,基膜远离脑室,顶膜朝向脑室腔与CSF直接接触。AQP1在人类和其他动物的脉络丛上皮细胞顶膜中高度表达,在侧膜、基膜中逐渐降低[306-308],这种分布模式提示AQP1可能与CSF的产生具有功能联系[309]。Brown等[310]和Praetorius[311]的研究认为,AQP1和Na^+-K^+-ATP酶在脉络丛上皮细胞膜的不同区域呈差异分布,从基膜、侧膜到顶膜逐渐递增,呈现出一定的梯度变化,这种差异分布模式可以解释水分子是如何由脉络丛上皮细胞进入脑室系统的,这是CSF产生的最重要的分子机制[306-308,310-312]。Oshio等[313]比较了野生型小鼠和*AQP1*敲除小鼠CSF产量和颅内压的差异,结果发现,*AQP1*敲除小鼠脉络膜上皮细胞的渗透性降低为原来的20%,而CSF的产量减少了20%~25%,颅内压则降低了56%。

2. 在脑积水的发生发展过程中AQP1的表达变化 Owler等[314]在小鼠枕大池中注射高岭土制作脑积水模型,并观察建模后脉络丛上皮细胞中AQP1的表达和定位:在造模后的第3天和第5天,与对照组相比,脑积水小鼠AQP1蛋白表达没有明显差异。但在第3天,脑积水小鼠的AQP1 mRNA表达降低,在第5天,AQP1 mRNA恢复到正常水平。H-Tx大鼠是一种常染色体隐性遗传的先天性脑积水模式动物[315],该类大鼠在妊娠早期通过复杂的遗传模式自发形成脑积水,其发病率为30%~50%[316]。Paul等[317]发现,在H-Tx大鼠出生后第5天和第10天脉络丛上皮细胞中的AQP1蛋白表达降低,在出生后第26天恢复到正常水平。此外,Smith等[318]报道了一例15个月大女婴脉络丛增生病例,这是一种罕见的非梗阻性脑积水,与脉络丛CSF过度生成有关。当对脉络丛标本进行免疫组化检测后,发现AQP1较对照组显著低表达。

综上所述,推测在脑积水早期阶段,机体可能通过下调AQP1的表达,以降低CSF的产量,从而延缓脑积水进程。

(二) AQP4

1. AQP4在中枢神经系统的定位分布及功能　AQP4是中枢神经系统中最主要的一种水通道蛋白。据报道[100,124]，在生理情况下，AQP4密集表达于室管膜上皮细胞顶膜、与软脑膜直接接触的星形胶质细胞细胞膜，以及血管周围的星形胶质细胞终足膜，而在终足膜上的表达尤为丰富，学者们将这一分布模式称为AQP4的极性表达[319]。AQP4通过与肌营养不良蛋白（dystrophin）复合体中的α-syntrophin相互作用，锚定到细胞膜上，从而发挥其正常的生理功能，起到调节水运输的作用[107]。AQP4通过介导水分子顺渗透压、流体静力压梯度跨膜转运，从而维持细胞内外的水和电解质平衡[18]。除此之外，AQP4还可以与内向整流钾通道Kir4.1偶联，进而介导中枢神经系统中水和钾离子的转运，维持组织细胞内外水和电解质的平衡[165-167]。

2. 在脑积水的发生发展过程中AQP4的表达变化　Skjolding等[320]的研究表明，用高岭土穿刺注射枕大池制造的交通性脑积水模型中，AQP4的表达具有动态变化趋势：建模后第2天，AQP4在皮质和侧脑室周围区域下降，1周后恢复正常，2周后在侧脑室周围区域表达上升。在另一项研究中，Mao等[321]发现在注射高岭土3～4周后AQP4表达上调，其表达主要集中在血管周围、顶叶、海马和室管膜内侧等区域的星形胶质细胞膜上。在H-Tx大鼠模型中，AQP4的表达水平在7周龄时显著升高，这一趋势一直持续到出生后的第9个月。

Bloch等[322]报道：利用高岭土制作的脑积水模型中，*AQP4*敲除小鼠较野生型小鼠的脑室明显扩张，颅内压和脑含水量均增加2%～3%；造模5天内野生型小鼠的存活率为84%，而*AQP4*敲除鼠的存活率仅为66%，表明*AQP4*敲除鼠的脑积水更加严重。Bloch等[322]基于上述结果建立多室集中参数模型（a multi-compartment lumped parameter model），以模拟梗阻性脑积水的动态变化与AQP4表达量之间的关系；其结果表明，AQP4的表达量和脑积水的严重程度呈负相关，当AQP4表达量增加时，脑积水的严重程度将降低，提示AQP4对脑积水具有缓解作用。

Feng等[323]对612只*AQP4*基因敲除小鼠研究后发现，梗阻性脑积水发生率为9.6%，其梗阻位于中脑导水管，该部位存在明显的室管膜上皮细胞紊乱，提示*AQP4*敲除对于导水管狭窄的发展有一定的作用。该研究认为原发性室管膜增厚、剥脱等形态异常可能导致中脑导水管粘连，从而引起阻塞性脑积水。目前有关室管膜AQP4的功能尚不十分清楚，据推测，AQP4可能促进了脑室和脑实质之间的水分子转运，也可能参与了维持室管膜结构的完整性，故而*AQP4*敲除导致CSF与脑实质间的水分子转运减少可能是室管膜异常的诱因之一。Feng等的研究观察到*AQP4*敲除小鼠室管膜上皮呈现非连续性，因此提出室管膜剥脱是一个关键的病理事件，其可能造成导水管闭塞，并促进梗阻性脑积水的发展。AQP4功能异常在人类导水管狭窄中的作用及其机制值得进一步研究。其研究结果可能为儿童先天性脑积水的临床治疗提供新的思路。

目前基于AQP1和AQP4的研究结果，可以做出推断：在脑积水早期阶段，机体可能通过下调AQP1的表达，以降低CSF的产量，从而延缓脑积水进程；而在脑积水后期，机体可能通过上调AQP4的表达，发挥水分调节作用，加强CSF的吸收，清除多余水分，并延缓脑积水的进程。

（三）AQP1 和 AQP4 与脑积水防治的转化研究和展望

Long 等[324]采用枕大池注射自体血造成大鼠蛛网膜下腔出血模型，经组织学确诊 SAH 大鼠形成脑积水后，检测侧脑室的 AQP1 和 AQP4 蛋白表达量。结果表明，AQP1/AQP4 值与侧脑室相对面积呈正相关，当 AQP1 对脑积水的促进作用大于 AQP4 对脑积水的缓解作用时，便可能发生脑积水。该研究以 AQP1/AQP4 值 0.864 作为脑积水发生的临界值，当高于该值时，便会出现脑积水，而低于该值时，则比较安全。同时，上述研究结果也表明，AQP1 在脑积水的发生发展过程中起促进作用，而 AQP4 则主要参与脑积水的缓解。

如前所述，有关 AQP1 和 AQP4 对脑积水的影响都是基于动物实验的结果获得的。如果这些结果能够通过人体研究得到印证，那么 AQP1 抑制剂和 AQP4 激活剂有可能成为治疗脑积水的潜在药物[325-327]。Tanimura 等[328]将纯化的人 AQP1 重组成脂质囊泡，并应用蛋白脂质体结合流光散射进行测量，用以研究乙酰唑胺对 AQP1 透水性的影响，结果表明，乙酰唑胺对 AQP1 有明显的抑制作用，可以抑制 AQP1 的功能；Niemietz 等[194]从当地血库获取人类红细胞，可在细胞膜表面检测出 AQP1，利用 PBM 囊泡进行抑制剂实验，发现 $AgNO_3$ 和磺胺嘧啶是有效的水通道蛋白 AQP1 的抑制剂；Tang 等[329]报道，应用 50nmol/L 浓度的凝血酶可以最大限度地抑制 AQP4 的表达，但并不影响体外培养的星形胶质细胞的活性，而这种作用可能是由 PKC 途径介导的，PKC 抑制剂 H-7 或 TPA 可以有效拮抗凝血酶对 AQP4 的抑制作用，上调 AQP4 的表达，成为 AQP4 的激活剂。

对于 AQP 抑制剂的研究结果，应当谨慎对待：首先，许多抑制剂具有生物毒性（$AgNO_3$ 等）；其次，这些潜在抑制剂的作用在脑积水模型中并没有得到证实；最后，候选药物应该具有足够的血脑屏障通透性，这在临床应用中非常关键，而上述潜在药物的血脑屏障通透性尚未见报道[330]。总之，AQP1 和 AQP4 在脑积水的发展过程中发挥了关键的作用，对 AQP1 和 AQP4 的深入研究可能为在临床上寻求治疗脑积水的新方式提供启发。

（黄　娟　张兴业　邱国平　骆世芳　刘　辉　万珊珊　何骏驰　程崇杰　孙晓川

甘胜伟　陈　鸿）

参 考 文 献

[1] Barthels D, Das H. Current advances in ischemic stroke research and therapies[J]. Biochim Biophys Acta Mol Basis Dis, 2020, 1866(4): 165260.

[2] Lin MP, Liebeskind DS. Imaging of ischemic stroke[J]. Continuum (Minneap Minn), 2016, 22(5): 1399-1423.

[3] Yang CJ, Hawkins KE, Doré S, et al. Neuroinflammatory mechanisms of blood-brain barrier damage in ischemic stroke[J]. Am J Physiol Cell Physiol, 2019, 316(2): C135-C153.

[4] Radak D, Katsiki N, Resanovic I, et al. Apoptosis and acute brain ischemia in ischemic stroke[J]. Curr Vasc Pharmacol, 2017, 15(2): 115-122.

[5] Zador Z, Stiver S, Wang V, et al. Role of aquaporin-4 in cerebral edema and stroke[J]. Handb Exp Pharmacol, 2009, (190): 159-170.

[6] Clément T, Rodriguez-Grande B, Badaut J. Aquaporins in brain edema[J]. J Neurosci Res, 2020, 98(1): 9-18.

[7] Marmarou A. A review of progress in understanding the pathophysiology and treatment of brain edema[J]. Neurosurg Focus, 2007, 22(5): E1.

[8] Vella J, Zammit C, Di Giovanni G, et al. The central role of aquaporins in the pathophysiology of ischemic stroke[J]. Front Cell Neurosci, 2015, 9: 108.

[9] O'Connor ER, Kimelberg HK. Role of calcium in astrocyte volume regulation and in the release of ions and amino acids[J]. J Neurosci, 1993, 13(6): 2638-2650.

[10] Pantoni L, Garcia JH, Gutierrez JA. Cerebral white matter is highly vulnerable to ischemia[J]. Stroke, 1996, 27(9): 1641-1646; discussion1647.

[11] Nielsen S, Nagelhus EA, Amiry-Moghaddam M, et al. Specialized membrane domains for water transport in glial cells: high-resolution immunogold cytochemistry of aquaporin-4 in rat brain[J]. J Neurosci, 1997, 17(1): 171-180.

[12] Ames A 3rd, Wright RL, Kowada M, et al. Cerebral ischemia. II. The no-reflow phenomenon[J]. Am J Pathol, 1968, 52(2): 437-453.

[13] Choi DW, Rothman SM. The role of glutamate neurotoxicity in hypoxic-ischemic neuronal death[J]. Annu Rev Neurosci, 1990, 13: 171-182.

[14] Wang CX, Shuaib A. Critical role of microvasculature basal lamina in ischemic brain injury[J]. Prog Neurobiol, 2007, 83(3): 140-148.

[15] Papadopoulos MC, Verkman AS. Aquaporin-4 and brain edema[J]. Pediatr Nephrol, 2007, 22(6): 778-784.

[16] Verkman AS, Binder DK, Bloch O, et al. Three distinct roles of aquaporin-4 in brain function revealed by knockout mice[J]. Biochim Biophys Acta, 2006, 1758(8): 1085-1093.

[17] Ishibashi K, Hara S, Kondo S. Aquaporin water channels in mammals[J]. Clin Exp Nephrol, 2009, 13(2): 107-117.

[18] Nagelhus EA, Ottersen OP. Physiological roles of aquaporin-4 in brain[J]. Physiol Rev, 2013, 93(4): 1543-1562.

[19] Badaut J, Ashwal S, Obenaus A. Aquaporins in cerebrovascular disease: a target for treatment of brain edema?[J]. Cerebrovasc Dis, 2011, 31(6): 521-531.

[20] Ribeiro Mde C, Hirt L, Bogousslavsky J, et al. Time course of aquaporin expression after transient focal cerebral ischemia in mice[J]. J Neurosci Res, 2006, 83(7): 1231-1240.

[21] Yang M, Gao F, Liu H, et al. Temporal changes in expression of aquaporin-3, -4, -5 and -8 in rat brains after permanent focal cerebral ischemia[J]. Brain Res, 2009, 1290: 121-132.

[22] Meng S, Qiao M, Foniok T, et al. White matter damage precedes that in gray matter despite similar magnetic resonance imaging changes following cerebral hypoxia-ischemia in neonatal rats[J]. Exp Brain Res, 2005, 166(1): 56-60.

[23] Badaut J, Lasbennes F, Magistretti PJ, et al. Aquaporins in brain: distribution, physiology, and pathophysiology[J]. J Cereb Blood Flow Metab, 2002, 22(4): 367-378.

[24] Li Q, Li Z, Mei Y, et al. Neuregulin attenuated cerebral ischemia—Creperfusion injury via inhibiting apoptosis and upregulating aquaporin-4[J]. Neurosci Lett, 2008, 443(3): 155-159.

[25] Rash JE, Yasumura T, Hudson CS, et al. Direct immunogold labeling of aquaporin-4 in square arrays of astrocyte and ependymocyte plasma membranes in rat brain and spinal cord[J]. Proc Natl Acad Sci USA, 1998, 95(20): 11981-11986.

[26] Huang J, Sun SQ, Lu WT, et al. The internalization and lysosomal degradation of brain AQP4 after ischemic injury[J]. Brain Res, 2013, 1539: 61-72.

[27] Gonda A, Kabagwira J, Senthil GN, et al. Internalization of exosomes through receptor-mediated endocytosis[J]. Mol Cancer Res, 2019, 17(2): 337-347.

[28] Ginn FL, Hochstein P, Trump BF. Membrane alterations in hemolysis: internalization of plasmalemma induced by primaquine[J]. Science, 1969, 164(3881): 843-845.

[29] Endo M, Jain RK, Witwer B, et al. Water channel (aquaporin 1) expression and distribution in mammary carcinomas and glioblastomas[J]. Microvasc Res, 1999, 58(2): 89-98.

[30] Papadopoulos MC, Krishna S, Verkman AS. Aquaporin water channels and brain edema[J]. Mt Sinai J Med, 2002, 69(4): 242-248.

[31] Badaut J, Brunet JF, Grollimund L, et al. Aquaporin 1 and aquaporin 4 expression in human brain after subarachnoid hemorrhage and in peritumoral tissue[J]. Acta Neurochir Suppl, 2003, 86: 495-498.

[32] Oshio K, Binder DK, Bollen A, et al. Aquaporin-1 expression in human glial tumors suggests a potential novel therapeutic target for tumor-associated edema[J]. Acta Neurochir Suppl, 2003, 86: 499-502.

[33] Huber VJ, Tsujita M, Yamazaki M, et al. Identification of arylsulfonamides as aquaporin 4 inhibitors[J]. Bioorg Med Chem Lett, 2007, 17(5): 1270-1273.

[34] Huber VJ, Tsujita M, Kwee IL, et al. Inhibition of aquaporin 4 by antiepileptic drugs[J]. Bioorg Med Chem, 2009, 17(1): 418-424.

[35] Igarashi H, Huber VJ, Tsujita M, et al. Pretreatment with a novel aquaporin 4 inhibitor, TGN-020, significantly reduces ischemic cerebral edema[J]. Neurol Sci, 2011, 32(1): 113-116.

[36] 段升强, 段虎斌. 乙酰唑胺对缺血性脑卒中大鼠脑组织水肿及水通道蛋白4表达的影响[J]. 中华实验外科杂志, 2018, 35(10): 1963.

[37] Wang R, Wu X, Zhao H, et al. Effects of erythropoietin combined with tissue plasminogen activator on the rats following cerebral ischemia and reperfusion[J]. Brain Circ, 2016, 2(1): 54-60.

[38] Higashida T, Kreipke CW, Rafols JA, et al. The role of hypoxia-inducible factor-1α, aquaporin-4, and matrix metalloproteinase-9 in blood-brain barrier disruption and brain edema after traumatic brain injury[J]. J Neurosurg, 2011, 114(1): 92-101.

[39] Wang Z, Meng CJ, Shen XM, et al. Potential contribution of hypoxia-inducible factor-1α, aquaporin-4, and matrix metalloproteinase-9 to blood-brain barrier disruption and brain edema after experimental subarachnoid hemorrhage[J]. J Mol Neurosci, 2012, 48(1): 273-280.

[40] Higashida T, Peng C, Li J, et al. Hypoxia-inducible factor-1α contributes to brain edema after stroke by regulating aquaporins and glycerol distribution in brain[J]. Curr Neurovasc Res, 2011, 8(1): 44-51.

[41] Yang Y, Estrada EY, Thompson JF, et al. Matrix metalloproteinase-mediated disruption of tight junction proteins in cerebral vessels is reversed by synthetic matrix metalloproteinase inhibitor in focal ischemia in rat[J]. J Cereb Blood Flow Metab, 2007, 27(4): 697-709.

[42] Lee JH, Cui HS, Shin SK, et al. Effect of propofol post-treatment on blood-brain barrier integrity and cerebral edema after transient cerebral ischemia in rats[J]. Neurochem Res, 2013, 38(11): 2276-2286.

[43] Tomás-Camardiel M, Venero JL, de Pablos RM, et al. In vivo expression of aquaporin-4 by reactive microglia[J]. J Neurochem, 2004, 91(4): 891-899.

[44] Ikeshima-Kataoka H, Abe Y, Abe T, et al. Immunological function of aquaporin-4 in stab-wounded mouse brain in concert with a pro-inflammatory cytokine inducer, osteopontin[J]. Mol Cell Neurosci, 2013, 56: 65-75.

[45] Bhattacharya P, Pandey AK, Paul S, et al. Aquaporin-4 inhibition mediates piroxicam-induced neuroprotection against focal cerebral ischemia/reperfusion injury in rodents[J]. PLoS One, 2013, 8(9): e73481.

[46] Chen YY, Gong ZC, Zhang MM, et al. Brain-targeting emodin mitigates ischemic stroke via inhibiting

AQP4-mediated swelling and neuroinflammation[J]. Transl Stroke Res, 2023, 15(4): 818-830.

[47] Liu Y, Tang GH, Sun YH, et al. The protective role of Tongxinluo on blood-brain barrier after ischemia-reperfusion brain injury[J]. J Ethnopharmacol, 2013, 148(2): 632-639.

[48] Tang G, Liu Y, Zhang Z, et al. Mesenchymal stem cells maintain blood-brain barrier integrity by inhibiting aquaporin-4 upregulation after cerebral ischemia[J]. Stem Cells, 2014, 32(12): 3150-3162.

[49] Xie B, Gu P, Wang W, et al. Therapeutic effects of human umbilical cord mesenchymal stem cells transplantation on hypoxic ischemic encephalopathy[J]. Am J Transl Res, 2016, 8(7): 3241-3250.

[50] van Kralingen JC, McFall A, Ord ENJ, et al. Altered extracellular vesicle microRNA expression in ischemic stroke and small vessel disease[J]. Transl Stroke Res, 2019, 10(5): 495-508.

[51] Wang Y, Huang J, Ma Y, et al. MicroRNA-29b is a therapeutic target in cerebral ischemia associated with aquaporin 4[J]. J Cereb Blood Flow Metab, 2015, 35(12): 1977-1984.

[52] Gan CS, Wang CW, Tan KS. Circulatory microRNA-145 expression is increased in cerebral ischemia[J]. Genet Mol Res, 2012, 11(1): 147-152.

[53] Zheng L, Cheng W, Wang X, et al. Overexpression of microRNA-145 ameliorates astrocyte injury by targeting aquaporin 4 in cerebral ischemic stroke[J]. Biomed Res Int, 2017, 2017: 9530951.

[54] Sepramaniam S, Armugam A, Lim KY, et al. MicroRNA 320a functions as a novel endogenous modulator of aquaporins 1 and 4 as well as a potential therapeutic target in cerebral ischemia[J]. J Biol Chem, 2010, 285(38): 29223-29230.

[55] Zheng Y, Wang L, Chen L, et al. Upregulation of miR-130b protects against cerebral ischemic injury by targeting water channel protein aquaporin 4(AQP4)[J]. Am J Transl Res, 2017, 9(7): 3452-3461.

[56] Gunnarson E, Zelenina M, Aperia A. Regulation of brain aquaporins[J]. Neuroscience, 2004, 129(4): 947-955.

[57] Moeller HB, Fenton RA, Zeuthen T, et al. Vasopressin-dependent short-term regulation of aquaporin 4 expressed in *Xenopus* oocytes[J]. Neuroscience, 2009, 164(4): 1674-1684.

[58] Madrid R, Le Maout S, Barrault MB, et al. Polarized trafficking and surface expression of the AQP4 water channel are coordinated by serial and regulated interactions with different clathrin-adaptor complexes[J]. EMBO J, 2001, 20(24): 7008-7021.

[59] Bhattacharya P, Pandey AK, Paul S, et al. Melatonin renders neuroprotection by protein kinase C mediated aquaporin-4 inhibition in animal model of focal cerebral ischemia[J]. Life Sci, 2014, 100(2): 97-109.

[60] Wei X, Zhang B, Cheng L, et al. Hydrogen sulfide induces neuroprotection against experimental stroke in rats by down-regulation of AQP4 via activating PKC[J]. Brain Res, 2015, 1622: 292-299.

[61] Chu H, Yang X, Huang C, et al. Apelin-13 protects against ischemic blood-brain barrier damage through the effects of aquaporin-4[J]. Cerebrovasc Dis, 2017, 44(1-2): 10-25.

[62] Nito C, Kamada H, Endo H, et al. Involvement of mitogen-activated protein kinase pathways in expression of the water channel protein aquaporin-4 after ischemia in rat cortical astrocytes[J]. J Neurotrauma, 2012, 29(14): 2404-2412.

[63] 王建枝，殷莲华. 病理生理学[M]. 8版. 北京：人民卫生出版社，2016.

[64] Jennings RB, Steenbergen C Jr, Reimer KA. Myocardial ischemia and reperfusion[J]. Monogr Pathol, 1995, 37: 47-80.

[65] Verkman AS. Aquaporin water channels and endothelial cell function[J]. J Anat, 2002, 200(6): 617-627.

[66] 蒲传强，郎森和，吴卫平. 脑血管病学[M]. 北京：人民军医出版社，1999.

[67] Taoufik E, Probert L. Ischemic neuronal damage[J]. Curr Pharm Des, 2008, 14(33): 3565-3573.

[68] White BC, Sullivan JM, DeGracia DJ, et al. Brain ischemia and reperfusion: molecular mechanisms of

neuronal injury[J]. Neurol Sci, 2000, 179(1): 1-33.

[69] 张兴业, 孙善全, 刘辉, 等. 脑局灶性缺血再灌大鼠AQP4与Kir4.1表达与脑水肿正相关[J]. 基础医学与临床, 2010, 30(8): 5.

[70] Preston GM, Carroll TP, Guggino WB, et al. Appearance of water channels in *Xenopus* oocytes expressing red cell CHIP28 protein[J]. Science, 1992, 256(5055): 385-387.

[71] Benga G. Birth of water channel proteins——the aquaporins[J]. Cell Biol Int, 2003, 27(9): 701-709.

[72] Miao R, Li Cy. The research progress of AQP-dependent cell migration[J]. Chin Pharmacol Bull, 2011, 27(5): 601-605.

[73] Dóczi T, Schwarcz A, Gallyas F, et al. Regulation of water transport in brain oedema[J]. Ideggyogy Sz, 2005, 58(9-10): 298-304.

[74] Rama Rao KV, Chen M, Simard JM, et al. Increased aquaporin-4 expression in ammonia-treated cultured astrocytes[J]. Neuroreport, 2003, 14(18): 2379-2382.

[75] Amiry-Moghaddam M, Williamson A, Palomba M, et al. Delayed K^+ clearance associated with aquaporin-4 mislocalization: phenotypic defects in brains of alpha-syntrophin-null mice[J]. Proc Natl Acad Sci USA, 2003, 100(23): 13615-13620.

[76] Warth A, Mittelbronn M, Wolburg H. Redistribution of the water channel protein aquaporin-4 and the K^+ channel protein Kir4.1 differs in low- and high-grade human brain tumors[J]. Acta Neuropathol, 2005, 109(4): 418-426.

[77] Pannicke T, Iandiev I, Uckermann O, et al. A potassium channel-linked mechanism of glial cell swelling in the postischemic retina[J]. Mol Cell Neurosci, 2004, 26(4): 493-502.

[78] Ding T, Zhou Y, Sun K, et al. Knockdown a water channel protein, aquaporin-4, induced glioblastoma cell apoptosis[J]. PLoS One, 2013, 8(8): e66751.

[79] Badaut J, Regli L. Distribution and possible roles of aquaporin 9 in the brain[J]. Neuroscience, 2004, 129(4): 971-981.

[80] Kuriyama H, Kawamoto S, Ishida N, et al. Molecular cloning and expression of a novel human aquaporin from adipose tissue with glycerol permeability[J]. Biochem Biophys Res Commun, 1997, 241(1): 53-58.

[81] Schrier RW, Chen YC, Cadnapaphornchai MA. From finch to fish to man: role of aquaporins in body fluid and brain water regulation[J]. Neuroscience, 2004, 129(4): 897-904.

[82] Warth A, Mittelbronn M, Hülper P, et al. Expression of the water channel protein aquaporin-9 in malignant brain tumors[J]. Appl Immunohistochem Mol Morphol, 2007, 15(2): 193-198.

[83] Saadoun S, Papadopoulos MC, Krishna S. Water transport becomes uncoupled from K^+ siphoning in brain contusion, bacterial meningitis, and brain tumours: immunohistochemical case review[J]. J Clin Pathol, 2003, 56(12): 972-975.

[84] Belayev L, Alonso OF, Busto R, et al. Middle cerebral artery occlusion in the rat by intraluminal suture[J]. Stroke, 1996, 27(9): 1616-1622.

[85] Belayer L, Busto R, Zhao W, et al. Quantitative evaluation of blood-brain barrier permeability following middle cerebral artery occlusion in rats[J]. Brain Res, 1996, 739(1-2): 88-96.

[86] 王耀辉, 李国辉. p38 MAPK在大鼠脑缺血再灌注脑组织AQP4表达变化及脑水肿形成中的作用[J]. 中风与神经疾病杂志, 2013, 30(8): 700-702.

[87] 陈俊, 陈寿权, 李章平, 等. 七叶皂苷对心肺复苏后大鼠脑水肿期脑水通道蛋白4 mRNA变化的作用[J]. 中国中西医结合急救杂志, 2007, (4): 245-249.

[88] 刘凤琴, 牛小媛, 牛文华. 大鼠脑缺血再灌注后脑组织水通道蛋白4的表达及丹参川芎嗪干预作用的研究[J]. 中华老年心脑血管病杂志, 2012, 14(12): 1321-1324.

[89] Qureshi AI, Mendelow AD, Hanley DF. Intracerebral haemorrhage[J]. Lancet, 2009, 373(9675): 1632-1644.

[90] Sangha N, Gonzales NR. Treatment targets in intracerebral hemorrhage[J]. Neurotherapeutics, 2011, 8(3): 374-387.

[91] Bordone MP, Salman MM, Titus HE, et al. The energetic brain——a review from students to students[J]. J Neurochem, 2019, 151(2): 139-165.

[92] Gross BA, Jankowitz BT, Friedlander RM. Cerebral intraparenchymal hemorrhage: a review[J]. JAMA, 2019, 321(13): 1295-1303.

[93] Amiry-Moghaddam M, Ottersen OP. The molecular basis of water transport in the brain[J]. Nat Rev Neurosci, 2003, 4(12): 991-1001.

[94] Assentoft M, Larsen BR, MacAulay N. Regulation and function of AQP4 in the central nervous system[J]. Neurochem Res, 2015, 40(12): 2615-2627.

[95] Vandebroek A, Yasui M. Regulation of AQP4 in the central nervous system[J]. Int J Mol Sci, 2020, 21(5): 1603.

[96] Takata K, Matsuzaki T, Tajika Y. Aquaporins: water channel proteins of the cell membrane[J]. Prog Histochem Cytochem, 2004, 39(1): 1-83.

[97] Filippidis AS, Carozza RB, Rekate HL. Aquaporins in brain edema and neuropathological conditions[J]. Int J Mol Sci, 2016, 18(1): 55.

[98] Arciénega II, Brunet JF, Bloch J, et al. Cell locations for AQP1, AQP4 and 9 in the non-human primate brain[J]. Neuroscience, 2010, 167(4): 1103-1114.

[99] Papadopoulos MC, Verkman AS. Aquaporin water channels in the nervous system[J]. Nat Rev Neurosci, 2013, 14(4): 265-277.

[100] Iacovetta C, Rudloff E, Kirby R. The role of aquaporin 4 in the brain[J]. Physiol Rev, 2012, 41(1): 32-44.

[101] Qiu GP, Xu J, Zhuo F, et al. Loss of AQP4 polarized localization with loss of β-dystroglycan immunoreactivity may induce brain edema following intracerebral hemorrhage[J]. Neurosci Lett, 2015, 588: 42-48.

[102] Wolburg H, Noell S, Fallier-Becker P, et al. The disturbed blood-brain barrier in human glioblastoma[J]. Mol Aspects Med, 2012, 33(5-6): 579-589.

[103] Noell S, Wolburg-Buchholz K, Mack AF, et al. Evidence for a role of dystroglycan regulating the membrane architecture of astroglial endfeet[J]. Eur J Neurosci, 2011, 33(12): 2179-2186.

[104] 邱国平, 孙善全, 刘辉, 等. 脑出血后水通道蛋白4与内向整流性钾通道4.1的再分布与脑水肿形成的关系[J]. 解剖学杂志, 2009, 32: 765-769.

[105] 邱国平, 孙善全, 卓飞, 等. 脑出血脑组织中水通道蛋白4的极性表达缺失[J]. 重庆医科大学学报, 2015, 40(5): 666-670.

[106] 邱国平, 孙善全, 卓飞, 等. 脑出血后脑组织中水通道蛋白4的定位分布变化[J]. 中国组织化学与细胞化学杂志, 2016, 25: 23-29.

[107] Amiry-Moghaddam M, Frydenlund DS, Ottersen OP. Anchoring of aquaporin-4 in brain: molecular mechanisms and implications for the physiology and pathophysiology of water transport[J]. Neuroscience, 2004, 129(4): 999-1010.

[108] Steiner E, Enzmann GU, Lin S, et al. Loss of astrocyte polarization upon transient focal brain ischemia as a possible mechanism to counteract early edema formation[J]. Glia, 2012, 60(11): 1646-1659.

[109] Constantin B. Dystrophin complex functions as a scaffold for signalling proteins[J]. Biochim Biophys Acta, 2014, 1838(2): 635-642.

[110] Carmosino M, Procino G, Nicchia GP, et al. Histamine treatment induces rearrangements of orthogonal arrays of particles(OAPs)in human AQP4-expressing gastric cells[J]. J Cell Biol, 2001, 154(6): 1235-1243.

[111] Buffoli B. Aquaporin biology and nervous system[J]. Curr Neuropharmacol, 2010, 8(2): 97-104.

[112] Badaut J, Petit JM, Brunet JF, et al. Distribution of aquaporin 9 in the adult rat brain: preferential expression in catecholaminergic neurons and in glial cells[J]. Neuroscience, 2004, 128(1): 27-38.

[113] Amiry-Moghaddam M, Xue R, Haug FM, et al. Alpha-syntrophin deletion removes the perivascular but not endothelial pool of aquaporin-4 at the blood-brain barrier and delays the development of brain edema in an experimental model of acute hyponatremia[J]. FASEB J, 2004, 18(3): 542-544.

[114] 邱国平，孙善全，刘辉，等. AQP9在脑出血大鼠脑组织中的表达变化[J]. 第三军医大学学报, 2009, 31(5): 402-405.

[115] 杨美，孙善全，汪克建，等. 水通道蛋白5在大鼠大脑组织中的分布及表达[J]. 解剖学杂志, 2008, (1): 5-7+142.

[116] Chai RC, Jiang JH, Wong AK, et al. AQP5 is differentially regulated in astrocytes during metabolic and traumatic injuries[J]. Glia, 2013, 61(10): 1748-1765.

[117] Yamamoto N, Yoneda K, Asai K, et al. Alterations in the expression of the AQP family in cultured rat astrocytes during hypoxia and reoxygenation[J]. Brain Res Mol Brain Res, 2001, 90(1): 26-38.

[118] Koike S, Tanaka Y, Matsuzaki T, et al. Aquaporin-11(AQP11)expression in the mouse brain[J]. Int J Mol Sci, 2016, 17(6): 861.

[119] Gorelick DA, Praetorius J, Tsunenari T, et al. Aquaporin-11: a channel protein lacking apparent transport function expressed in brain[J]. BMC Biochem, 2006, 7: 14.

[120] Soejima K, Schmalstieg FC, Sakurai H, et al. Pathophysiological analysis of combined burn and smoke inhalation injuries in sheep[J]. Am J Physiol Lung Cell Mol Physiol, 2001, 280(6): L1233-L1241.

[121] 杨宗城. 烧伤后血管内皮细胞损伤及其在早期脏器损害中的作用[J]. 中华烧伤杂志, 2001, (3): 4-6.

[122] Barone CM, Jimenez DF, Huxley VH, et al. In vivo visualization of cerebral microcirculation in systemic thermal injury[J]. J Burn Care Rehabil, 2000, 21(1 Pt 1): 20-25.

[123] King LS, Agre P. Pathophysiology of the aquaporin water channels[J]. Annu Rev Physiol, 1996, 58: 619-648.

[124] Papadopoulos MC, Verkman AS. Aquaporin water channels in the nervous system[J]. Nat Rev Neurosci, 2013, 14(4): 265-277.

[125] Venero JL, Vizuete ML, Ilundáin AA, et al. Detailed localization of aquaporin-4 messenger RNA in the CNS: preferential expression in periventricular organs[J]. Neuroscience, 1999, 94(1): 239-250.

[126] 李燕华，孙善全. 低渗液对星形胶质细胞水通道蛋白-4表达的影响[J]. 中华医学杂志, 2004, (6): 60-65.

[127] 李燕华，孙善全. 高渗液对体外培养星形胶质细胞中水通道蛋白4 mRNA表达的影响[J]. 中国危重病急救医学, 2004, (4): 210-213+258.

[128] Tang Y, Wu P, Su J, et al. Effects of aquaporin-4 on edema formation following intracerebral hemorrhage[J]. Exp Neurol, 2010, 223(2): 485-495.

[129] Manley GT, Fujimura M, Ma T, et al. Aquaporin-4 deletion in mice reduces brain edema after acute water intoxication and ischemic stroke[J]. Nat Med, 2000, 6(2): 159-163.

[130] 张艳伟，黎海涛，胡俊，等. 烫伤早期脑水肿家兔脑组织水通道蛋白4表达与磁共振成像结果相关性研究[J]. 中华烧伤杂志, 2011, (6): 441-445.

[131] 骆世芳，孙善全，汪克建. 严重烧伤后脑组织水通道蛋白-4的表达变化及意义[J]. 中华创伤杂志, 2007, 23(6): 465-468.

[132] Verkman AS, Smith AJ, Phuan PW, et al. The aquaporin-4 water channel as a potential drug target in neurological disorders[J]. Expert Opin Ther Targets, 2017, 21(12): 1161-1170.

[133] Ko SB, Uchida S, Naruse S, et al. Cloning and functional expression of rAOP9L a new member of aquaporin family from rat liver[J]. Biochem Mol Biol Int, 1999, 47(2): 309-318.

[134] Hwang IK, Yoo KY, Li H, et al. Aquaporin 9 changes in pyramidal cells before and is expressed in astrocytes after delayed neuronal death in the ischemic hippocampal CA1 region of the gerbil[J]. J Neurosci Res, 2007, 85(11): 2470-2479.

[135] Liu H, Yang M, Qiu GP, et al. Aquaporin 9 in rat brain after severe traumatic brain injury[J]. Arq Neuropsiquiatr, 2012, 70(3): 214-220.

[136] Badaut J, Hirt L, Granziera C, et al. Astrocyte-specific expression of aquaporin-9 in mouse brain is increased after transient focal cerebral ischemia[J]. J Cereb Blood Flow Metab, 2001, 21(5): 477-482.

[137] Ji W, Wang J, Xu J, et al. Lack of aquaporin 9 reduces brain angiogenesis and exaggerates neuronal loss in the hippocampus following intracranial hemorrhage in mice[J]. J Mol Neurosci, 2017, 61(3): 351-358.

[138] Wang BF, Cui ZW, Zhong ZH, et al. Curcumin attenuates brain edema in mice with intracerebral hemorrhage through inhibition of AQP4 and AQP9 expression[J]. Acta Pharmacol Sin, 2015, 36(8): 939-948.

[139] Yang M, Gao F, Liu H, et al. Hyperosmotic induction of aquaporin expression in rat astrocytes through a different MAPK pathway[J]. J Cell Biochem, 2013, 114(1): 111-119.

[140] Wei X, Ren X, Jiang R, et al. Phosphorylation of p38 MAPK mediates aquaporin 9 expression in rat brains during permanent focal cerebral ischaemia[J]. J Mol Histol, 2015, 46(3): 273-281.

[141] Verney EB. The antidiuretic hormone and the factors which determine its release[J]. Proc R Soc Lond B Biol Sci, 1947, 135(878): 25-106.

[142] Rosenberg GA, Scremin O, Estrada E, et al. Arginine vasopressin V1-antagonist and atrial natriuretic peptide reduce hemorrhagic brain edema in rats[J]. Stroke, 1992, 23(12): 1767-1773; discussion 1773-1774.

[143] 佚名.《神经科学纲要》即将发行[J].生理科学进展, 1993, 1: 30.

[144] Armstead WM. Role of endothelin in pial artery vasoconstriction and altered responses to vasopressin after brain injury[J]. J Neurosurg, 1996, 85(5): 901-907.

[145] Koshimizu TA, Nasa Y, Tanoue A, et al. V1a vasopressin receptors maintain normal blood pressure by regulating circulating blood volume and baroreflex sensitivity[J]. Proc Natl Acad Sci USA, 2006, 103(20): 7807-7812.

[146] Zeynalov E, Jones SM, Elliott JP. Vasopressin and vasopressin receptors in brain edema[J]. Vitam Horm, 2020, 113: 291-312.

[147] Koshimizu TA, Nakamura K, Egashira N, et al. Vasopressin V1a and V1b receptors: from molecules to physiological systems[J]. Physiol Rev, 2012, 92(4): 1813-1864.

[148] Baylis PH. The syndrome of inappropriate antidiuretic hormone secretion[J]. Int J Biochem Cell Biol, 2003, 35(11): 1495-1499.

[149] Zeynalov E, Jones SM, Seo JW, et al. Arginine-vasopressin receptor blocker conivaptan reduces brain edema and blood-brain barrier disruption after experimental stroke in mice[J]. PLoS One, 2015, 10(8): e0136121.

[150] Kleindienst A, Fazzina G, Dunbar JG, et al. Protective effect of the V1a receptor antagonist SR49059 on brain edema formation following middle cerebral artery occlusion in the rat[J]. Acta Neurochir Suppl, 2006, 96: 303-306.

[151] Kleindienst A, Dunbar JG, Glisson R, et al. The role of vasopressin V1A receptors in cytotoxic brain

edema formation following brain injury[J]. Acta Neurochir（Wien）, 2013, 155（1）: 151-164.

[152] Liu X, Nakayama S, Amiry-Moghaddam M, et al. Arginine-vasopressin V1 but not V2 receptor antagonism modulates infarct volume, brain water content, and aquaporin-4 expression following experimental stroke[J]. Neurocrit Care, 2010, 12（1）: 124-131.

[153] Yuan ZH, Zhu JY, Huang WD, et al. Early change of plasma and cerebrospinal fluid arginine vasopressin in traumatic subarachnoid hemorrhage[J]. Chin J Traumatol, 2010, 13（1）: 42-45.

[154] 骆世芳, 孙善全. 严重烧伤后血浆AVP水平与脑含水量的相关性研究[J]. 重庆医科大学学报, 2005,（2）: 246-248.

[155] Capizzi A, Woo J, Verduzco-Gutierrez M. Traumatic brain injury: an overview of epidemiology, pathophysiology, and medical management[J]. Med Clin North Am, 2020, 104（2）: 213-238.

[156] Dixon KJ. Pathophysiology of traumatic brain injury[J]. Phys Med Rehabil Clin N Am, 2017, 28（2）: 215-225.

[157] Browne KD, Chen XH, Meaney DF, et al. Mild traumatic brain injury and diffuse axonal injury in swine[J]. J Neurotrauma, 2011, 28（9）: 1747-1755.

[158] Tang-Schomer MD, Johnson VE, Baas PW, et al. Partial interruption of axonal transport due to microtubule breakage accounts for the formation of periodic varicosities after traumatic axonal injury[J]. Exp Neurol, 2012, 233（1）: 364-372.

[159] Sorby-Adams AJ, Marcoionni AM, Dempsey ER, et al. The role of neurogenic inflammation in blood-brain barrier disruption and development of cerebral oedema following acute central nervous system（CNS）injury[J]. Int J Mol Sci, 2017, 18（8）: 1788.

[160] Nimmo AJ, Cernak I, Heath DL, et al. Neurogenic inflammation is associated with development of edema and functional deficits following traumatic brain injury in rats[J]. Neuropeptides, 2004, 38（1）: 40-47.

[161] Jha RM, Kochanek PM, Simard JM. Pathophysiology and treatment of cerebral edema in traumatic brain injury[J]. Neuropharmacology, 2019, 145（Pt B）: 230-246.

[162] Katada R, Nishitani Y, Honmou O, et al. Expression of aquaporin-4 augments cytotoxic brain edema after traumatic brain injury during acute ethanol exposure[J]. Am J Pathol, 2012, 180（1）: 17-23.

[163] Badaut J, Nehlig A, Verbavatz J, et al. Hypervascularization in the magnocellular nuclei of the rat hypothalamus: relationship with the distribution of aquaporin-4 and markers of energy metabolism[J]. J Neuroendocrinol, 2000, 12（10）: 960-969.

[164] Verkman AS, Ratelade J, Rossi A, et al. Aquaporin-4: orthogonal array assembly, CNS functions, and role in neuromyelitis optica[J]. Acta Pharmacol Sin, 2011, 32（6）: 702-710.

[165] Nagelhus EA, Horio Y, Inanobe A, et al. Immunogold evidence suggests that coupling of K^+ siphoning and water transport in rat retinal Müller cells is mediated by a coenrichment of Kir4.1 and AQP4 in specific membrane domains[J]. Glia, 1999, 26（1）: 47-54.

[166] Strohschein S, Hüttmann K, Gabriel S, et al. Impact of aquaporin-4 channels on K^+ buffering and gap junction coupling in the hippocampus[J]. Glia, 2011, 59（6）: 973-980.

[167] Nagelhus EA, Mathiisen TM, Ottersen OP. Aquaporin-4 in the central nervous system: cellular and subcellular distribution and coexpression with KIR4.1[J]. Neuroscience, 2004, 129（4）: 905-913.

[168] Yan JH, Khatibi NH, Han HB, et al. p53-induced uncoupling expression of aquaporin-4 and inwardly rectifying $K^+4.1$ channels in cytotoxic edema after subarachnoid hemorrhage[J]. CNS Neurosci Ther, 2012, 18（4）: 334-342.

[169] Liu XQ, Kobayashi H, Jin ZB, et al. Differential expression of Kir4.1 and aquaporin 4 in the retina from endotoxin-induced uveitis rat[J]. Mol Vis, 2007, 13: 309-317.

[170] Fukuda AM, Adami A, Pop V, et al. Posttraumatic reduction of edema with aquaporin-4 RNA interference improves acute and chronic functional recovery[J]. J Cereb Blood Flow Metab, 2013, 33(10): 1621-1632.

[171] Fukuda AM, Pop V, Spagnoli D, et al. Delayed increase of astrocytic aquaporin 4 after juvenile traumatic brain injury: possible role in edema resolution?[J]. Neuroscience, 2012, 222: 366-378.

[172] Sun MC, Honey CR, Berk C, et al. Regulation of aquaporin-4 in a traumatic brain injury model in rats[J]. J Neurosurg, 2003, 98(3): 565-569.

[173] Gu M, Kawoos U, McCarron R, et al. Protection against blast-induced traumatic brain injury by increase in brain volume[J]. Biomed Res Int, 2017, 2017: 2075463.

[174] Xiong A, Xiong R, Yu J, et al. Aquaporin-4 is a potential drug target for traumatic brain injury via aggravating the severity of brain edema[J]. Burns Trauma, 2021, 9: tkaa050.

[175] Guo Q, Sayeed I, Baronne LM, et al. Progesterone administration modulates AQP4 expression and edema after traumatic brain injury in male rats[J]. Exp Neurol, 2006, 198(2): 469-478.

[176] Zhao J, Moore AN, Clifton GL, et al. Sulforaphane enhances aquaporin-4 expression and decreases cerebral edema following traumatic brain injury[J]. J Neurosci Res, 2005, 82(4): 499-506.

[177] Kiening KL, van Landeghem FK, Schreiber S, et al. Decreased hemispheric aquaporin-4 is linked to evolving brain edema following controlled cortical impact injury in rats[J]. Neurosci Lett, 2002, 324(2): 105-108.

[178] Zhao ZA, Li P, Ye SY, et al. Perivascular AQP4 dysregulation in the hippocampal CA1 area after traumatic brain injury is alleviated by adenosine A2A receptor inactivation[J]. Sci Rep, 2017, 7(1): 2254.

[179] Saadoun S, Papadopoulos MC, Watanabe H, et al. Involvement of aquaporin-4 in astroglial cell migration and glial scar formation[J]. J Cell Sci, 2005, 118(Pt 24): 5691-5698.

[180] Cheng X, Wang J, Sun X, et al. Morphological and functional alterations of astrocytes responding to traumatic brain injury[J]. J Integr Neurosci, 2019, 18(2): 203-215.

[181] Auguste KI, Jin S, Uchida K, et al. Greatly impaired migration of implanted aquaporin-4-deficient astroglial cells in mouse brain toward a site of injury[J]. FASEB J, 2007, 21(1): 108-116.

[182] Laird MD, Shields JS, Sukumari-Ramesh S, et al. High mobility group box protein-1 promotes cerebral edema after traumatic brain injury via activation of Toll-like receptor 4[J]. Glia, 2014, 62(1): 26-38.

[183] Kapoor S, Kim SM, Farook JM, et al. Foxo3a transcriptionally upregulates AQP4 and induces cerebral edema following traumatic brain injury[J]. J Neurosci, 2013, 33(44): 17398-17403.

[184] Li A H, Sun X, Ni Y, et al. HIF-1α involves in neuronal apoptosis after traumatic brain injury in adult rats[J]. J Mol Neurosci, 2013, 51(3): 1052-1062.

[185] Dai Y, Xu M, Wang Y, et al. HIF-1alpha induced-VEGF overexpression in bone marrow stem cells protects cardiomyocytes against ischemia[J]. J Mol Cell Cardiol, 2007, 42(6): 1036-1044.

[186] Suzuki R, Okuda M, Asai J, et al. Astrocytes co-express aquaporin-1, -4, and vascular endothelial growth factor in brain edema tissue associated with brain contusion[J]. Acta Neurochir Suppl, 2006, 96: 398-401.

[187] Neri M, Frati A, Turillazzi E, et al. Immunohistochemical evaluation of aquaporin-4 and its correlation with CD68, IBA-1, HIF-1α, GFAP, and CD15 expressions in fatal traumatic brain injury[J]. Int J Mol Sci, 2018, 19(11): 3544.

[188] Yuan F, Xu ZM, Lu LY, et al. SIRT2 inhibition exacerbates neuroinflammation and blood-brain barrier disruption in experimental traumatic brain injury by enhancing NF-κB p65 acetylation and activation[J]. J Neurochem, 2016, 136(3): 581-593.

[189] DeKosky ST, Goss JR, Miller PD, et al. Upregulation of nerve growth factor following cortical trauma[J]. Exp Neurol, 1994, 130(2): 173-177.

[190] Dixon CE, Flinn P, Bao J, et al. Nerve growth factor attenuates cholinergic deficits following traumatic brain injury in rats[J]. Exp Neurol, 1997, 146(2): 479-490.

[191] Philips MF, Mattiasson G, Wieloch T, et al. Neuroprotective and behavioral efficacy of nerve growth factor-transfected hippocampal progenitor cell transplants after experimental traumatic brain injury[J]. J Neurosurg, 2001, 94(5): 765-774.

[192] Lv Q, Fan X, Xu G, et al. Intranasal delivery of nerve growth factor attenuates aquaporins-4-induced edema following traumatic brain injury in rats[J]. Brain Res, 2013, 1493: 80-89.

[193] Verkman AS, Anderson MO, Papadopoulos MC. Aquaporins: important but elusive drug targets[J]. Nat Rev Drug Discov, 2014, 13(4): 259-277.

[194] Niemietz CM, Tyerman SD. New potent inhibitors of aquaporins: silver and gold compounds inhibit aquaporins of plant and human origin[J]. FEBS Lett, 2002, 531(3): 443-447.

[195] Martins AP, Ciancetta A, de Almeida A, et al. Aquaporin inhibition by gold(Ⅲ) compounds: new insights[J]. ChemMedChem, 2013, 8(7): 1086-1092.

[196] Detmers FJ, de Groot BL, Müller EM, et al. Quaternary ammonium compounds as water channel blockers. Specificity, potency, and site of action[J]. J Biol Chem, 2006, 281(20): 14207-14214.

[197] Migliati E, Meurice N, DuBois P, et al. Inhibition of aquaporin-1 and aquaporin-4 water permeability by a derivative of the loop diuretic bumetanide acting at an internal pore-occluding binding site[J]. Mol Pharmacol, 2009, 76(1): 105-112.

[198] Ma B, Xiang Y, Mu SM, et al. Effects of acetazolamide and anordiol on osmotic water permeability in AQP1-cRNA injected *Xenopus* oocyte[J]. Acta Pharmacol Sin, 2004, 25(1): 90-97.

[199] Tradtrantip L, Zhang H, Saadoun S, et al. Anti-aquaporin-4 monoclonal antibody blocker therapy for neuromyelitis optica[J]. Ann Neurol, 2012, 71(3): 314-322.

[200] Sun C, Lin L, Yin L, et al. Acutely inhibiting AQP4 with TGN-020 improves functional outcome by attenuating edema and peri-infarct astrogliosis after cerebral ischemia[J]. Front Immunol, 2022, 13: 870029.

[201] Cui D, Jia S, Li T, et al. Alleviation of brain injury by applying TGN-020 in the supraoptic nucleus via inhibiting vasopressin neurons in rats of focal ischemic stroke[J]. Life Sci, 2021, 264: 118683.

[202] Lu Q, Xiong J, Yuan Y, et al. Minocycline improves the functional recovery after traumatic brain injury via inhibition of aquaporin-4[J]. Int J Biol Sci, 2022, 18(1): 441-458.

[203] Ding Z, Zhang J, Xu J, et al. Propofol administration modulates AQP-4 expression and brain edema after traumatic brain injury[J]. Cell Biochem Biophys, 2013, 67(2): 615-622.

[204] Yin J, Zhang H, Chen H, et al. Hypertonic saline alleviates brain edema after traumatic brain injury via downregulation of aquaporin 4 in rats[J]. Med Sci Monit, 2018, 24: 1863-1870.

[205] Zhang H, Liu J, Liu Y, et al. Hypertonic saline improves brain edema resulting from traumatic brain injury by suppressing the NF-κB/IL-1β signaling pathway and AQP4[J]. Exp Ther Med, 2020, 20(5): 71.

[206] Lopez NE, Krzyzaniak MJ, Blow C, et al. Ghrelin prevents disruption of the blood-brain barrier after traumatic brain injury[J]. J Neurotrauma, 2012, 29(2): 385-393.

[207] Ding JY, Kreipke CW, Speirs SL, et al. Hypoxia-inducible factor-1alpha signaling in aquaporin upregulation after traumatic brain injury[J]. Neurosci Lett, 2009, 453(1): 68-72.

[208] Xiong A, Li J, Xiong R, et al. Inhibition of HIF-1α-AQP4 axis ameliorates brain edema and neurological functional deficits in a rat controlled cortical injury (CCI) model[J]. Sci Rep, 2022,

12(1): 2701.

[209] Chen B, Kong X, Li Z, et al. Downregulation of NF-κB by Shp-1 alleviates cerebral venous sinus thrombosis-induced brain edema via suppression of AQP4[J]. J Stroke Cerebrovasc Dis, 2022, 31(8): 106570.

[210] Zhang Y, Wang J, Zhang Y, et al. Overexpression of long noncoding RNA Malat1 ameliorates traumatic brain injury induced brain edema by inhibiting AQP4 and the NF-κB/IL-6 pathway[J]. J Cell Biochem, 2019, 120(10): 17584-17592.

[211] Tomura S, Nawashiro H, Otani N, et al. Effect of decompressive craniectomy on aquaporin-4 expression after lateral fluid percussion injury in rats[J]. J Neurotrauma, 2011, 28(2): 237-243.

[212] Badaut J, Fukuda AM, Jullienne A, et al. Aquaporin and brain diseases[J]. Biochim Biophys Acta, 2014, 1840(5): 1554-1565.

[213] 刘辉，孙善全，杨美，等. 大鼠轻度脑外伤后水通道蛋白9在大脑皮质及海马的表达变化[J]. 第三军医大学学报，2008，(3): 220-222.

[214] Verkman A. More than just water channels: unexpected cellular roles of aquaporins[J]. J Cell Sci, 2005, 118(Pt 15): 3225-3232.

[215] Iandiev I, Biedermann B, Reichenbach A, et al. Expression of aquaporin-9 immunoreactivity by catecholaminergic amacrine cells in the rat retina[J]. Neurosci Lett, 2006, 398(3): 264-267.

[216] Rojek AM, Skowronski MT, Füchtbauer EM, et al. Defective glycerol metabolism in aquaporin 9 (AQP9) knockout mice[J]. Proc Natl Acad Sci USA, 2007, 104(9): 3609-3614.

[217] Dibas A, Yang MH, Bobich J, et al. Stress-induced changes in neuronal Aquaporin-9 (AQP9) in a retinal ganglion cell-line[J]. Pharmacol Res, 2007, 55(5): 378-384.

[218] 姚泰. 人体生理学[M]. 3版. 北京：人民卫生出版社，2001.

[219] 余维华，孙善全，汪克建，等. 大鼠脑外伤后蓝斑去甲肾上腺素能神经元变化的研究[J]. 中国急救医学，2007，27(3): 232-234.

[220] Singhi P, Saini AG. Fungal and parasitic CNS infections[J]. Indian J Pediatr, 2019, 86(1): 83-90.

[221] Roos KL, 2015. Bacterial infections of the central nervous system[J]. Continuum (Minneap Minn), 21(6 Neuroinfectious Disease): 1679-1691.

[222] Norbury AJ, Calvert JK, Al-Shujairi WH, et al. Dengue virus infects the mouse eye following systemic or intracranial infection and induces inflammatory responses[J]. J Gen Virol, 2020, 101(1): 79-85.

[223] Wu W, Jia F, Wang W, et al. Antiparasitic treatment of cerebral cysticercosis: lessons and experiences from China[J]. Parasitol Res, 2013, 112(8): 2879-2890.

[224] Gerber J, Nau R. Mechanisms of injury in bacterial meningitis[J]. Curr Opin Neuro, 2010, 23(3): 312-318.

[225] Holm A, Karlsson T, Vikström E. Pseudomonas aeruginosa lasI/rhlI quorum sensing genes promote phagocytosis and aquaporin 9 redistribution to the leading and trailing regions in macrophages[J]. Front Microbiol, 2015, 6: 915.

[226] Liu S, Mao J, Wang T, et al. Downregulation of aquaporin-4 protects brain against hypoxia ischemia via anti-inflammatory mechanism[J]. Mol Neurobiol, 2017, 54(8): 6426-6435.

[227] Dai W, Yan J, Chen G, et al. AQP4-knockout alleviates the lipopolysaccharide-induced inflammatory response in astrocytes via SPHK1/MAPK/AKT signaling[J]. Int J Mol Med, 2018, 42(3): 1716-1722.

[228] Steiner I, Kennedy PG. Herpes simplex virus latent infection in the nervous system[J]. J Neurovirol, 1995, 1(1): 19-29.

[229] Meyding-Lamadé U, Lamadé W, Kehm R, et al. Herpes simplex virus encephalitis: chronic progressive cerebral MRI changes despite good clinical recovery and low viral load—an experimental mouse study[J].

Eur J Neurol, 1999, 6(5): 531-538.
[230] McGrath N, Anderson NE, Croxson MC, et al. Herpes simplex encephalitis treated with acyclovir: diagnosis and long term outcome[J]. J Neurol Neurosurg Psychiatry, 1997, 63(3): 321-326.
[231] St Hillaire C, Vargas D, Pardo CA, et al. Aquaporin 4 is increased in association with human immunodeficiency virus dementia: implications for disease pathogenesis[J]. J NeuroVirol, 2005, 11(6): 535-543.
[232] Martínez-Torres FJ, Wagner S, Haas J, et al. Increased presence of matrix metalloproteinases 2 and 9 in short- and long-term experimental herpes simplex virus encephalitis[J]. Neurosci Lett, 2004, 368(3): 274-278.
[233] Jennische E, Eriksson CE, Lange S, et al. The anterior commissure is a pathway for contralateral spread of herpes simplex virus type 1 after olfactory tract infection[J]. J Neurovirol, 2015, 21(2): 129-147.
[234] Elkjaer M, Vajda Z, Nejsum LN, et al. Immunolocalization of AQP9 in liver, epididymis, testis, spleen, and brain[J]. Biochem Biophys Res Commun, 2000, 276(3): 1118-1128.
[235] Ishibashi K, Kuwahara M, Gu Y, et al. Cloning and functional expression of a new aquaporin (AQP9) abundantly expressed in the peripheral leukocytes permeable to water and urea, but not to glycerol[J]. Biochem Biophys Res Commun, 1998, 244(1): 268-274.
[236] Xing HQ, Zhang Y, Izumo K, et al. Decrease of aquaporin-4 and excitatory amino acid transporter-2 indicate astrocyte dysfunction for pathogenesis of cortical degeneration in HIV-associated neurocognitive disorders[J]. Neuropathology, 2017, 37(1): 25-34.
[237] Papadopoulos MC, Verkman AS. Aquaporin-4 gene disruption in mice reduces brain swelling and mortality in pneumococcal meningitis[J]. J Biol Chem, 2005, 280(14): 13906-13912.
[238] Huang J, Lu WT, Sun SQ, et al. Upregulation and lysosomal degradation of AQP4 in rat brains with bacterial meningitis[J]. Neurosci Lett, 2014, 566: 156-161.
[239] Gao M, Lu W, Shu Y, et al. Poldip2 mediates blood-brain barrier disruption and cerebral edema by inducing AQP4 polarity loss in mouse bacterial meningitis model[J]. CNS Neurosci Ther, 2020, 26(12): 1288-1302.
[240] Warth A, Mittelbronn M, Wolburg H. Redistribution of the water channel protein aquaporin-4 and the K^+ channel protein Kir4.1 differs in low- and high-grade human brain tumors[J]. Acta Neuropathol, 2005, 109(4): 418-426.
[241] Connors NC, Adams ME, Froehner SC, et al. The potassium channel Kir4.1 associates with the dystrophin-glycoprotein complex via alpha-syntrophin in glia[J]. J Biol Chem, 2004, 279(27): 28387-28392.
[242] Neely JD, Amiry-Moghaddam M, Ottersen OP, et al. Syntrophin-dependent expression and localization of Aquaporin-4 water channel protein[J]. Proc Natl Acad Sci USA, 2001, 98(24): 14108-14113.
[243] Frigeri A, Nicchia GP, Nico B, et al. Aquaporin-4 deficiency in skeletal muscle and brain of dystrophic mdx mice[J]. FASEB J, 2001, 15(1): 90-98.
[244] Nico B, Tamma R, Annese T, et al. Glial dystrophin-associated proteins, laminin and agrin, are downregulated in the brain of mdx mouse[J]. Lab Invest, 2010, 90(11): 1645-1660.
[245] Dalloz C, Sarig R, Fort P, et al. Targeted inactivation of dystrophin gene product Dp71: phenotypic impact in mouse retina[J]. Hum Mol Genet, 2003, 12(13): 1543-1554.
[246] Hernandes MS, Lassègue B, Griendling KK. Polymerase δ-interacting protein 2: a multifunctional protein[J]. J Cardiovasc Pharmacol, 2017, 69(6): 335-342.
[247] Mathisen GE, Johnson JP. Brain abscess[J]. Clin Infect Dis, 1997, 25(4): 763-779; quiz 780-781.
[248] Townsend GC, Scheld WM, 1998. Infections of the central nervous system[J]. Adv Intern Med, 43: 403-447.

[249] Bloch O, Papadopoulos MC, Manley GT, et al. Aquaporin-4 gene deletion in mice increases focal edema associated with staphylococcal brain abscess[J]. J Neurochem, 2005, 95（1）: 254-262.

[250] Connors NC, Kofuji P. Dystrophin Dp71 is critical for the clustered localization of potassium channels in retinal glial cells[J]. J Neurosci, 2002, 22（11）: 4321-4327.

[251] Blocher J, Eckert I, Elster J, et al. Aquaporins AQP1 and AQP4 in the cerebrospinal fluid of bacterial meningitis patients[J]. Neurosci Lett, 2011, 504（1）: 23-27.

[252] Verkman AS, Ruiz-Ederra J, Levin MH. Functions of aquaporins in the eye[J]. Prog Retin Eye Res, 2008, 27（4）: 420-433.

[253] Goodyear MJ, Crewther SG, Junghans BM. A role for aquaporin-4 in fluid regulation in the inner retina[J]. Vis Neurosci, 2009, 26（2）: 159-165.

[254] Hamann S, Zeuthen T, La Cour M, et al. Aquaporins in complex tissues: distribution of aquaporins 1-5 in human and rat eye[J]. Am J Physiol, 1998, 274（5）: C1332-C1345.

[255] Thiagarajah JR, Verkman AS. Aquaporin deletion in mice reduces corneal water permeability and delays restoration of transparency after swelling[J]. J Biol Chem, 2002, 277（21）: 19139-19144.

[256] Bonanno JA. Molecular mechanisms underlying the corneal endothelial pump[J]. Exp Eye Res, 2012, 95（1）: 2-7.

[257] Naka M, Kanamori A, Negi A, et al. Reduced expression of aquaporin-9 in rat optic nerve head and retina following elevated intraocular pressure[J]. Invest Ophthalmol Vis Sci, 2010, 51（9）: 4618-4626.

[258] Stamer WD, Peppel K, O'Donnell ME, et al. Expression of aquaporin-1 in human trabecular meshwork cells: role in resting cell volume[J]. Invest Ophthalmol Vis Sci, 2001, 42（8）: 1803-1811.

[259] Baetz NW, Hoffman EA, Yool AJ, et al. Role of aquaporin-1 in trabecular meshwork cell homeostasis during mechanical strain[J]. Exp Eye Res, 2009, 89（1）: 95-100.

[260] Yamaguchi Y, Watanabe T, Hirakata A, et al. Localization and ontogeny of aquaporin-1 and-4 expression in iris and ciliary epithelial cells in rats[J]. Cell Tissue Res, 2006, 325（1）: 101-109.

[261] Chepelinsky AB. Structural function of MIP/aquaporin 0 in the eye lens; genetic defects lead to congenital inherited cataracts[J]. Handb Exp Pharmacol, 2009,（190）: 265-297.

[262] Varadaraj K, Kumari S, Shiels A, et al. Regulation of aquaporin water permeability in the lens[J]. Invest Ophthalmol Vis Sci, 2005, 46（4）: 1393.

[263] Iandiev I, annicke T, Härtig W, et al. Localization of aquaporin-0 immunoreactivity in the rat retina[J]. Neurosci Lett, 2007, 426（2）: 81-86.

[264] Kim IB, Oh SJ, Nielsen S, et al. Immunocytochemical localization of aquaporin 1 in the rat retina[J]. Neurosci Lett, 1998, 244（1）: 52-54.

[265] Iandiev I, Pannicke T, Reichel MB, et al. Expression of aquaporin-1 immunoreactivity by photoreceptor cells in the mouse retina[J]. Neurosci Lett, 2005, 388（2）: 96-99.

[266] Levin MH, Verkman AS. Aquaporin-dependent water permeation at the mouse ocular surface: in vivo microfluorimetric measurements in cornea and conjunctiva[J]. Invest Ophthalmol Vis Sci, 2004, 45（12）: 4423-4432.

[267] Da T, Verkman AS. Aquaporin-4 gene disruption in mice protects against impaired retinal function and cell death after ischemia[J]. Invest Ophthalmol Vis Sci, 2004, 45（12）: 4477-4483.

[268] Tanihara H, Hangai M, Sawaguchi S, et al. Up-regulation of glial fibrillary acidic protein in the retina of primate eyes with experimental glaucoma[J]. Arch Ophthalmol, 1997, 115（6）: 752-756.

[269] Jiang Y, Zhang C, Ma J, et al. Expression of matrix metalloproteinases-2 and aquaporin-1 in corneoscleral junction after angle-closure in rabbits[J]. BMC Ophthalmol, 2019, 19（1）: 43.

[270] Hollborn M, Rehak M, Iandiev I, et al. Transcriptional regulation of aquaporins in the ischemic rat retina: upregulation of aquaporin-9[J]. Curr Eye Res, 2012, 37（6）: 524-531.

[271] 车成业，赵桂秋，张丽丽. 慢性高眼压状态下水通道蛋白-1在家兔角膜内皮细胞的表达变化[J]. 眼科研究，2010，28（9）：832-835.

[272] Huang OS，Seet LF，Ho HW，et al. Altered iris aquaporin expression and aqueous humor osmolality in glaucoma[J]. Invest Ophthalmol Vis Sci，2021，62（2）：34.

[273] Zhang D，Vetrivel L，Verkman AS. Aquaporin deletion in mice reduces intraocular pressure and aqueous fluid production[J]. J Gen Physiol，2002，119（6）：561-569.

[274] Wu J，Bell OH，Copland DA，et al. Gene therapy for glaucoma by ciliary body aquaporin 1 disruption using CRISPR-Cas9[J]. Mol Ther，2020，28（3）：820-829.

[275] Bogner B，Schroedl F，Trost A，et al. Aquaporin expression and localization in the rabbit eye[J]. Exp Eye Res，2016，147：20-30.

[276] Hatakeyama S，Yoshida Y，Tani T，et al. Cloning of a new aquaporin（AQP10）abundantly expressed in duodenum and jejunum[J]. Biochem Biophys Res Commun，2001，287（4）：814-819.

[277] 冉建华，孙善全. 水通道蛋白4在大鼠眼球中的表达[J]. 解剖学报，2004，（6）：656-659.

[278] 陈海，孙善全，汪克建，等. 水通道蛋白-4表达变化与急性高眼压视网膜损伤的相关性研究[J]. 解剖学报，2008，（5）：677-682.

[279] 罗莎莎，陈琴，于东毅，等. 慢性高眼压状态下小鼠水通道蛋白4基因对视网膜胶质细胞活化的影响[J]. 中华眼科杂志，2012，（7）：598-603.

[280] Wang YF，Parpura V. Central role of maladapted astrocytic plasticity in ischemic brain edema formation[J]. Front Cell Neurosci，2016，10：129.

[281] Nicchia GP，Pisani F，Simone L，et al. Glio-vascular modifications caused by aquaporin-4 deletion in the mouse retina[J]. Exp Eye Res，2016，146：259-268.

[282] 张瑶，张星慧，佘美华，等. Neu-P11对急性高眼压大鼠视网膜AQP4表达及眼内压的影响[J]. 中国药理学通报，2016，32（11）：1624-1624，1625.

[283] Chen C，Fan P，Zhang L，et al. Bumetanide rescues aquaporin-4 depolarization via suppressing β-dystroglycan cleavage and provides neuroprotection in rat retinal ischemia-reperfusion injury[J]. Neuroscience，2023，510：95-108.

[284] Funaki H，Yamamoto T，Koyama Y，et al. Localization and expression of AQP5 in cornea，serous salivary glands，and pulmonary epithelial cells[J]. Am J Physiol，1998，275（4）：C1151-C1157.

[285] Patil RV，Saito I，Yang X，et al. Expression of aquaporins in the rat ocular tissue[J]. Exp Eye Res，1997，64（2）：203-209.

[286] Tran TL，Bek T，la Cour M，et al. Altered aquaporin expression in glaucoma eyes[J]. APMIS，2014，122（9）：772-780.

[287] Iandiev I，Dukic-Stefanovic S，Hollborn M，et al. Immunolocalization of aquaporin-6 in the rat retina[J]. Neurosci Lett，2011，490（2）：130-134.

[288] Tran TL，Bek T，Holm L，et al. Aquaporins 6-12 in the human eye[J]. Acta Ophthalmol，2013，91（6）：557-563.

[289] Yang MH，Dibas A，Tyan YC. Changes in retinal aquaporin-9（AQP9）expression in glaucoma[J]. Biosci Rep，2013，33（2）：e00035.

[290] Mizokami J，Kanamori A，Negi A，et al. A preliminary study of reduced expression of aquaporin-9 in the optic nerve of primate and human eyes with glaucoma[J]. Curr Eye Res，2011，36（11）：1064-1067.

[291] Binhammer RT. CSF anatomy with emphasis on relations to nasal cavity and labyrinthine fluids[J]. Ear Nose Throat，1992，71（7）：292-294，297-299.

[292] Sato O，Bering EA. Extra-ventricular formation of cerebrospinal fluid[J]. Brain Nerv，1967，19（9）：883-885.

[293] Lorenzo AV, Page LK, Watters GV. Relationship between cerebrospinal fluid formation, absorption and pressure in human hydrocephalus[J]. Brain, 1970, 93(4): 679-692.

[294] Bering EA Jr, Sato O. Hydrocephalus: changes in formation and absorption of cerebrospinal fluid within the cerebral ventricles[J]. Ncurosurg, 1963, 20: 1050-1063.

[295] Shapiro WR. Cerebrospinal fluid circulation and the blood-brain barrier[J]. Ann N Y Acad Sci, 1988, 531: 9-14.

[296] Bhattathiri PS, Gregson B, Prasad KS, et al. Intraventricular hemorrhage and hydrocephalus after spontaneous intracerebral hemorrhage: results from the STICH trial[J]. Acta Neurochir Suppl, 2006, 96: 65-68.

[297] Hemphill JC 3rd, Greenberg SM, Anderson CS, et al. Guidelines for the management of spontaneous intracerebral hemorrhage: a guideline for healthcare professionals from the American Heart Association/American Stroke Association[J]. Stroke, 2015, 46(7): 2032-2060.

[298] Chen QW, Feng Z, Tan Q, et al. Post-hemorrhagic hydrocephalus: recent advances and new therapeutic insights[J]. J Neurol Sci, 2017, 375: 220-230.

[299] Wright Z, Larrew TW, Eskandari R. Pediatric hydrocephalus: current state of diagnosis and treatment[J]. Pediatr Rev, 2016, 37(11): 478-490.

[300] Kahle KT, Kulkarni AV, Limbrick DD Jr, et al. Hydrocephalus in children[J]. Lancet, 2016, 387(10020): 788-799.

[301] 王忠诚. 王忠诚神经外科学[M]. 2版. 武汉: 湖北科学技术出版社, 2015.

[302] 赵玉沛, 陈孝平. 外科学[M]. 3版. 北京: 人民卫生出版社, 2015.

[303] Hashimoto PH. Tracer in cisternal cerebrospinal fluid is soon detected in choroid plexus capillaries[J]. Brain Res, 1988, 440(1): 149-152.

[304] Orešković D, Radoš M, Klarica M. New concepts of cerebrospinal fluid physiology and development of Hydrocephalus[J]. Pediatr Neurosurg, 2017, 52(6): 417-425.

[305] Xu H. New concept of the pathogenesis and therapeutic orientation of acquired communicating hydrocephalus[J]. Neurol Sci, 2016, 37(9): 1387-1391.

[306] Praetorius J, Nielsen S. Distribution of sodium transporters and aquaporin-1 in the human choroid plexus[J]. Am J Physiol Cell Physiol, 2006, 291(1): C59-C67.

[307] Longatti PL, Basaldella L, Orvieto E, et al. Choroid plexus and aquaporin-1: a novel explanation of cerebrospinal fluid production[J]. Pediatr Neurosurg, 2004, 40(6): 277-283.

[308] Speake T, Freeman LJ, Brown PD. Expression of aquaporin 1 and aquaporin 4 water channels in rat choroid plexus[J]. Biochim Biophys Acta, 2003, 1609(1): 80-86.

[309] Mobasheri A, Marples D. Expression of the AQP-1 water channel in normal human tissues: a semiquantitative study using tissue microarray technology[J]. Am J Physiol Cell Physiol, 2004, 286(3): C529-C537.

[310] Brown PD, Davies SL, Speake T, et al. Molecular mechanisms of cerebrospinal fluid production[J]. Neuroscience, 2004, 129(4): 957-970.

[311] Praetorius J. Water and solute secretion by the choroid plexus[J]. Pflugers Arch, 2007, 454(1): 1-18.

[312] Johansson PA, Dziegielewska KM, Ek CJ, et al. Aquaporin-1 in the choroid plexuses of developing mammalian brain[J]. Cell Tissue Res, 2005, 322(3): 353-364.

[313] Oshio K, Song Y, Verkman AS, et al. Aquaporin-1 deletion reduces osmotic water permeability and cerebrospinal fluid production[J]. Acta Neurochir Suppl, 2003, 86: 525-528.

[314] Owler BK, Pitham T, Wang D. Aquaporins: relevance to cerebrospinal fluid physiology and therapeutic potential in hydrocephalus[J]. Cerebrospinal Fluid Re, 2010, 7: 15.

[315] Cai X, McGraw G, Pattisapu JV, et al. Hydrocephalus in the H-Tx rat: a monogenic disease?[J]. Exp Neurol, 2000, 163（1）: 131-135.

[316] Kohn DF, Chinookoswong N, Chou SM. A new model of congenital hydrocephalus in the rat[J]. Acta Neuropathol, 1981, 54（3）: 211-218.

[317] Paul L, Madan M, Rammling M, et al. The altered expression of aquaporin 1 and 4 in choroid plexus of congenital hydrocephalus[J]. Cerebrospinal Fluid Res, 2009, 6（1）: S7.

[318] Smith ZA, Moftakhar P, Malkasian D, et al. Choroid plexus hyperplasia: surgical treatment and immunohistochemical results. Case report[J]. J Neurosurg, 2007, 107（3 Suppl）: 255-262.

[319] Verkman AS. Novel roles of aquaporins revealed by phenotype analysis of knockout mice[J]. Rev Physiol Biochem Pharmacol, 2005, 155: 31-55.

[320] Skjolding AD, Rowland IJ, Søgaard LV, et al. Hydrocephalus induces dynamic spatiotemporal regulation of aquaporin-4 expression in the rat brain[J]. Cerebrospinal Fluid Res, 2010, 7: 20.

[321] Mao X, Enno TL, Del Bigio MR. Aquaporin 4 changes in rat brain with severe hydrocephalus[J]. Eur J Neurosci, 2006, 23（11）: 2929-2936.

[322] Bloch O, Auguste KI, Manley GT, et al. Accelerated progression of kaolin-induced hydrocephalus in aquaporin-4-deficient mice[J]. J Cereb Blood Flow Metab, 2006, 26（12）: 1527-1537.

[323] Feng XC, Papadopoulos MC, Liu J, et al. Sporadic obstructive hydrocephalus in Aqp4 null mice[J]. J Neurosci Res, 2009, 87（5）: 1150-1155.

[324] Long CY, Huang GQ, Du Q, et al. The dynamic expression of aquaporins 1 and 4 in rats with hydrocephalus induced by subarachnoid haemorrhage[J]. Folia Neuropathol, 2019, 57（2）: 182-195.

[325] Frigeri A, Nicchia GP, Svelto M. Aquaporins as targets for drug discovery[J]. Curr Pharm Des, 2007, 13（23）: 2421-2427.

[326] Gunnarson E, Zelenina M, Aperia A. Regulation of brain aquaporins[J]. Neuroscience, 2004, 129（4）: 947-955.

[327] Papadopoulos MC, Verkman AS. Potential utility of aquaporin modulators for therapy of brain disorders[J]. Prog Brain Res, 2008, 170: 589-601.

[328] Tanimura Y, Hiroaki Y, Fujiyoshi Y. Acetazolamide reversibly inhibits water conduction by aquaporin-4[J]. J Struct Biol, 2009, 166（1）: 16-21.

[329] Tang Y, Cai D, Chen Y. Thrombin inhibits aquaporin 4 expression through protein kinase C-dependent pathway in cultured astrocytes[J]. J Mol Neurosci, 2007, 31（1）: 83-93.

[330] Filippidis AS, Kalani MY, Rekate HL. Hydrocephalus and aquaporins: lessons learned from the bench[J]. Childs Nerv Syst, 2011, 27（1）: 27-33.

第四节　水通道蛋白与脑胶质瘤

胶质瘤是最常见的脑肿瘤，起源于神经上皮组织，占颅内肿瘤的40%～50%。近年来，AQP在胶质瘤发生发展中的作用也在不断地被发现。研究表明，AQP能影响胶质瘤细胞的存活与生长[1,2]。一方面，水的快速转运是肿瘤细胞存活与生长的重要保障，而水转运能力依赖于AQP的表达[3]。另一方面，AQP可以调控肿瘤细胞中与生长相关的信号通路，从而影响肿瘤细胞的存活与生长[4,5]。除此以外，AQP在胶质瘤引发的脑水肿中也可能发挥重要的作用。AQP在胶质瘤中的作用及机制研究为胶质瘤的治疗提供了新方向。

（一）AQP1与胶质瘤

在中枢神经系统中，AQP1主要分布于脉络膜上皮细胞，通过水转运影响脑脊液的产生。胶质瘤中AQP1的表达相较于正常脑组织有所上调，且这种上调与胶质瘤的恶性程度呈正相关[6, 7]。进一步研究显示，尽管AQP1在整个胶质瘤组织中高表达，但肿瘤组织核心区域AQP1的表达远低于富含新生血管的瘤周区域，提示AQP1的表达上调与肿瘤组织新生血管的形成和肿瘤细胞的侵袭能力关系密切[8]。沉默*AQP1*表达后，胶质瘤细胞A172和U251的迁移和侵袭能力降低；裸鼠成瘤实验结果表明，与对照组细胞相比，沉默*AQP1*的表达后，肿瘤体积变小且重量减轻[9]。过表达AQP1可以促进胶质瘤细胞A172和U251的侵袭与迁移，AQP1还可以通过下调血栓反应蛋白1型结构域7A来参与新生血管的形成[10]。

除了对胶质瘤细胞本身表达的AQP1的研究外，也有研究关注胶质瘤与周围环境的关系，胶质瘤中的胶质瘤干细胞（glioma stem cell，GSC）在胶质瘤的生长、侵袭、血管生成和免疫逃逸中发挥关键作用，也是产生化疗耐药的主要细胞[11]。胶质瘤组织常被小胶质细胞和外周巨噬细胞浸润，这种浸润促进了胶质瘤的进展[12]。研究发现，GSC通过TLR4信号通路促进小胶质细胞分泌IL-6，IL-6能够与GSC膜受体结合，促进胶质瘤干细胞生长相关信号通路的激活。GSC还可激活胶质瘤相关的小胶质细胞/巨噬细胞（glioma-associated microglia and macrophages，GAM），从而促进胶质瘤的生长并产生免疫抑制作用。将外源性胶质瘤细胞GL261植入*AQP1*敲除鼠与对照组小鼠，AQP1的缺失显著改变了胶质瘤相关的GAM比例，同时也影响了小胶质细胞与巨噬细胞的功能，AQP1缺失小鼠体内的胶质瘤生长明显减慢，存活率却显著上升[6, 13-15]。另外，AQP1的表达也与糖酵解水平有关，胶质瘤中AQP1、乳酸脱氢酶（lactate dehydrogenase，LDH）和组织蛋白酶B的上调有助于细胞外环境的酸化，这提高了胶质瘤的有氧糖酵解效率，保障了胶质瘤细胞在增殖中的能量需求[16]。胶质瘤相关致癌基因1（*Gli1*）结合*AQP1*启动子可在胶质瘤中促进AQP1的表达[17]。虽然针对AQP1的特异性抑制药物尚在研究之中，但在啮齿动物的脑水肿过程中，褪黑素和胍丁胺可以有效降低AQP1的表达，使用褪黑素和胍丁胺的激动剂可延缓脑水肿的发展[18]，这为寻找降低胶质瘤中AQP1表达的药物提供了一些参考。

（二）AQP4与胶质瘤

中枢神经系统中，AQP4在血管周围星形胶质细胞终足端、室管膜下星形胶质细胞和室管膜细胞基底外侧膜广泛表达，AQP4可以形成同型和异型四聚体，异型四聚体组装成粒子正交阵列（orthogonal arrays of particles，OAP）。OAP可增加水的渗透性，调节AQP4在细胞膜上的分布；OAP还参与了血脑屏障的发育[19, 20]。星形胶质细胞中AQP4与Kir4.1共定位，AQP4可能通过转运水并协同Kir4.1调节神经元细胞外钾离子浓度，影响神经元的兴奋性，所以AQP4被认为是大脑水稳态和离子平衡的重要组成部分[21]。

相较于正常脑组织，胶质瘤患者脑组织中AQP4表达上调。与其他AQP不同的是，AQP4在胶质瘤中的表达与胶质瘤的恶性程度之间没有明显的正相关关系，胶质瘤恶性程度Ⅰ级的毛细胞型星形细胞瘤和Ⅳ级胶质母细胞瘤的AQP4表达显著高于Ⅱ级的星形细胞瘤[22]。在人正常大脑中，星形胶质细胞终足端的AQP4参与了血脑屏障的组成。在胶质瘤患者中，星形胶质细胞的AQP4表达量上调，但星形胶质细胞AQP4上调的作用还存在争议。有学者认为AQP4表达上调可清除在胶质瘤进展中血脑屏障破坏所致的毛细血管渗透液的增多；另有研究报道，胶质瘤进展中胶质瘤细胞主动引发星形胶质细胞AQP4的上调[23-25]。对AQP4与胶质瘤引发脑水肿之间的关系也有一些共识：在胶质瘤引发的血管源性脑水肿中，AQP4可以清除毛细血管渗透液而发挥保护大脑的作用，但在胶质瘤引发的细胞毒性脑水肿中，AQP4则是星形细胞发生肿胀的主要原因，它加剧了细胞水肿[26-28]。鉴于此，部分学者认为，胶质瘤细胞可通过特异性诱导AQP4来预防或减轻血管源性脑水肿，但诱导AQP4作为预防和治疗胶质瘤引发的血管源性脑水肿的方法和适用性仍被质疑，特别是在小鼠胶质瘤引发的脑水肿手术模型中，AQP4缺失的小鼠表现出预后更好的现象[29, 30]。因此，AQP4在胶质瘤导致的脑水肿中的具体作用还有待进一步探讨。

目前普遍认为，胶质瘤相关水肿是血管源性的，由于血脑屏障功能紊乱而导致血管通透性增加，脑肿瘤血管结构的改变导致屏障功能的丧失，并使血浆和蛋白质发生渗漏引发水肿[31, 32]。此外，胶质瘤细胞产生各种细胞因子作用于肿瘤内部及周围的内皮细胞，其中最重要的细胞因子是血管内皮生长因子（vascular endothelial growth factor，VEGF）。在胶质瘤的星形胶质细胞中存在AQP4与VEGF的共定位现象，VEGF可以激活MAPK通路，进而调节AQP4的表达，使血管周围细胞AQP4蛋白水平上升，内皮细胞通透性增加，引发血脑屏障的紊乱和相关水肿[33-36]。

使用小干扰RNA敲除或药物抑制胶质瘤AQP4可造成胶质瘤细胞的侵袭和迁移能力受到抑制[37, 38]。研究显示，在从胶质瘤组织中分离培养的胶质瘤细胞中过表达AQP1和AQP4，过表达AQP1的胶质瘤细胞表现出更强的迁移与侵袭能力，但过表达AQP4的胶质瘤细胞的迁移能力却降低[39]。不同的课题组在研究AQP4对胶质瘤细胞迁移能力的影响方面得出了相反的结论。因此，关于AQP4在胶质瘤细胞生物学行为中的作用，还需要更多的研究加以证实。

总之，AQP4在胶质瘤的进展中，尤其是在胶质瘤引发的脑水肿过程中可能发挥重要作用，这为进一步阐明胶质瘤的水肿发生机制提供了理论支撑。

（三）AQP5与胶质瘤

AQP5在中枢神经系统中主要表达于脉络丛、梨状皮质、海马、背侧丘脑等区域的星形胶质细胞和神经元[40]，AQP5表达水平受氧含量和蛋白激酶A（PKA）的调节[41, 42]。

最近的研究显示，AQP5的表达也与胶质瘤患者瘤周水肿的发展和强度有关。敲低AQP5的表达后，EGFR/ERK/p38 MAPK信号通路被抑制，胶质瘤细胞增殖、迁移能力降低，细胞凋亡增加[43]。AQP5影响胶质瘤发展的具体机制仍在进一步研究中。

（四）AQP8与胶质瘤

AQP8在哺乳动物大脑中表达较弱，对AQP8的敲除不会明显影响正常小鼠的生存[44]。

在人胶质瘤细胞中，AQP8表达量较高，且AQP8的表达变化与胶质瘤组织的病理级别呈正相关，提示AQP8在胶质瘤生长过程中可能发挥重要作用[45]。Yang等[46]证明，AQP8在脑缺血后表达上调，并且参与了脑缺血引发的早期水肿的形成，但AQP8是否参与了胶质瘤引发的脑水肿仍未可知，需要更多的研究揭示AQP8在胶质瘤中的具体作用。

（五）AQP9与胶质瘤

AQP9主要存在于脑室周围的室管膜细胞和下丘脑的神经元细胞，也可表达于星形胶质细胞和软脑膜血管内皮细胞的细胞膜[47,48]；AQP9还分布于胞内线粒体内膜。生理条件下，AQP9可以通过水转运协同线粒体进行氧化磷酸化产生ATP，为细胞提供能量[49]。

AQP9在恶性星形胶质细胞中高表达[50]；在胶质瘤来源的分化细胞中，与其他AQP相比较，AQP9表达量也相对较高，并与胶质瘤干细胞密切相关[51]。长链非编码RNA 00320（LINC00320）可以通过抑制NFKB1在*AQP9*启动子区域募集从而下调AQP9，进而抑制胶质瘤细胞的增殖和微血管内皮细胞（microvascular endothelial cell，MVEC）血管生成[52]。除此之外，AQP9与能量代谢也有关联。由于胶质瘤恶性增殖需要进行大量的能量代谢，导致胶质瘤细胞周围产生甘油和乳酸等代谢废物，而AQP9可以促进细胞外甘油和乳酸的清除，以减轻胶质瘤相关的乳酸酸中毒[53]，维持胶质瘤细胞的正常能量代谢。

目前暂无水通道蛋白与其他脑肿瘤相关方面的研究。

（刘　茜　朱淑娟）

参 考 文 献

[1] Beitz E, Golldack A, Rothert M, et al. Challenges and achievements in the therapeutic modulation of aquaporin functionality[J]. Pharmacol Ther, 2015, 155：22-35.

[2] Papadopoulos MC, Verkman AS. Aquaporin water channels in the nervous system[J]. Nat Rev Neurosci, 2013, 14（4）：265-277.

[3] Wang J, Feng L, Zhu Z, et al. Aquaporins as diagnostic and therapeutic targets in cancer：how far we are?[J]. J Transl Med, 2015, 13：96.

[4] Alkhalifa H, Mohammed F, Taurin S, et al. Inhibition of aquaporins as a potential adjunct to breast cancer cryotherapy[J]. Oncol Lett, 2021, 21（6）：458.

[5] Wang L, Huo D, Zhu H, et al. Deciphering the structure, function, expression and regulation of aquaporin-5 in cancer evolution[J]. Oncol Lett, 2021, 21（4）：309.

[6] El Hindy N, Rump K, Lambertz N, et al. The functional aquaporin 1-783G/C-polymorphism is associated with survival in patients with glioblastoma multiforme[J]. J Surg Oncol, 2013, 108（7）：492-498.

[7] Saadoun S, Papadopoulos MC, Davies DC, et al. Increased aquaporin 1 water channel expression inhuman brain tumours[J]. Br J Cancer, 2002, 87（6）：621-623.

[8] El Hindy N, Bankfalvi A, Herring A, et al. Correlation of aquaporin-1 water channel protein expression with tumor angiogenesis in human astrocytoma[J]. Anticancer Res, 2013, 33（2）：609-613.

[9] Yang WY, Tan ZF, Dong DW, et al. Association of aquaporin-1 with tumor migration, invasion and vasculogenic mimicry in glioblastoma multiforme[J]. Mol Med Rep, 2018, 17（2）：3206-3211.

[10] Oishi M, Munesue S, Harashima A, et al. Aquaporin 1 elicits cell motility and coordinates vascular bed

formation by downregulating thrombospondin type-1 domain-containing 7A in glioblastoma[J]. Cancer Med, 2020, 9(11): 3904-3917.

[11] Hambardzumyan D, Gutmann DH, Kettenmann H. The role of microglia and macrophages in glioma maintenance and progression[J]. Nat Neurosci, 2016, 19(1): 20-27.

[12] Yang I, Han SJ, Kaur G, et al. The role of microglia in central nervous system immunity and glioma immunology[J]. J Clin Neurosci, 2010, 17(1): 6-10.

[13] Dzaye O, Hu F, Derkow K, et al. Glioma stem cells but not bulk glioma cells upregulate IL-6 secretion in microglia/brain macrophages via toll-like receptor 4 signaling[J]. J Neuropathol Exp Neurol, 2016, 75(5): 429-440.

[14] Hu F, Huang Y, Semtner M, et al. Down-regulation of aquaporin-1 mediates a microglial phenotype switch affecting glioma growth[J]. Exp Cell Res, 2020, 396(2): 112323.

[15] Vinnakota K, Hu F, Ku MC, et al. Toll-like receptor 2 mediates microglia/brain macrophage MT1-MMP expression and glioma expansion[J]. Neuro-oncol, 2013, 15(11): 1457-1468.

[16] Hayashi Y, Edwards NA, Proescholdt MA, et al. Regulation and function of aquaporin-1 in glioma cells[J]. Neoplasia, 2007, 9(9): 777-787.

[17] Liao ZQ, Ye M, Yu PG, et al. Glioma-associated oncogene homolog1(Gli1)-Aquaporin1 pathway promotes glioma cell metastasis[J]. BMB Rep, 2016, 49(7): 394-399.

[18] Kim JH, Lee YW, Park KA, et al. Agmatine attenuates brain edema through reducing the expression of aquaporin-1 after cerebral ischemia[J]. J Cereb Blood Flow Metab, 2010, 30(5): 943-949.

[19] Nicchia GP, Nico B, Camassa LM, et al. The role of aquaporin-4 in the blood-brain barrier development and integrity: studies in animal and cell culture models[J]. Neuroscience, 2004, 129(4): 935-945.

[20] Potokar M, Jorgačevski J, Zorec R. Astrocyte aquaporin dynamics in health and disease[J]. Int J Mol Sci, 2016, 17(7): 1121.

[21] Nagelhus EA, Mathiisen TM, Ottersen OP4. Aquaporin-4 in the central nervous system: cellular and subcellular distribution and coexpression with KIR4. 1[J]. Neuroscience, 200, 129(4): 905-913.

[22] Warth A, Simon P, Capper D, et al. Expression pattern of the water channel aquaporin-4 in human gliomas is associated with blood–brain barrier disturbance but not with patient survival[J]. J Neurosci Res, 2007, 85(6): 1336-1346.

[23] Tani K, Hiroaki Y, Fujiyoshi Y. Aquaporin-4[J]. Clinical Neurology, 2008, 48(11): 941-944.

[24] Ding T, Zhou Y, Sun K, et al. Knockdown a water channel protein, aquaporin-4, induced glioblastoma cell apoptosis[J]. PLoS One, 2013, 8(8): e66751.

[25] Lan YL, Wang X, Lou JC, et al. The potential roles of aquaporin 4 in malignant gliomas[J]. Oncotarget, 2017, 8(19): 32345-32355.

[26] Manley GT, Binder DK, Papadopoulos MC, et al. New insights into water transport and edema in the central nervous system from phenotype analysis of aquaporin-4 null mice[J]. Neuroscience, 2004, 129(4): 983-991.

[27] Papadopoulos MC, Manley GT, Krishna S, et al. Aquaporin-4 facilitates reabsorption of excess fluid in vasogenic brain edema[J]. FASEB J, 2004, 18(11): 1291-1293.

[28] Papadopoulos MC, Saadoun S, Binder DK, et al. Molecular mechanisms of brain tumor edema[J]. Neuroscience, 2004, 129(4): 1011-1020.

[29] Saadoun S, Papadopoulos MC, Davies DC, et al. Aquaporin-4 expression is increased in oedematous human brain tumours[J]. J Neurol Neurosurg Psychiatry, 2002, 72(2): 262-265.

[30] Warth A, Kröger S, Wolburg H. Redistribution of aquaporin-4 in human glioblastoma correlates with loss of agrin immunoreactivity from brain capillary basal laminae[J]. Acta Neuropathol, 2004, 107(4): 311-318.

[31] Hirano A, Kawanami T, Llena JF. Electron microscopy of the blood-brain barrier in disease[J]. Microsc Res Tech, 1994, 27(6): 543-556.

[32] Roth P, Regli L, Tonder M, et al. Tumor-associated edema in brain cancer patients: pathogenesis and management. Expert Rev Anticancer Ther, 2013, 13(11): 1319-1325.

[33] 黄婷, 陈波, 曾昭明. VEGF基因修饰人骨髓间充质干细胞对脑出血大鼠早期脑水肿的影响[J]. 四川大学学报（医学版）, 2020, 51(5): 622-629.

[34] Park H, Choi SH, Kong MJ, et al. Dysfunction of 67-kDa laminin receptor disrupts BBB integrity via impaired dystrophin/AQP4 complex and p38 MAPK/VEGF activation following status epilepticus[J]. Front Cell Neurosci, 2019, 13: 236.

[35] Pietsch T, Valter MM, Wolf HK, et al. Expression and distribution of vascular endothelial growth factor protein in human brain tumors[J]. Acta Neuropathol, 1997, 93(2): 109-117.

[36] Rite I, Machado A, Cano J, et al. Intracerebral VEGF injection highly upregulates AQP4 mRNA and protein in the perivascular space and glia limitans externa[J]. Neurochem Int, 2008, 52(4-5): 897-903.

[37] Ding T, Gu F, Fu L, et al. Aquaporin-4 in glioma invasion and an analysis of molecular mechanisms[J]. J Clin Neurosci, 2010, 17(11): 1359-1361.

[38] Ding T, Ma Y, Li W, et al. Role of aquaporin-4 in the regulation of migration and invasion of human glioma cells[J]. Int J Oncol, 2011, 38(6): 1521-1531.

[39] McCoy E, Sontheimer H. Expression and function of water channels (aquaporins) in migrating malignant astrocytes[J]. Glia, 2007, 55(10): 1034-1043.

[40] Yang M, Gao F, Liu H, et al. Immunolocalization of aquaporins in rat brain[J]. Anat Histol Embryol, 2011, 40(4): 299-306.

[41] Yamamoto N, Sobue K, Fujita M, et al. Differential regulation of aquaporin-5 and -9 expression in astrocytes by protein kinase A[J]. Brain Res Mol Brain Res, 2002, 104(1): 96-102.

[42] Yamamoto N, Yoneda K, Asai K, et al. Alterations in the expression of the AQP family in cultured rat astrocytes during hypoxia and reoxygenation[J]. Brain Res Mol Brain Res, 2001, 90(1): 26-38.

[43] Yang J, Zhang JN, Chen WL, et al. Effects of AQP5 gene silencing on proliferation, migration and apoptosis of human glioma cells through regulating EGFR/ERK/p38 MAPK signaling pathway[J]. Oncotarget, 2017, 8(24): 38444-38455.

[44] Yang M, Gao F, Liu H, et al. Temporal changes in expression of aquaporin-3, -4, -5 and -8 in rat brains after permanent focal cerebral ischemia[J]. Brain Res, 2009, 1290: 121-132.

[45] Zhu SJ, Wang KJ, Gan SW, et al. Expression of aquaporin 8 in human astrocytomas: correlation with pathologic grade[J]. Biochem Biophys Res Commun, 2013, 440(1): 168-172.

[46] Yang B, Song Y, Zhao D, et al. Phenotype analysis of aquaporin-8 null mice[J]. Am J Physiol Cell Physiol, 2005, 288(5): C1161-C1170.

[47] Badaut J, Hirt L, Granziera C, et al. Astrocyte-specific expression of aquaporin-9 in mouse brain is increased after transient focal cerebral ischemia[J]. J Cereb Blood Flow Metab, 2001, 21(5): 477-482.

[48] Elkjaer M, Vajda Z, Nejsum LN, et al. Immunolocalization of AQP9 in liver, epididymis, testis, spleen, and brain[J]. Biochem Biophys Res Commun, 2000, 276(3): 1118-1128.

[49] Amiry-Moghaddam M, Lindland H, Zelenin S, et al. Brain mitochondria contain aquaporin water channels: evidence for the expression of a short AQP9 isoform in the inner mitochondrial membrane[J]. FASEB J, 2005, 19(11): 1459-1467.

[50] Jelen S, Parm Ulhøi B, Larsen A, et al. AQP9 expression in glioblastoma multiforme tumors is limited to a small population of astrocytic cells and CD15+/CalB(+)leukocytes[J]. PLoS One, 2013, 8(9): e75764.

[51] Fossdal G, Vik-Mo EO, Sandberg C, et al. Aqp 9 and brain tumour stem cells[J]. The Scientific World Journal, 2012, 2012: 915176.

[52] Chang L, Bian Z, Xiong X, et al. Long non-coding RNA LINC00320 inhibits tumorigenicity of glioma

cells and angiogenesis through downregulation of NFKB1-mediated AQP9[J]. Front Cell Neurosci, 2020, 14: 542552.

[53] Warth A, Mittelbronn M, Hülper P, et al. Expression of the water channel protein aquaporin-9 in malignant brain tumors[J]. Appl Immunohistochem Mol Morphol, 2007, 15(2): 193-198.

第五节　水通道蛋白与神经退行性疾病

一、水通道蛋白与阿尔茨海默病

阿尔茨海默病（Alzheimer's disease，AD）是一种常见的中枢神经系统退行性疾病，其病理改变包括淀粉样β蛋白（amyloid β-protein，Aβ）沉积形成的老年斑、神经原纤维缠结、神经元丢失和突触损伤等，以进行性记忆减退和认知功能障碍为主要临床症状[1]。尽管AD的病因和发病机制尚未明确，但有研究表明AD患者和AD动物模型脑内AQP的表达发生改变，提示AQP表达变化可能参与了AD的发病过程。本部分将总结AQP在AD中的表达变化，以及AQP在AD发病中的作用及机制。

（一）AQP1与AD

Moftakhar等的研究发现AD患者脑内AQP1的表达水平与对照者无显著差异[2]。但是，Pérez等和Hoshi等则发现，AD患者脑内AQP1的表达有所增加。Pérez等发现在散发性AD早期患者的额叶皮质中，AQP1的表达水平上调，星形胶质细胞中AQP1的免疫反应增强[3]。Hoshi等的研究显示[4]，与对照组相比，散发性和家族性AD患者大脑皮质中AQP1阳性细胞均更为丰富。在AD患者脑内，大量AQP1阳性细胞主要位于锥体细胞层，具有反应性星形胶质细胞胞体大、突起分支多的形态学特征。Hoshi等利用免疫荧光标记对AD患者脑内斑块及血管淀粉样变周围AQP1表达特征进行了研究[4]，结果显示，在家族性和散发性AD患者脑内，AQP1阳性的星形胶质细胞很少出现在Aβ42斑块密集的区域，在Aβ42斑块稀疏的区域则有大量的AQP1阳性细胞，即AQP1与Aβ42在不同区域的表达水平呈负相关。

有学者指出，AQP1表达增强对减轻AD患者大脑中的Aβ沉积具有重要作用[5]。在人类AD患者大脑中，激活的星形胶质细胞中存在大量的Aβ[6-9]，这些细胞可表达Aβ降解酶，用以清除Aβ[6, 9, 10]。在Aβ斑块附近存在表达AQP1的星形胶质细胞，将有利于Aβ的吸收和降解，从而减少局部老年斑的形成。此外，有学者发现AQP1在星形胶质细胞的迁移过程中发挥关键作用。在仓鼠卵巢细胞（CHO细胞）中，外源性AQP1表达于细胞突起部位，可加速细胞迁移[11]。因此，AD患者脑中星形胶质细胞AQP1的表达对增强星形胶质细胞的迁移极为重要[4]。Hoshi认为，AD患者中AQP1的表达上调可能在增强星形胶质细胞向Aβ斑块迁移的过程中起到作用。Aβ42与AQP1的表达水平呈负相关，这意味着缺乏AQP1表达但Aβ42超负荷的星形胶质细胞可能与老年斑的形成有关[12, 13]。

（二）AQP4 与 AD

目前，对于 AD 患者脑内 AQP4 表达变化的研究报道不一。Pérez 等和 Nagele 等[3, 13]的研究发现，与对照组相比，AD 患者额叶皮质内的 AQP4 表达并无显著改变。但另有研究则发现[14, 15]，AD 患者脑内反应性星形胶质细胞 AQP4 的表达水平增高。

1. AQP4 与 AD 中淀粉样蛋白沉积　AD 患者脑内血管壁上出现淀粉样蛋白沉积物，此种情况称为脑淀粉样蛋白血管病（cerebral amyloid angiopathy，CAA）。研究发现，CAA 患者脑内 AQP4 免疫反应性呈现两种截然不同的现象，有的患者增加[2]，而有的患者降低[16]。

Wilcock 等发现，随着 CAA 病情加重，AQP4 和 dystrophin 蛋白、mRNA 水平以及血管周围 AQP4 染色水平均有所降低[7]。进一步研究发现，在伴有 CAA 的 AD 小鼠模型中[15, 17, 18]，AQP4 在星形胶质细胞终足膜上的定位发生改变，极性表达缺失。尽管转基因小鼠中 AQP4 蛋白或 mRNA 的整体表达水平未见明显变化，但在额叶皮质中，血管淀粉样蛋白沉积增加，AQP4 阳性血管减少，dystrophin 蛋白/mRNA 表达水平降低。因此，推测 CAA 可能与 AQP4 错误定位和极性表达缺失有关。

大脑内 β 淀粉样蛋白斑块沉积是 AD 的病理标志之一。Aβ42 主要存在于 AD 患者脑内，而 Aβ40 则常见于受 CAA 影响的小动脉壁中。AQP4 在 Aβ42 斑块致密核心区无表达，但在边缘区域免疫反应增强[4]。在出现 CAA 的大血管和毛细血管（Aβ40 阳性血管）中，AQP4 表达程度各异；与 Aβ40 阴性血管相比，轻度和中度 Aβ40 沉积的大血管周围可见强烈的 AQP4 免疫反应[4]。在 AD 转基因小鼠模型中[15, 17, 18]，淀粉样蛋白沉积血管周围的星形细胞终足膜上 AQP4 的表达下调，而周围的神经鞘膜 AQP4 表达上调，即 AQP4 的极性表达形式发生改变，且 AQP4 的极性表达改变发生在斑块出现之后。因此，Aβ 沉积可能会影响 AD 发展过程中 AQP4 的表达模式。

2. AQP4 与 AD 中神经血管单元受损　神经血管单元（neurovascular unit，NVU）是由多种类型细胞组成的一个多细胞通信部位，包括毛细血管内皮细胞、神经元、星形胶质细胞终足和周细胞。Aβ 斑块插入 NVU 并在空间上破坏星形胶质细胞膜结构[17]，从而改变 AQP4 的极性分布。在卵巢切除术后注射右旋半乳糖的 AD 模型中，海马 AQP4 蛋白表达上升，星形胶质细胞增生，NVU 被破坏[19, 20]。在 $AQP4^{-/-}$ 小鼠中，Aβ 产生和积累增加[18, 20]，这可能是 $AQP4^{-/-}$ 小鼠对细胞间质中包括 Aβ 在内的溶质清除能力减少的直接后果[21]。

3. AQP4 与 AD 中记忆障碍和认知功能受损　在注射右旋半乳糖的 AD 模型小鼠中，小鼠脑内出现氧化应激、认知障碍和突触相关蛋白丢失。敲除此模型的 $AQP4$ 后，小鼠认知功能受损等症状加剧，表现为 Aβ 聚集、血管淀粉样变、胶质增生、认知障碍、皮质和海马中脑源性神经生长因子（brain-derived neurotrophic factor，BDNF）缺乏以及炎症因子 IL-1β 产生等[18, 20]。

AD 动物模型的认知功能受损与突触前囊泡蛋白突触素（synaptophysin，SYP）和突触后密度蛋白-95（postsynaptic density protein-95，PSD-95）的表达降低有关[18, 20]。在卵巢切除并敲除 $AQP4$ 的小鼠中，SYP 和 PSD-95 的表达下降，同时伴随空间学习和记忆功能下降[20]。APP/PS1小鼠在敲除 *AQP4* 基因后，海马和皮质中的BDNF水平降低。因此，AQP4在AD认知功能损伤中的作用可能与缺乏PSD-95导致突触对BDNF的反应性减弱有关[22]。

（三）AQP9与AD

AQP9是一种负责转运水、甘油、乳酸和尿素等小分子溶质的膜蛋白通道。Liu等[23]的研究报道，APP/PS1转基因AD小鼠模型在3月龄、6月龄和10月龄时，海马和大脑皮质中AQP9 mRNA和蛋白表达水平呈下降趋势。采用Aβ1-40处理后，PC12细胞中AQP9的表达呈剂量依赖性降低。抑制AQP9的表达会导致PC12细胞对Aβ1-40产生更严重的神经毒性反应。以上研究结果表明了AQP9在维持神经元正常活动中的重要作用。

孙善全研究团队在APP/PS1 AD转基因小鼠模型中得到的研究结果[24]与Liu等的结果[23]存在差异。免疫组化、免疫印迹和定量反转录PCR（qRT-PCR）结果显示：与对照组相比，AD小鼠脑内AQP9的吸光度值在海马不同区域均有所增高；AQP9蛋白和mRNA表达水平也呈现增高趋势[24]。Badaut等发现，脑缺血后缺血周边区域AQP9表达增高，他们认为缺血后脑内乳酸含量增加会促进软脑膜和微血管周围的内皮细胞中AQP9表达增高，从而有利于清除聚集在细胞外间隙的乳酸和甘油等物质[25]。孙善全研究团队推测[24]，AD小鼠模型中，AQP9表达增高将增加乳酸和甘油等能量底物从脑实质向血液和脑脊液的转运，导致神经元可利用的能量底物减少，从而引起脑内能量代谢下降。因此，AQP9作为乳酸和甘油等神经元能量底物的转运通道，在AD模型小鼠中表达增高，既有可能是引起脑内能量代谢下降的原因，也可能是脑内能量代谢下降的结果。

有文献报道，AD动物脑内与能量代谢相关的胰岛素信号通路下调[26]，*AQP9*基因在−496/−502的启动子中包含一个负性胰岛素反应元件TGTTTTC[27]。因此，AQP9表达水平增高可能是由下调的胰岛素信号通路对*AQP9*启动子区域的负性调控作用减弱所致。无论AQP9表达增高是AD能量代谢障碍的原因还是结果，均提示AQP9的表达变化与AD模型动物脑内能量代谢异常有关。AD模型小鼠脑内AQP9的表达变化为从能量代谢角度探究AD的发病机制提供了新的切入点。

二、水通道蛋白与帕金森病

帕金森病（Parkinson's disease，PD）是一种常见的神经系统退行性疾病，患者可出现静止性震颤和运动迟缓等特征性运动障碍，并伴有嗅觉障碍和精神、认知障碍等非运动症状[28]。病理特征为黑质致密区多巴胺能神经元丢失和残留神经元胞质内出现嗜酸性的包涵体，即路易小体（Lewy neurite，LN或Lewy body，LB）[29]。这些病理改变与α-突触核蛋白（α-synuclein，α-syn）磷酸化沉积有关[30]。本部分将总结AQP在PD中的表达变化、作用和可能机制。

（一）AQP1与PD

α-syn的异常沉积是PD的病理标志物之一[31]。α-syn可通过激活caspase诱导细胞凋亡[32]、促进活性氧（reactive oxygen species，ROS）的产生[33]、激活小胶质细胞导致促炎分子的表达和释放，以及促进多巴胺能神经元退行性变[34, 35]等多种途径参与PD的疾病进程。降低α-syn表达或聚集，以及增加α-syn的清除作为治疗帕金森病的一种策略[36]。

Hoshi等[37]探讨了PD患者颞叶新皮质AQP1的表达变化及其与α-syn沉积的关系。结果显示，PD患者新皮质各层均存在AQP1表达，特别是锥体细胞层中有大量AQP1阳性细胞。通过形态学观察和免疫荧光双标技术，发现AQP1在反应性星形胶质细胞中表达最为明显。进一步分析发现，AQP1和α-syn在Ⅱ～Ⅲ层皮质中表达呈负相关，即α-syn聚积区域缺乏AQP1阳性细胞；而在AQP1阳性表达增强的区域，α-syn表达则相应减弱。研究已表明，反应性星形胶质细胞具有吞噬和降解α-syn的能力[38, 39]，因此，表达AQP1的反应性星形胶质细胞可能在减少Ⅱ～Ⅲ层皮质α-syn的局部沉积中发挥作用。此外，AQP1与α-syn之间的负相关，可能源于星形胶质细胞对α-syn的摄取会抑制星形胶质细胞的活性，从而影响AQP1的表达[40]。

（二）AQP4与PD

PD的发生和进展涉及诸多细胞和分子机制，包括氧化应激、神经元凋亡、炎症途径、胶质细胞激活等。本部分将围绕AQP4在PD中的作用，从其与胶质细胞、α-syn、炎症等方面的关系展开探讨。

1. PD中的AQP4和胶质细胞 小胶质细胞是脑内的主要免疫细胞，在中枢神经系统的病理生理过程中发挥重要作用。已有较多研究关注到PD中小胶质细胞激活与疾病发生之间的关系。在PD动物模型中，脑内小胶质细胞被激活，导致NF-κB通路和NADPH氧化酶的激活，促进炎症介质的释放[41, 42]。有研究表明，α-syn可通过诱导基质金属蛋白酶表达及激活蛋白酶依赖的受体-1（protease-activated receptor-1，PAR）来激活小胶质细胞[43]，从而使小胶质细胞反应性增强，增加促炎性细胞因子的释放[44]。另一方面，小胶质细胞的激活可通过多种促炎介质和NO导致神经元死亡[45, 46]。据报道，多巴胺能神经元的丢失与黑质中激活的小胶质细胞数量增加有关[47, 48]。Lee等的研究显示，小胶质细胞激活可导致多巴胺能神经元损伤[49]。

星形胶质细胞是中枢神经系统的稳态调节细胞，参与分泌神经递质和营养因子、K^+缓冲、突触传递、突触形成和突触可塑性等过程。星形胶质细胞是PD发生和进展过程中的重要介质[50]，其活化在PD的病理生理过程中发挥重要作用[51]。有证据表明，α-syn可引发星形胶质细胞的活化[52]。神经元中的α-syn可通过连续内吞和胞吐转移至星形胶质细胞，进而刺激星形胶质细胞的炎症反应[39]。

有关AQP4参与PD疾病中小胶质细胞和星形胶质细胞激活的研究较少。在PD动物模型中，AQP4的缺失会导致小胶质细胞激活增加[53]。在另一项研究中发现，PD动物模型脑内AQP4缺失可导致腹侧被盖区（ventral tegmental area，VTA）和黑质（substantia nigra，SN）的小胶质细胞和星形胶质细胞被激活[54]。

2. PD中的AQP4和炎症因子 炎症因子在PD的病理生理过程中发挥重要作用。例如，在PD患者和动物模型中，白细胞介素-1β（interleukin-1β，IL-1β）和肿瘤坏死因子-α （tumor necrosis factor-α，TNF-α）水平显著升高[55, 56]，且与疾病的严重程度和进展密切相关[57, 58]。众多炎症因子参与了PD中多巴胺能神经元变性和凋亡信号通路[59, 60]。有报道称，AQP4缺失可促进PD动物模型中脑内TNF-α和IL-1β的产生。AQP4的表达降低激活了IKK/NF-κB通路，星形胶质细胞中TNF-α和IL-1β的产生增加[53]。在AQP4缺乏的PD动物模型中，腹侧被盖区和黑质致密部（SNc）中的促炎性细胞因子（包括TNF-α和IL-1）水平升高。此外，有研究报道在AQP4缺陷的PD小鼠中，$CD4^+CD25^+$调节性T细胞减少[61]。

3. PD中的AQP4和α-syn　Hoshi等[37]对PD患者颞叶新皮质中AQP4的表达改变及其与α-syn沉积的关系进行了探讨。结果显示，与前述PD脑内AQP1的表达状况类似，多数GFAP阳性细胞表现出不同程度的AQP4免疫反应性。在Ⅴ～Ⅵ层皮质中，AQP4与α-syn表达呈负相关，即α-syn聚集区域缺乏AQP4阳性细胞；而在AQP4阳性增强的区域，α-syn表达则相应减弱。反应性星形胶质细胞具有吸收和降解α-syn的能力[38, 39]，表达AQP4的反应性星形胶质细胞可能在降低Ⅴ～Ⅵ层皮质α-syn的沉积中发挥作用。另外，AQP4与α-syn表达呈负相关，也可能是因为星形胶质细胞对α-syn的摄取抑制了星形胶质细胞的活性，从而影响了AQP4的表达[40]。

（三）AQP9与PD

中脑黑质中多巴胺能神经元丢失被视为PD的主要病理改变之一，而AQP9在中脑多巴胺能神经元中表达丰富。因此，AQP9在PD疾病过程中的作用引起了学者的关注[62]。

研究发现，在散发性PD中，多巴胺能神经元死亡的原因与接触除草剂、亚砷酸盐和其他环境毒素有关[63-65]。Stahl利用在体和离体实验表明，多巴胺能神经元对这些毒素的选择易感性与AQP9密切相关[66]。例如，在注射AQP9 cDNA的爪蟾卵母细胞中，PD毒素MPP^+（1-甲基-4-苯基-吡啶离子）可渗透进细胞；稳定表达AQP9可增加HEK细胞对MPP^+和亚砷酸盐的易感性；相反，靶向敲除*AQP9*可保护黑质多巴胺能神经元免受MPP^+毒性的影响。可见，作为水-甘油跨膜输送的蛋白通道，AQP9除了可供包括嘌呤、嘧啶和乳酸等在内的广泛底物通过外，还与毒素更易通过细胞膜从而损害多巴胺能神经元的生存能力有关。在亚细胞水平上，AQP9存在于质膜和线粒体内膜[67, 68]，因此，AQP9可能允许MPP^+和其他毒素渗透至线粒体基质，导致电子传递链被破坏，从而与帕金森病的发生有关[69, 70]。

AQP9具备转运乳酸和β-羟基丁酸等能量代谢底物的功能，线粒体具有AQP9的多巴胺能神经元能够利用乳酸和β-羟基丁酸等替代底物产生能量，从而抵抗缺血带来的损伤[71]。然而，关于AQP9表达上调能否对帕金森病起保护作用，尚需进一步研究确定。

三、水通道蛋白与肌萎缩侧索硬化

肌萎缩侧索硬化（amyotrophic lateral sclerosis，ALS），亦称为Lou Gehrig病，是一种

侵犯运动神经元的神经退行性疾病，最终导致患者肌无力和瘫痪[72]。现将AQP在ALS中的表达、作用及其潜在的神经生物学机制予以概述。

（一）AQP1与ALS

Daniela等学者对*AQP1*基因在运动皮质和脊髓中的表达、调控及其与ALS风险的关联进行了系统研究，提出AQP1表达上调可能是一种与ALS风险相关的全新生物标志物[73]。该研究推测，AQP1在神经元中可直接与蛋白激酶C相互作用蛋白1（protein interacting with C kinase 1，PICK1）作用。AQP1表达的增强会降低或抑制ALS患者脊髓运动神经元中PICK1的表达，导致氧化应激水平增高，触发兴奋性毒性和运动神经元死亡，从而加重损伤并促进ALS的病理过程。

（二）AQP4与ALS

AQP4作为脑内主导的AQP成员，特异性分布于星形胶质细胞在毛细血管周围形成的终足膜、软脑膜内表面的外胶质界膜和室管膜细胞附近的内胶质界膜等部位[74]，其功能主要包括调节水和离子的转运[75]、钾的释放和摄取[76]、胶质瘢痕形成[77]、星形胶质细胞迁移[78,79]、神经信号传导[80]、促炎性细胞因子分泌[81]和星形胶质细胞间通信[82]等神经活动。运动神经元的功能发挥高度依赖于星形胶质细胞，因此，位于星形胶质细胞膜上的AQP4在ALS发生发展中发挥重要作用。目前，AQP4和ALS相互关系的研究有了显著的进展。

1. ALS中AQP4表达的改变　利用超氧化物歧化酶1（superoxide dismutase 1，SOD1）突变的转基因ALS小鼠模型进行研究，结果显示，脊髓中星形胶质细胞AQP4蛋白表达水平升高。Nicaise[83]在大鼠ALS模型中也发现，脊髓灰质中AQP4 mRNA和蛋白含量增高，但白质中AQP4的表达无变化。另有报道[84]发现，在ALS晚期的*SOD1*转基因小鼠模型中，脑干和皮质中AQP4的表达水平亦呈现升高趋势。

在ALS模型中，脑和脊髓内的AQP4不仅表达量发生改变，其分布形式亦有所不同。Dai等[85]发现，AQP4在ALS发病前呈极性分布于血管周围星形胶质细胞的终足膜。然而，在ALS的发病过程中，血管周围的AQP4缺失，呈去极性分布，且随着疾病进展，发生去极化的AQP4表达量逐渐增高。以上研究提示，AQP4去极化可能是ALS发生和进展的关键病理特征。AQP4持续去极化可能导致脊髓内水稳态功能失调，进而引发ALS患者神经元肿胀和脊髓功能受损。

2. AQP4表达变化引起ALS中K^+稳态失衡　ALS的一个重要病理生理学特征是轴突中的离子通道（如钾通道和钠通道）功能失调[86-88]。轴突离子功能障碍导致运动性轴突过度兴奋，从而引起运动神经元退行性变和肌肉震颤等临床表现[88]。

Bataveljić等[84]对ALS转基因小鼠脑内AQP4和内向性钾通道Kir4.1的表达水平进行了检测，结果显示，AQP4表达水平增加，而Kir4.1表达水平减少。在体外培养的ALS星形胶质细胞中，AQP4和Kir4.1的表达变化与体内一致，同时还检测到Kir4.1的电生理学特性发生了改变。AQP4和Kir4.1共定位，均分布于沿血管分布的星形胶质细胞终足膜[89]，且两者均可通过与α-syntrophin结合在特定的细胞膜区域[90]，提示两者可能协同维持脑内渗透压调节。AQP4的表达变化可通过影响与Kir4.1的协同作用，引起细胞外K⁺稳态失衡，进而干扰神经元正常的微环境，对运动神经元的存活和功能产生负面影响，导致ALS疾病发生和进展[88]。

3. AQP4功能障碍引起ALS中血脑屏障损伤和炎症反应　研究发现，ALS患者血脑屏障的完整性受到严重破坏[91, 92]，导致外周炎症细胞和细胞因子侵入中枢神经系统，并触发运动神经元损伤[93]。

S100B是由星形胶质细胞分泌的一种具有细胞因子样活性的主要产物，在神经退行性疾病中发挥作用[94]。研究表明，包括rs1058424的T等位基因、rs335929的A等位基因、rs3763043的A等位基因和TAA单倍型等*AQP4*基因的变异，均与S100B表达水平升高相关。受多态性调控的*AQP4*基因表达变化可能会通过S100B影响ALS的炎症过程[95]。此外，AQP4表达变化还可通过影响血流和水的平衡，进一步影响神经炎症的发展[88]。因此，在ALS疾病进程中，AQP4表达变化至关重要。

高亲和力的AQP4特异性抗体（AQP4-specific antibodies，AQP4-abs）可单独进入中枢神经系统与星形胶质细胞结合，导致AQP4反应性缺失，干扰星形胶质细胞和内皮细胞之间的相互作用，诱导血脑屏障损伤，促进ALS的进展[96, 97]。针对AQP4的TAA单倍型抑制剂有望保护ALS患者大脑免受持续性神经炎症和血脑屏障损伤的影响[88, 95]。

4. AQP4功能障碍导致ALS中的连接蛋白功能失调　连接蛋白（connexin，Cx）在相邻星形胶质细胞之间或星形胶质细胞和少突胶质细胞[64]之间构成缝隙连接[98]。目前研究证实，连接蛋白功能失调与ALS的疾病进展相关。Cui等[99]发现，在ALS模型小鼠脊髓前角，Cx32和Cx47表达受到抑制，星形胶质细胞和少突胶质细胞上的缝隙连接蛋白在ALS的晚期被破坏尤为明显。Díaz-Amarilla等[100]从*SOD1*转基因的ALS模型小鼠中提取异常星形胶质细胞，发现这些细胞中Cx43的免疫反应性增强，将此类异常的星形胶质细胞与健康神经元共培养，会诱导健康运动神经元死亡。由此可见，星形胶质细胞上的Cx43表达增加可以引起胶质细胞活化，进而引发后续兴奋性毒性神经退行性变。

研究证实，AQP4和连接蛋白共定位表达[101]。在海马区域，Cx30和Cx42作为主要的缝隙连接蛋白，与AQP4相似，均分布于血管周围的足细胞膜[102]。除了细胞定位相似之外，连接蛋白亦具有与AQP4相似的功能。Katoozi发现，在靶向敲除*AQP4*后，海马星形胶质细胞的缝隙连接蛋白表达升高，血管周围Cx43阳性细胞的数量减少[103]。这些结果与Strohschein的结果相吻合[76]，表明在AQP4缺失的情况下，连接蛋白（尤其是Cx43）的表达发生上调。此外，有学者推测Cx43可能是AQP4引发脑水肿的下游效应因子[104]。对AQP4与Cx43相互作用的进一步探索，将有助于优化ALS的治疗策略。

5. AQP4表达改变通过NGF参与ALS疾病发生　神经生长因子（nerve growth factor，NGF）在维持神经元健康、生长和生存等方面发挥重要作用。对于神经退行性疾病患者，NGF具有神经保护效应，NGF缺乏将诱发神经退行性疾病[105, 106]。有研究发现，ALS患者脊髓背侧标本NGF表达水平降低[107]。Chen等[108]发现，通过慢病毒介导的RNAi抑制AQP4表达，可以上调NGF表达，促进运动神经元的功能恢复。以上研究为探索AQP4和

ALS疾病之间的关系开辟了新的途径。

6. AQP4表达改变通过谷氨酸稳态失调参与ALS疾病发生 谷氨酸功能稳态失调所导致的兴奋性毒性是运动神经元变性的一个潜在因素[51]。谷氨酸转运体1（glutamate transporter 1，GLT-1）是星形胶质细胞上的一种重要的谷氨酸转运体，负责摄取中枢神经系统中大部分的谷氨酸[109]。在ALS患者和转基因ALS大鼠模型中，脊髓和运动皮质区的星形胶质细胞均出现GLT-1功能受损[110, 111]，导致谷氨酸摄取失调。研究证实，AQP4缺失会下调GLT-1表达，减少星形胶质细胞对谷氨酸的摄取[112]，AQP4亦参与了GLT-1功能和表达的调控[113]。AQP4和GLT-1可能作为星形胶质细胞膜中的一个复合物，共同维持神经元的完整性[112, 114, 115]。关于GLT-1和AQP4在ALS中相互作用及机制，仍有待进一步研究。

（陆蔚天　廖玉慧）

参 考 文 献

[1] Forlenza OV, Diniz BS, Gattaz WF. Diagnosis and biomarkers of predementia in Alzheimer's disease[J]. BMC Med, 2010, 8: 89.

[2] Moftakhar P, Lynch MD, Pomakian JL, et al. Aquaporin expression in the brains of patients with or without cerebral amyloid angiopathy[J]. J Neuropathol Exp Neurol, 2010, 69(12): 1201-1209.

[3] Pérez E, Barrachina M, Rodríguez A, et al. Aquaporin expression in the cerebral cortex is increased at early stages of Alzheimer disease[J]. Brain Res, 2007, 1128(1): 164-174.

[4] Hoshi A, Yamamoto T, Shimizu K, et al. Characteristics of aquaporin expression surrounding senile plaques and cerebral amyloid angiopathy in Alzheimer disease[J]. J Neuropathol Exp Neurol, 2012, 71(8): 750-759.

[5] Park J, Madan M, Chigurupati S, et al. Neuronal aquaporin 1 inhibits amyloidogenesis by suppressing the interaction between beta-secretase and amyloid precursor protein[J]. J Gerontol A Biol Sci Med Sci, 2021, 76(1): 23-31.

[6] Wyss-Coray T, Loike JD, Brionne TC, et al. Adult mouse astrocytes degrade amyloid-beta in vitro and in situ[J]. Nat Med, 2003, 9(4): 453-457.

[7] Verkhratsky A, Olabarria M, Noristani HN, et al. Astrocytes in Alzheimer's disease[J]. Neuro, 2010, 7(4): 399-412.

[8] Allaman I, Bélanger M, Magistretti PJ. Astrocyte-neuron metabolic relationships: for better and for worse[J]. Trends Neurosci, 2011, 34(2): 76-87.

[9] Koistinaho M, Lin S, Wu X, et al. Apolipoprotein E promotes astrocyte colocalization and degradation of deposited amyloid-beta peptides[J]. Nature Med, 2004, 10(7): 719-726.

[10] Dorfman VB, Pasquini L, Riudavets M, et al. Differential cerebral deposition of IDE and NEP in sporadic and familial Alzheimer's disease[J]. Neurobiol Aging, 2010, 31(10): 1743-1757.

[11] Verkman AS. More than just water channels: unexpected cellular roles of aquaporins[J]. J Cell Sci, 2005, 118(Pt 15): 3225-3232.

[12] Nuutinen T, Huuskonen J, Suuronen T, et al. Amyloid-beta 1-42 induced endocytosis and clusterin/apoJ protein accumulation in cultured human astrocytes[J]. Neurochem Int, 2007, 50(3): 540-547.

[13] Nagele RG, Wegiel J, Venkataraman V, et al. Contribution of glial cells to the development of amyloid

plaques in Alzheimer's disease[J]. Neurobiol Aging, 2004, 25(5): 663-674.

[14] Rodríguez A, Pérez-Gracia E, Espinosa JC, et al. Increased expression of water channel aquaporin 1 and aquaporin 4 in Creutzfeldt-Jakob disease and in bovine spongiform encephalopathy-infected bovine-PrP transgenic mice[J]. Acta Neuropathol, 2006, 112(5): 573-585.

[15] Yang J, Lunde LK, Nuntagij P, et al. Loss of astrocyte polarization in the tg-ArcSwe mouse model of Alzheimer's disease[J]. J Alzheimers Dis, 2011, 27(4): 711-722.

[16] Wilcock DM, Vitek MP, Colton CA. Vascular amyloid alters astrocytic water and potassium channels in mouse models and humans with Alzheimer's disease[J]. Neuroscience, 2009, 159(3): 1055-1069.

[17] Zago W, Schroeter S, Guido T, et al. Vascular alterations in PDAPP mice after anti-Aβ immunotherapy: implications for amyloid-related imaging abnormalities[J]. Alzheimers Dement, 2013, 9(5 Suppl): S105-S115.

[18] Xu Z, Xiao N, Chen Y, et al. Deletion of aquaporin-4 in APP/PS1 mice exacerbates brain Aβ accumulation and memory deficits[J]. Mol Neurodegener, 2015, 10: 58.

[19] Liu L, Su Y, Yang W, et al. Disruption of neuronal-glial-vascular units in the hippocampus of ovariectomized mice injected with D-galactose[J]. Neuroscience, 2010, 169(2): 596-608.

[20] Liu L, Lu Y, Kong H, et al. Aquaporin-4 deficiency exacerbates brain oxidative damage and memory deficits induced by long-term ovarian hormone deprivation and D-galactose injection[J]. Int J Neuropsychopharmacol, 2012, 15(1): 55-68.

[21] Iliff JJ, Wang M, Liao Y, et al. A paravascular pathway facilitates CSF flow through the brain parenchyma and the clearance of interstitial solutes, including amyloid β[J]. Sci Transl Med, 2012, 4(147): 147ra111.

[22] Hubbard JA, Szu JI, Binder DK. The role of aquaporin-4 in synaptic plasticity, memory and disease[J]. Brain Res Bull, 2018, 136: 118-129.

[23] Liu JY, Chen XX, Chen HY, et al. Downregulation of aquaporin 9 exacerbates beta-amyloid-induced neurotoxicity in Alzheimer's disease models in vitro and in vivo[J]. Neuroscience, 2018, 394: 72-82.

[24] 陆蔚天, 黄娟, 李昱, 等. AQP9在APP/PS1双转基因小鼠海马中的表达变化[J]. 重庆医科大学学报, 2017, 42(1): 120-124.

[25] Badaut J, Regli L. Distribution and possible roles of aquaporin 9 in the brain[J]. Neuroscience, 2004, 129(4): 971-981.

[26] Rönnemaa E, Zethelius B, Sundelöf J, et al. Impaired insulin secretion increases the risk of Alzheimer disease[J]. Neurology, 2008, 71(14): 1065-1071.

[27] Kuriyama H, Shimomura I, Kishida K, et al. Coordinated regulation of fat-specific and liver-specific glycerol channels, aquaporin adipose and aquaporin 9[J]. Diabetes, 2002, 51(10): 2915-2921.

[28] Jellinger KA. Neuropathology of sporadic Parkinson's disease: evaluation and changes of concepts[J]. Mov Disord, 2012, 27(1): 8-30.

[29] Schulz-Schaeffer WJ. Neurodegeneration in Parkinson disease: moving Lewy bodies out of focus[J]. Neurology, 2012, 79(24): 2298-2299.

[30] Mavroeidi P, Xilouri M. Neurons and glia interplay in α-synucleinopathies[J]. Int J Mol Sci, 2021, 22(9): 4994.

[31] Stefanis L. α-Synuclein in Parkinson's disease[J]. Cold Spring Harb Perspect Med, 2012, 2(2): a009399.

[32] Flower TR, Chesnokova LS, Froelich CA, et al. Heat shock prevents alpha-synuclein-induced apoptosis in a yeast model of Parkinson's disease[J]. J Mol Biol, 2005, 351(5): 1081-1100.

[33] Stefanova N, Schanda K, Klimaschewski L, et al. Tumor necrosis factor-alpha-induced cell death in

U373 cells overexpressing alpha-synuclein[J]. J Neurosci Res, 2003, 73(3): 334-340.

[34] Masliah E, Rockenstein E, Veinbergs I, et al. Dopaminergic loss and inclusion body formation in alpha-synuclein mice: implications for neurodegenerative disorders[J]. Science, 2000, 287(5456): 1265-1269.

[35] Saha AR, Ninkina NN, Hanger DP, et al. Induction of neuronal death by alpha-synuclein[J]. Eur J Neurosci, 2000, 12(8): 3073-3077.

[36] Lashuel HA, Overk CR, Oueslati A, et al. The many faces of α-synuclein: from structure and toxicity to therapeutic target[J]. Nat Rev Neurosci, 2013, 14(1): 38-48.

[37] Hoshi A, Tsunoda A, Tada M, et al. Expression of aquaporin 1 and aquaporin 4 in the temporal neocortex of patients with Parkinson's disease[J]. Brain Pathol, 2017, 27(2): 160-168.

[38] Braak H, Sastre M, Del Tredici K. Development of alpha-synuclein immunoreactive astrocytes in the forebrain parallels stages of intraneuronal pathology in sporadic Parkinson's disease[J]. Acta Neuropathol, 2007, 114(3): 231-241.

[39] Lee HJ, Suk JE, Patrick C, et al. Direct transfer of alpha-synuclein from neuron to astroglia causes inflammatory responses in synucleinopathies[J]. J Biol Chem, 2010, 285(12): 9262-9272.

[40] Tong J, Ang LC, Williams B, et al. Low levels of astroglial markers in Parkinson's disease: relationship to α-synuclein accumulation[J]. Neurobiol Dis, 2015, 82: 243-253.

[41] Bordt EA, Polster BM. NADPH oxidase- and mitochondria-derived reactive oxygen species in proinflammatory microglial activation: a bipartisan affair?[J]. Free Radic Biol Med, 2014, 76: 34-46.

[42] Wu DC, Teismann P, Tieu K, et al. NADPH oxidase mediates oxidative stress in the 1-methyl-4-phenyl-1, 2, 3, 6-tetrahydropyridine model of Parkinson's disease[J]. Proc Natl Acad Sci USA, 2003, 100(10): 6145-6150.

[43] Lee EJ, Woo MS, Moon PG, et al. Alpha-synuclein activates microglia by inducing the expressions of matrix metalloproteinases and the subsequent activation of protease-activated receptor-1[J]. J Immunol, 2010, 185(1): 615-623.

[44] Rojanathammanee L, Murphy EJ, Combs CK. Expression of mutant alpha-synuclein modulates microglial phenotype in vitro[J]. J Neuroinflammation, 2011, 8: 44.

[45] Boje KM, Arora PK. Microglial-produced nitric oxide and reactive nitrogen oxides mediate neuronal cell death[J]. Brain Res, 1992, 587(2): 250-256.

[46] Ghoshal A, Das S, Ghosh S, et al. Proinflammatory mediators released by activated microglia induces neuronal death in Japanese encephalitis[J]. Glia, 2007, 55(5): 483-496.

[47] Gao HM, Jiang J, Wilson B, et al. Microglial activation-mediated delayed and progressive degeneration of rat nigral dopaminergic neurons: relevance to Parkinson's disease[J]. J Neurochem, 2002, 81(6): 1285-1297.

[48] McGeer PL, Itagaki S, Boyes BE, et al. Reactive microglia are positive for HLA-DR in the substantia nigra of Parkinson's and Alzheimer's disease brains[J]. Neurology, 1988, 38(8): 1285-1291.

[49] Le W, Rowe D, Xie W, et al. Microglial activation and dopaminergic cell injury: an in vitro model relevant to Parkinson's disease[J]. J Neurosci, 2001, 21(21): 8447-8455.

[50] Fellner L, Jellinger KA, Wenning GK, et al. Glial dysfunction in the pathogenesis of α-synucleinopathies: emerging concepts[J]. Acta Neuropathol, 2011, 121(6): 675-693.

[51] Maragakis NJ, Rothstein JD. Mechanisms of disease: astrocytes in neurodegenerative disease[J]. Nat Clin Pract Neurol, 2006, 2(12): 679-689.

[52] Fellner L, Irschick R, Schanda K, et al. Toll-like receptor 4 is required for α-synuclein dependent activation of microglia and astroglia[J]. Glia, 2013, 61(3): 349-360.

[53] Sun H, Liang R, Yang B, et al. Aquaporin-4 mediates communication between astrocyte and microglia: implications of neuroinflammation in experimental Parkinson's disease[J]. Neuroscience, 2016, 317: 65-75.

[54] Zhang J, Yang B, Sun H, et al. Aquaporin-4 deficiency diminishes the differential degeneration of midbrain dopaminergic neurons in experimental Parkinson's disease[J]. Neurosci Lett, 2016, 614: 7-15.

[55] Bessler H, Djaldetti R, Salman H, et al. IL-1 beta, IL-2, IL-6 and TNF-alpha production by peripheral blood mononuclear cells from patients with Parkinson's disease[J]. Biomed Pharmacother, 1999, 53(3): 141-145.

[56] Mogi M, Togari A, Tanaka K, et al. Increase in level of tumor necrosis factor(TNF)-alpha in 6-hydroxydopamine-lesioned striatum in rats without influence of systemic L-DOPA on the TNF-alpha induction[J]. Neurosci Lett, 1999, 268(2): 101-104.

[57] Kouchaki E, Daneshvar Kakhaki R, Tamtaji OR, et al. Correlation of serum levels and gene expression of tumor necrosis factor-α-induced protein-8 like-2 with Parkinson disease severity[J]. Metab Brain Dis, 2018, 33(6): 1955-1959.

[58] Kouchaki E, Kakhaki RD, Tamtaji OR, et al. Increased serum levels of TNF-α and decreased serum levels of IL-27 in patients with Parkinson disease and their correlation with disease severity[J]. Clin Neurol Neurosurg, 2018, 166: 76-79.

[59] Burguillos MA, Hajji N, Englund E, et al. Apoptosis-inducing factor mediates dopaminergic cell death in response to LPS-induced inflammatory stimulus: evidence in Parkinson's disease patients[J]. Neurobiol Dis, 2011, 41(1): 177-188.

[60] Venderova K, Park DS. Programmed cell death in Parkinson's disease[J]. Cold Spring Harb Perspect Med, 2012, 2(8): a009365.

[61] Chi Y, Fan Y, He L, et al. Novel role of aquaporin-4 in CD4+ CD25+ T regulatory cell development and severity of Parkinson's disease[J]. Aging Cell, 2011, 10(3): 368-382.

[62] Yool AJ. Aquaporins: multiple roles in the central nervous system[J]. Neuroscientist, 2007, 13(5): 470-485.

[63] Pezzoli G, Cereda E. Exposure to pesticides or solvents and risk of Parkinson disease[J]. Neurology, 2013, 80(22): 2035-2041.

[64] Heikkila RE, Nicklas WJ, Vyas I, et al. Dopaminergic toxicity of rotenone and the 1-methyl-4-phenylpyridinium ion after their stereotaxic administration to rats: implication for the mechanism of 1-methyl-4-phenyl-1, 2, 3, 6-tetrahydropyridine toxicity[J]. Neurosci Lett, 1985, 62(3): 389-394.

[65] Fan SF, Chao PL, Lin AM. Arsenite induces oxidative injury in rat brain: synergistic effect of iron[J]. Ann N Y Acad Sci, 2010, 1199: 27-35.

[66] Stahl K, Rahmani S, Prydz A, et al. Targeted deletion of the aquaglyceroporin AQP9 is protective in a mouse model of Parkinson's disease[J]. PLoS One, 2018, 13(3): e0194896.

[67] Amiry-Moghaddam M, Lindland H, Zelenin S, et al. Brain mitochondria contain aquaporin water channels: evidence for the expression of a short AQP9 isoform in the inner mitochondrial membrane[J]. FASEB J, 2005, 19(11): 1459-1467.

[68] Mylonakou MN, Petersen PH, Rinvik E, et al. Analysis of mice with targeted deletion of AQP9 gene provides conclusive evidence for expression of AQP9 in neurons[J]. J Neurosci Res, 2009, 87(6): 1310-1322.

[69] Betarbet R, Sherer TB, MacKenzie G, et al. Chronic systemic pesticide exposure reproduces features of Parkinson's disease[J]. Nat Neurosci, 2000, 3(12): 1301-1306.

[70] Nicklas WJ, Youngster SK, Kindt MV, et al. MPTP, MPP$^+$ and mitochondrial function[J]. Life Sci, 1987, 40(8): 721-729.

[71] Badaut J. Aquaglyceroporin 9 in brain pathologies[J]. Neuroscience, 2010, 168(4): 1047-1057.

[72] van Es MA, Hardiman O, Chio A, et al. Amyotrophic lateral sclerosis[J]. Lancet, 2017, 390(10107): 2084-2098.

[73] Recabarren-Leiva D, Alarcón M. New insights into the gene expression associated to amyotrophic lateral sclerosis[J]. Life Sci, 2018, 193: 110-123.

[74] Nagelhus EA, Ottersen OP. Physiological roles of aquaporin-4 in brain[J]. Physiol Rev, 2013, 93(4): 1543-1562.

[75] MacVicar BA, Feighan D, Brown A, et al. Intrinsic optical signals in the rat optic nerve: role for K(+) uptake via NKCC1 and swelling of astrocytes[J]. Glia, 2002, 37(2): 114-123.

[76] Strohschein S, Hüttmann K, Gabriel S, et al. Impact of aquaporin-4 channels on K^+ buffering and gap junction coupling in the hippocampus[J]. Glia, 2011, 59(6): 973-980.

[77] Saadoun S, Papadopoulos MC, Watanabe H, et al. Involvement of aquaporin-4 in astroglial cell migration and glial scar formation[J]. J Cell Sci, 2005, 118(Pt 24): 5691-5698.

[78] Ikeshima-Kataoka H. Neuroimmunological implications of AQP4 in astrocytes[J]. Int J Mol Sci, 2016, 17(8): 1306.

[79] Saadoun S, Papadopoulos MC, Hara-Chikuma M, et al. Impairment of angiogenesis and cell migration by targeted aquaporin-1 gene disruption[J]. Nature, 2005, 434(7034): 786-792.

[80] Verkman AS, Ratelade J, Rossi A, et al. Aquaporin-4: orthogonal array assembly, CNS functions, and role in neuromyelitis optica[J]. Acta Pharmacol Sin, 2011, 32(6): 702-710.

[81] Li L, Zhang H, Varrin-Doyer M, et al. Proinflammatory role of aquaporin-4 in autoimmune neuroinflammation[J]. FASEB J, 2011, 25(5): 1556-1566.

[82] Thrane AS, Rappold PM, Fujita T, et al. Critical role of aquaporin-4(AQP4) in astrocytic Ca^{2+} signaling events elicited by cerebral edema[J]. Proc Natl Acad Sci USA, 2011, 108(2): 846-851.

[83] Nicaise C, Soyfoo MS, Authelet M, et al. Aquaporin-4 overexpression in rat ALS model[J]. Anat Rec (Hoboken), 2009, 292(2): 207-213.

[84] Bataveljić D, Nikolić L, Milosević M, et al. Changes in the astrocytic aquaporin-4 and inwardly rectifying potassium channel expression in the brain of the amyotrophic lateral sclerosis SOD1(G93A) rat model[J]. Glia, 2012, 60(12): 1991-2003.

[85] Dai J, Lin W, Zheng M, et al. Alterations in AQP4 expression and polarization in the course of motor neuron degeneration in SOD1G93A mice[J]. Molecular Medicine Reports, 2017, 16(2): 1739-1746.

[86] Forte G, Bocca B, Oggiano R, et al. Essential trace elements in amyotrophic lateral sclerosis(ALS): results in a population of a risk area of Italy[J]. Neurol Sci, 2017, 38(9): 1609-1615.

[87] Nakata M, Kuwabara S, Kanai K, et al. Distal excitability changes in motor axons in amyotrophic lateral sclerosis[J]. Clin Neurophysiol, 2006, 117(7): 1444-1448.

[88] Zou S, Lan YL, Wang H, et al. The potential roles of aquaporin 4 in amyotrophic lateral sclerosis[J]. Neurol Sci, 2019, 40(8): 1541-1549.

[89] Nagelhus EA, Mathiisen TM, Ottersen OP. Aquaporin-4 in the central nervous system: cellular and subcellular distribution and coexpression with KIR4.1[J]. Neuroscience, 2004, 129(4): 905-913.

[90] Masaki H, Wakayama Y, Hara H, et al. Immunocytochemical studies of aquaporin 4, Kir4.1, and α1-syntrophin in the astrocyte endfeet of mouse brain capillaries[J]. Acta Histochem Cytochem, 2010, 43(4): 99-105.

[91] Garbuzova-Davis S, Saporta S, Haller E, et al. Evidence of compromised blood-spinal cord barrier in early and late symptomatic SOD1 mice modeling ALS[J]. PLoS One, 2007, 2(11): e1205.

[92] Zhong Z, Deane R, Ali Z, et al. ALS-causing SOD1 mutants generate vascular changes prior to motor

neuron degeneration[J]. Nat Neurosci, 2008, 11(4): 420-422.

[93] Zlokovic BV. The blood-brain barrier in health and chronic neurodegenerative disorders[J]. Neuron, 2008, 57(2): 178-201.

[94] Steiner J, Bogerts B, Schroeter ML, et al. S100B protein in neurodegenerative disorders[J]. Clin Chem Lab Med, 2011, 49(3): 409-424.

[95] Wu YF, Sytwu HK, Lung FW. Human aquaporin 4 gene polymorphisms and haplotypes are associated with serum S100B level and negative symptoms of schizophrenia in a southern Chinese Han population[J]. Front Psychiatry, 2018, 9: 657.

[96] Howe CL, Kaptzan T, Magaña SM, et al. Neuromyelitis optica IgG stimulates an immunological response in rat astrocyte cultures[J]. Glia, 2014, 62(5): 692-708.

[97] Takeshita Y, Obermeier B, Cotleur AC, et al. Effects of neuromyelitis optica-IgG at the blood-brain barrier in vitro[J]. Neurol Neuroimmunol Neuroinflamm, 2016, 4(1): e311.

[98] Lin X, Xu Q, Veenstra RD. Functional formation of heterotypic gap junction channels by connexins-40 and-43[J]. Channels, 2014, 8(5): 433-443.

[99] Cui Y, Masaki K, Yamasaki R, et al. Extensive dysregulations of oligodendrocytic and astrocytic connexins are associated with disease progression in an amyotrophic lateral sclerosis mouse model[J]. J Neuroinflammation, 2014, 11: 42.

[100] Díaz-Amarilla P, Olivera-Bravo S, Trias E, et al. Phenotypically aberrant astrocytes that promote motoneuron damage in a model of inherited amyotrophic lateral sclerosis[J]. Proc Natl Acad Sci USA, 2011, 108(44): 18126-18131.

[101] Rash JE, Yasumura T, Dudek FE, et al. Cell-specific expression of connexins and evidence of restricted gap junctional coupling between glial cells and between neurons[J]. J Neurosci, 2001, 21(6): 1983-2000.

[102] Griemsmann S, Höft SP, Bedner P, et al. Characterization of panglial gap junction networks in the thalamus, neocortex, and hippocampus reveals a unique population of glial cells[J]. Cereb Cortex, 2015, 25(10): 3420-3433.

[103] Katoozi S, Skauli N, Rahmani S, et al. Targeted deletion of Aqp4 promotes the formation of astrocytic gap junctions[J]. Brain Struct Funct, 2017, 222(9): 3959-3972.

[104] Li G, Liu X, Liu Z, et al. Interactions of connexin 43 and aquaporin-4 in the formation of glioma-induced brain edema[J]. Mol Med Rep, 2015, 11(2): 1188-1194.

[105] Ferreira D, Westman E, Eyjolfsdottir H, et al. Brain changes in Alzheimer's disease patients with implanted encapsulated cells releasing nerve growth factor[J]. J Alzheimers Dis, 2015, 43(3): 1059-1072.

[106] Appel SH. A unifying hypothesis for the cause of amyotrophic lateral sclerosis, Parkinsonism, and Alzheimer disease[J]. Ann Neurol, 1981, 10(6): 499-505.

[107] Anand P, Parrett A, Martin J, et al. Regional changes of ciliary neurotrophic factor and nerve growth factor levels in post mortem spinal cord and cerebral cortex from patients with motor disease[J]. Nat Med, 1995, 1(2): 168-172.

[108] Chen J, Zeng X, Li S, et al. Lentivirus-mediated inhibition of AQP4 accelerates motor function recovery associated with NGF in spinal cord contusion rats[J]. Brain Res, 2017, 1669: 106-113.

[109] Maragakis NJ, Rothstein JD. Glutamate transporters: animal models to neurologic disease[J]. Neurobiol Dis, 2004, 15(3): 461-473.

[110] Rothstein JD, Martin LJ, Kuncl RW. Decreased glutamate transport by the brain and spinal cord in amyotrophic lateral sclerosis[J]. N Engl J Med, 1992, 326(22): 1464-1468.

[111] Rothstein JD, Van Kammen M, Levey AI, et al. Selective loss of glial glutamate transporter GLT-1 in

amyotrophic lateral sclerosis[J]. Ann Neurol, 1995, 38(1): 73-84.
[112] Lan YL, Chen JJ, Hu G, et al. Aquaporin 4 in astrocytes is a target for therapy in Alzheimer's disease[J]. Curr Pharm Des, 2017, 23(33): 4948-4957.
[113] Lan YL, Zou S, Chen JJ, et al. The neuroprotective effect of the association of aquaporin-4/glutamate transporter-1 against Alzheimer's disease[J]. Neural Plast, 2016, 2016: 4626593.
[114] Yang J, Li MX, Luo Y, et al. Chronic ceftriaxone treatment rescues hippocampal memory deficit in AQP4 knockout mice via activation of GLT-1[J]. Neuropharmacology, 2013, 75: 213-222.
[115] Hinson SR, Roemer SF, Lucchinetti CF, et al. Aquaporin-4-binding autoantibodies in patients with neuromyelitis optica impair glutamate transport by down-regulating EAAT2[J]. J Exp Med, 2008, 205(11): 2473-2481.

第六节　水通道蛋白与精神疾病

有研究在对自闭症及精神分裂症患者进行尸检时发现[1, 2]，其大脑前额叶皮质、扣带回及小脑等区域的AQP4表达水平明显降低，另有研究发现精神分裂症患者的基因组中存在罕见的AQP3重复序列[3]，这些证据都提示水通道蛋白家族中的某些成员与精神疾病之间可能存在联系。然而，在水通道蛋白的研究中，与精神疾病相关的报道却相对较少。因此，本节主要就水通道蛋白与自闭症、抑郁症及精神分裂症之间的关系进行简要说明。

一、AQP4与自闭症

自闭症是一种严重的神经发育障碍疾病，患者在社会交往方面存在明显障碍，常有刻板及重复的行为或语言，对周围环境刺激的感知过于敏感或迟钝，且常对特定的兴趣爱好表现出异乎寻常的专注度。

以往的研究发现自闭症的发生可能与遗传因素及孕早期的病毒感染有密切关系，但也有证据表明，AQP4可能也参与了自闭症的发生[4]。首先，在自闭症患者的尸检结果中发现，其小脑组织中的AQP4表达明显降低[4]。同时，研究人员还发现自闭症小鼠在孕期受人流感病毒H1N1感染后，其子代小鼠从出生后0～35天，AQP4的表达明显降低[1, 5]，该结果与自闭症患者的尸检结果相符，提示病毒感染引起的AQP4表达下调，可能在自闭症的发生中发挥了重要作用。

此外，通过体外培养星形胶质细胞，并利用RNA干扰技术抑制其AQP4的表达后，细胞会出现皱缩及渗透性降低[6]，提示AQP4对维持星形胶质细胞的形态及其功能有重要作用。由此推测，当AQP4的表达下调时，星形胶质细胞的形态及离子稳态的变化，使其功能出现障碍，继而引起星形胶质细胞与神经元之间的联络出现异常[1, 4, 7]，进而导致自闭症的发生。

二、AQP4与抑郁症

抑郁症是一类以显著而持久的情绪低落为主要临床表现的精神疾病,可影响患者的学习、工作、睡眠和饮食,部分患者可能出现自杀倾向。

抑郁症的发生可能与下丘脑-垂体-肾上腺轴的功能障碍、神经退行性变、神经炎症及神经再生受到抑制等多种因素有关[8]。从尸检结果来看[9,10],抑郁症患者的前额叶皮质、扣带回及海马的星形胶质细胞数量明显减少,同时伴随细胞体积变小。此外,实验证明通过选择性损伤前额叶皮质的星形胶质细胞,或者选择性阻断海马星形胶质细胞的功能,可引起小鼠出现抑郁表现[11,12]。由此可见,星形胶质细胞的损伤及其功能障碍可能在抑郁症的发病中起到重要作用[13]。

AQP4主要表达于星形胶质细胞,并与其功能的调节密切相关。当AQP4的表达下调时,可导致星形胶质细胞的功能出现障碍,如水和离子的平衡紊乱[14-16],影响细胞的迁移、胶质瘢痕的形成以及神经信息的传递,还可改变突触的化学可塑性[14,17-19]及影响促炎性细胞因子的分泌和释放等[20]。因此,AQP4表达下调引起的星形胶质细胞的功能障碍,可能在抑郁症中发挥重要作用。

Kong等[13]使用$AQP4^{-/-}$小鼠,通过皮下注射皮质酮构建抑郁症小鼠模型,发现$AQP4$敲除会加重小鼠的抑郁症状,分析其可能的原因如下。①$AQP4$的敲除可引起海马星形胶质细胞的兴奋性氨基酸转运体2(excitatory amino acid transporter 2,EAAT2)的表达下调,EAAT2的主要功能是摄取突触间隙内的谷氨酸并将其转运至星形胶质细胞内[21],而文献证实谷氨酸摄取和代谢异常与抑郁症的发生有密切联系[22,23]。在阻断前额叶皮质EAAT2的表达后,大鼠出现抑郁的症状[24]。因此,AQP4可能通过下调EAAT2的表达,使突触间隙的谷氨酸难以清除,导致神经元间信息的传递出现障碍,从而引起抑郁的表现。②$AQP4$的敲除可引起星形胶质细胞的胶质细胞源性神经营养因子(glia cell line-derived neurotrophic factor,GDNF)分泌减少,而GDNF分泌减少可影响神经元突触的形成,继而使synapse-1蛋白表达降低。synapse-1是一种在成熟神经元中含量最高的突触前囊泡蛋白[25]。研究证实[26,27],海马synapse-1的表达下调与抑郁症密切相关。因此,AQP4可能通过减少GDNF的分泌,间接下调synapse-1的表达而加重抑郁的表现。③海马区神经发生受阻是引起抑郁症的主要发病机制之一,研究发现无论是在体内还是离体环境中[13],$AQP4$的敲除均可阻碍海马神经发生,但是其具体机制仍不清楚。

综上所述,$AQP4$的敲除可导致星形胶质细胞脆性增加,加重海马星形胶质细胞的功能障碍,并阻碍其神经发生,而这些因素正是抑郁症发病机制中的重要分子机制,提示AQP4在抑郁症发生发展中发挥了重要作用。因此,AQP4有望成为抑郁症治疗的新靶点。

三、AQP3、AQP4与精神分裂症

精神分裂症是一种复杂的慢性精神疾病,病程大多迁延且反复发作,临床表现为认知、思维、情感及社会功能等多方面的障碍[28]。

尸检结果显示[2]，精神分裂症患者脑组织中AQP4及其mRNA的表达水平明显降低，且AQP4的表达下降有明显的区域特异性。AQP4的降低仅局限于前扣带回深层皮质的星形胶质细胞，而在浅层皮质及白质的星形胶质细胞中，AQP4的表达与对照组之间并无明显差异。AQP4表达异常的这种区域特异性，说明其可能在精神分裂症的发病中发挥了重要作用，但是其潜在的机制并不清楚。

除AQP4与精神分裂症可能存在联系外，AQP3也可能与精神分裂症相关。研究发现基因拷贝数变异是精神分裂症的重要病因之一[29]，Kunisawa等[3]在对精神分裂症患者的基因拷贝数变异进行分析时，发现了罕见的AQP3重复序列，说明AQP3可能是精神分裂症的致病基因之一。实验中还发现AQP3的过表达可能会导致大脑皮质的运动神经元出现肿胀，并引起caspase-3依赖的细胞凋亡[3]，最终导致神经元死亡。而在精神分裂症患者中也观察到caspase-3的水平增高且同时伴随神经元丢失[30]。以上证据表明AQP3可能也参与了精神分裂症的发生。

脑结构和功能的改变是精神分裂症的发病基础，研究发现精神分裂症患者的第一和第二躯体运动区均出现了皮质容量的减少[31-33]，而该区域的神经元活动改变与精神分裂症的阴性症状有密切联系[34]，提示大脑运动皮质区神经元的结构和功能的异常，可能在精神分裂症的发病中发挥着重要作用。由此推测，AQP3的异常表达可能通过影响运动皮质区神经元的存活，导致该区的神经功能失调，这一途径在精神分裂症中发挥着作用。

虽然从现有的文献报道来看，AQP3及AQP4与精神疾病之间可能存在联系，但大多都是基于观察到的实验现象而做出的推论，如阐明水通道蛋白在精神疾病发病中的具体机制，还有待更深入的研究来证实。

（卓 飞）

参 考 文 献

[1] Fatemi SH，Folsom TD，Reutiman TJ，et al. Expression of astrocytic markers aquaporin 4 and connexin 43 is altered in brains of subjects with autism[J]. Synapse，2008，62（7）：501-507.

[2] Katsel P，Byne W，Roussos P，et al. Astrocyte and glutamate markers in the superficial，deep，and white matter layers of the anterior cingulate gyrus in schizophrenia[J]. Neuropsychopharmacology，2011，36（6）：1171-1177.

[3] Kunisawa K，Shimizu T，Kushima I，et al. Dysregulation of schizophrenia-related aquaporin 3 through disruption of paranode influences neuronal viability[J]. J Neurochem，2018，147（3）：395-408.

[4] Fatemi SH，Pearce DA，Brooks AI，et al. Prenatal viral infection in mouse causes differential expression of genes in brains of mouse progeny：a potential animal model for schizophrenia and autism[J]. Synapse，2005，57（2）：91-99.

[5] Fatemi SH，Folsom TD，Reutiman TJ，et al. Viral regulation of aquaporin 4，connexin 43，microcephalin and nucleolin[J].Schizophr Res，2008，98（1-3）：163-177.

[6] Nicchia GP，Srinivas M，Li W，et al. New possible roles for aquaporin-4 in astrocytes：cell cytoskeleton and functional relationship with connexin43[J]. FASEB J，2005，19（12）：1674-1676.

[7] Albertini R，Bianchi R. Aquaporins and glia[J]. Curr Neuropharmacol，2010，8（2）：84-91.

[8] Zunszain PA, Anacker C, Cattaneo A, et al. Glucocorticoids, cytokines and brain abnormalities in depression[J]. Prog Neuropsychopharmacol Biol Psychiatry, 2011, 35(3): 722-729.

[9] Rajkowska G, Stockmeier CA. Astrocyte pathology in major depressive disorder: insights from human postmortem brain tissue[J]. Curr Drug Targets, 2013, 14(11): 1225-1236.

[10] Gosselin RD, Gibney S, O'Malley D, et al. Region specific decrease in glial fibrillary acidic protein immunoreactivity in the brain of a rat model of depression[J]. Neuroscience, 2009, 159(2): 915-925.

[11] Banasr M, Duman RS. Glial loss in the prefrontal cortex is sufficient to induce depressive-like behaviors[J]. Biol Psychiatry, 2008, 64(10): 863-870.

[12] Iwata M, Shirayama Y, Ishida H, et al. Hippocampal astrocytes are necessary for antidepressant treatment of learned helplessness rats[J]. Hippocampus, 2011, 21(8): 877-884.

[13] Kong H, Zeng XN, Fan Y, et al. Aquaporin-4 knockout exacerbates corticosterone-induced depression by inhibiting astrocyte function and hippocampal neurogenesis[J]. CNS Neurosci Ther, 2014, 20(5): 391-402.

[14] Kong H, Fan Y, Xie J, et al. AQP4 knockout impairs proliferation, migration and neuronal differentiation of adult neural stem cells[J]. J Cell Sci, 2008, 121(24): 4029-4036.

[15] Thrane AS, Rappold PM, Fujita T, et al. Critical role of aquaporin-4 (AQP4) in astrocytic Ca^{2+} signaling events elicited by cerebral edema[J]. Proc Natl Acad Sci USA, 2011, 108(2): 846-851.

[16] Li X, Gao J, Ding J, et al. Aquaporin-4 expression contributes to decreases in brain water content during mouse postnatal development[J]. Brain Res Bull, 2013, 94: 49-55.

[17] Auguste KI, Jin S, Uchida K, et al. Greatly impaired migration of implanted aquaporin-4-deficient astroglial cells in mouse brain toward a site of injury[J]. FASEB J, 2007, 21(1): 108-116.

[18] Scharfman HE, Binder DK. Aquaporin-4 water channels and synaptic plasticity in the hippocampus[J]. Neurochem Int, 2013, 63(7): 702-711.

[19] Fan Y, Zhang J, Sun XL, et al. Sex- and region-specific alterations of basal amino acid and monoamine metabolism in the brain of aquaporin-4 knockout mice[J]. J Neurosci Res, 2005, 82(4): 458-464.

[20] Shi WZ, Zhao CZ, Zhao B, et al. Aggravated inflammation and increased expression of cysteinyl leukotriene receptors in the brain after focal cerebral ischemia in AQP4-deficient mice[J]. Neurosci Bull, 2012, 28(6): 680-692.

[21] Sanacora G, Banasr M. From pathophysiology to novel antidepressant drugs: glial contributions to the pathology and treatment of mood disorders[J]. Biol Psychiatry, 2013, 73(12): 1172-1179.

[22] Chandley MJ, Szebeni K, Szebeni A, et al. Gene expression deficits in pontine locus coeruleus astrocytes in men with major depressive disorder[J]. J Psychiatry Neurosci, 2013, 38(4): 276-284.

[23] Zink M, Vollmayr B, Gebicke-Haerter PJ, et al. Reduced expression of glutamate transporters vGluT1, EAAT2 and EAAT4 in learned helpless rats, an animal model of depression[J]. Neuropharmacology, 2010, 58(2): 465-473.

[24] John CS, Smith KL, Van't Veer A, et al. Blockade of astrocytic glutamate uptake in the prefrontal cortex induces anhedonia[J]. Neuropsychopharmacology, 2012, 37(11): 2467-2475.

[25] Dagyte G, Luiten PG, De Jager T, et al. Chronic stress and antidepressant agomelatine induce region-specific changes in synapsin I expression in the rat brain[J]. J Neurosci Res, 2011, 89(10): 1646-1657.

[26] Bessa JM, Ferreira D, Melo I, et al. The mood-improving actions of antidepressants do not depend on neurogenesis but are associated with neuronal remodeling[J]. Mol Psychiatry, 2009, 14(8): 764-773, 739.

[27] Silva R, Mesquita AR, Bessa J, et al. Lithium blocks stress-induced changes in depressive-like behavior and hippocampal cell fate: the role of glycogen-synthase-kinase-3beta[J]. Neuroscience, 2008, 152(3):

656-669.

[28] Wu YF, Sytwu HK, Lung FW. Human aquaporin 4 gene polymorphisms and haplotypes are associated with serum S100B level and negative symptoms of schizophrenia in a southern Chinese Han population[J]. Front Psychiatry, 2018, 9: 657.
[29] Crespi BJ, Crofts HJ. Association testing of copy number variants in schizophrenia and autism spectrum disorders[J]. J Neurodev Disord, 2012, 4(1): 15.
[30] Gassó P, Mas S, Molina O, et al. Increased susceptibility to apoptosis in cultured fibroblasts from antipsychotic-naïve first-episode schizophrenia patients[J]. J Psychiatr Res, 2014, 48(1): 94-101.
[31] Walther S. Psychomotor symptoms of schizophrenia map on the cerebral motor circuit[J]. Psychiatry Res, 2015, 233(3): 293-298.
[32] Douaud G, Smith S, Jenkinson M, et al. Anatomically related grey and white matter abnormalities in adolescent-onset schizophrenia[J]. Brain, 2007, 130(Pt 9): 2375-2386.
[33] Wang YM, Zou LQ, Xie WL, et al. Altered grey matter volume and cortical thickness in patients with schizo-obsessive comorbidity[J]. Psychiatry Res Neuroimaging, 2018, 276: 65-72.
[34] Walther S, Stegmayer K, Federspiel A, et al. Aberrant hyperconnectivity in the motor system at rest is linked to motor abnormalities in schizophrenia spectrum disorders[J]. Schizophr Bull, 2017, 43(5): 982-992.

第七节　水通道蛋白与脊髓疾病

一、AQP4与视神经脊髓炎

视神经脊髓炎谱系疾病（neuromyelitis optica spectrum disorders，NMOSD）是一种免疫介导的中枢神经系统脱髓鞘疾病，常累及视神经和脊髓[1]，侵及视神经通常会导致失明，而侵及脊髓则可诱发广泛的横贯性脊髓炎，进而导致腿部和手臂肌肉无力或瘫痪[2]，该疾病的反复发作可导致视神经和脊髓出现严重的累积损伤[3]。长期以来，视神经脊髓炎因其临床表现与多发性硬化相似，且均具有反复发作的特点，二者容易混淆。直到2004年，水通道蛋白4（AQP4）特异性IgG抗体的发现，为这两种疾病的鉴别提供了新思路。研究表明，AQP4主要分布于毛细血管、心室壁和软脑膜-神经胶质界面的星形胶质细胞，AQP4的特异性抗体（AQP4-IgG，也称为NMO-IgG）可结合至AQP4的外域，该抗体被认为是视神经脊髓炎（optical neuromyelitis，NMO）的生物标志物[4]，这一关键发现可用于鉴别NMO与多发性硬化和其他炎症性中枢神经系统脱髓鞘疾病。有研究表明，约80%的NMOSD患者在血清中可检测出AQP4-IgG[5]。

经典的NMO表型包括单侧或双侧视神经炎和横贯性脊髓炎，然而在一些典型病例中还存在其他临床表现（如极后区综合征或急性脑干综合征），因此专家定义了更广泛的NMO谱系疾病（NMOSD）[6]。2015年，Wingerchuk等发表了关于NMOSD诊断标准的国际共识（表2-2）。据此，NMOSD可分为经典NMO[即视神经炎（optic neuritis，ON）合并纵向广泛横贯性脊髓炎（longitudinal extensive transverse myelitis，LETM）]，孤立性ON或LETM，ON和（或）LETM伴有自身免疫性系统疾病，ON和LETM伴有脑干、间脑或脑受累症状，以及亚洲眼脊髓型多发性硬化[7,8]。

表 2-2 NMOSD 诊断标准

核心临床症状	■视神经炎 ■急性脊髓炎 ■极后区综合征—不明原因的打嗝、恶心或呕吐 ■急性脑干综合征（动眼神经障碍、眼球综合征、呼吸衰竭） ■症状性发作性睡病或急性间脑综合征（冷漠或躁动、嗜睡、肥胖、自主神经功能障碍）伴 NMOSD 典型磁共振成像变化 ■症状性大脑综合征（意识模糊、癫痫发作）伴 NMOSD 典型脑损伤
水通道蛋白 4 抗体阳性的 NMOSD	■至少一种核心症状 ■水通道蛋白 4 抗体 IgG 试验阳性 ■排除其他诊断
水通道蛋白 4 抗体 IgG 试验阴性或未标记的 NMOSD	■至少 2 种核心临床症状为以下 1 种或多种临床发作： —至少 1 种核心临床症状必须是视神经炎伴 LETM 或极后区综合征 —空间播散（2 种或多种核心临床症状） —满足其他磁共振成像标准 ■水通道蛋白 4 抗体 IgG 试验阴性或未测试 ■排除任何其他诊断
无水通道蛋白 4 的 NMOSD 磁共振成像标准	■急性视神经炎： —大脑白质无变化或无非特异性变化 —视神经伴 T_2 高信号强度病变或者 T_1 加权钆增强损伤超过视神经长度一半或者累及视交叉 ■急性脊髓炎： —髓内磁共振成像显示病变累及 ≥ 3 个连续段 —急性脊髓炎的患者局灶性脊髓萎缩 ≥ 3 个相邻节段 ■极后区综合征 ■急性脑干综合征

AQP4 广泛分布于中枢神经系统的星形胶质细胞的足突，同时也高表达于视神经和脊髓，而 NMO-IgG 可与 AQP4 特异性结合，这也解释了 NMO 的病灶主要位于视神经及脊髓的缘由。当 AQP4 抗体经血脑屏障进入中枢神经系统，与星形胶质细胞足突上的 AQP4 结合后，将导致细胞依赖的毒性反应，而星形胶质细胞足突被 NMO-IgG 和补体降解，则将激活巨噬细胞、嗜酸性粒细胞及中性粒细胞产生细胞因子、氧自由基等，从而造成血管和脑实质的损伤，并最终导致包括轴索和少突胶质细胞在内的白质和灰质的损伤。然而，AQP4 的自身免疫攻击是如何启动的，机制尚不清楚，可能与 T 细胞对 AQP4 的不正确识别以及早期 B 细胞耐受性检查点的受损有关。在这一过程中，T 细胞和 B 细胞间复杂的相互作用，例如相关抗原的呈递、Th17 和 Treg 细胞之间的动态平衡（后者下调）、促炎性细胞因子的分泌以及 B 细胞扩增和增殖的刺激，都可能参与了 NMO 的发病。

二、水通道蛋白与脊髓损伤

脊髓损伤（spinal cord injury，SCI）是脊柱外科常见的危重症之一。随着现代交通和工矿行业的发展,其患病率呈逐年上升趋势。调查显示,发达国家每百万人脊髓损伤的发病率为13.1～163.4,而在非发达国家,发病率为每百万人13.0～220.0[9]。根据世界卫生组织的数据,脊髓损伤患者的抑郁症发病率较普通人群高,且其过早死亡的可能性是普通人群的55倍[10]。脊髓损伤不仅给患者带来身体上的痛苦,其高昂的治疗费用和个人护理费用等[11]还给患者的家庭乃至社会带来了巨大的经济负担。

现代医学研究将脊髓损伤分为原发性脊髓损伤和继发性脊髓损伤两种类型。原发性脊髓损伤主要指由机械、化学及生物等因素对脊髓造成的直接且不可逆的损伤,而继发性脊髓损伤是指在原发性损伤后几小时至几天内,继发于原发性损伤的组织缺血缺氧、水肿、炎症反应、神经细胞凋亡、自由基生成等一系列自身破坏性病变而导致的残余正常神经组织发生的慢性损伤。

(一)脊髓损伤与脊髓水肿

水肿是脊髓损伤进展中的一个重要临床特征[12]。脊髓水肿是在受损区域周围实质中出现的显著的水分子聚集,在继发性损伤的急性期可观察到脊髓水肿的发生[13]。水肿在原发性损伤后几分钟内即可发生,并在48小时由内向外发展为一个更大的液体填充腔,并可能持续至损伤后14天[14]。虽然促进水肿形成的确切分子机制尚不清楚,但可能与神经元肿胀和星形胶质细胞终足膜肿胀(即细胞毒性水肿)以及血-脊髓屏障破坏和渗漏(即血管源性水肿)有关。一方面水肿通过增加脊髓的鞘内压力,进一步加重原发性损伤。另一方面,水肿也可进一步造成血流减少,血-脊髓屏障破坏、出血等,从而加剧细胞死亡[15]。水肿是影响脊髓损伤病理进展的一个重要因素,水肿的严重程度(包括水肿区的大小和位置)与临床预后密切相关,尤其与运动功能的恢复关系密切[13, 16]。综上,脊髓损伤后水肿的形成与水转运失衡有关,提示参与水分子转运的蛋白可能参与了脊髓损伤后水肿形成的病理过程。

(二)水通道蛋白家族与脊髓水肿

水通道蛋白(AQP)是一类广泛存在于微生物、动植物以及人体组织内的膜通道蛋白,可快速介导水分子的跨膜转运[17-19]。目前在哺乳动物体内共发现了13种AQP,即AQP0～AQP12[20-24],其中与脊髓水肿相关的水通道蛋白包括AQP1、AQP4、AQP9。

1. AQP1 目前的研究显示,AQP1可能参与了脊髓损伤后不同的病理过程。在多种神经疾病所致的脊髓损伤中,如阿尔茨海默病[25]、克罗伊茨费尔特-雅各布病(Creutzfeldt-Jakob disease)[26]、脑外伤[27]和脊髓损伤[28]等,均观察到AQP1的表达水平显著增加。AQP1主要定位于正常脊髓组织的神经纤维和室管膜细胞[29-31]。而在脊髓损伤后,不仅分布于神经纤维和室管膜细胞上的AQP1的表达明显上调,而且星形胶质细胞胞体上表达的AQP1也明显增加,这种高表达甚至可持续至损伤后11个月[28]。多种类型细胞的AQP1表达持续性增高,提示AQP1可能参与了脊髓损伤后的病理过程。此外,氧化应激会显著上调AQP1蛋白的表达,而抗氧化剂褪黑素则会抑制AQP1的表达[28]。在缺氧状态下,内皮细胞上表达的AQP1在血管生成和细胞迁移中发挥了重要作用[32, 33],此外,由于AQP1与轴突生长标志物GAP43共定位,推测AQP1在脊髓星形胶质细胞或轴突延伸的水摄取中发挥了类似的作用[28]。脊髓损伤后，AQP1表达的持续上调可能会加重神经元或神经轴突的肿胀[34]，并导致脑室脉络膜上皮细胞过度产生脑脊液[28, 35]。在脊髓损伤部位周围的反应性星形胶质细胞上，也发现了AQP1的表达上调，提示AQP1可能促进了星形胶质细胞向病变部位的迁移。

2. AQP4　AQP4在脊髓灰质表层、后角白质以及中央管周围的星形胶质细胞终足膜和有髓神经纤维上表达[36, 37]。脊髓损伤后局部组织常有明显的水肿，推测脊髓组织中的AQP4可能参与了水肿的形成。研究发现[38, 39]，脊髓损伤后1天AQP4的表达明显增加，在损伤后3天达到高峰，此时水肿表现为细胞内水肿，提示AQP4表达上调可能促进了脊髓水肿的发生。但也有学者的研究得出了截然相反的结论。研究发现，在*AQP4*基因敲除小鼠脊髓半横断损伤动物模型中，损伤后3天，其损伤侧脊髓的含水量明显高于未损伤侧脊髓[40]，同时，*AQP4*基因敲除小鼠的运动功能恢复情况也明显差于野生型小鼠[41]。以上结果提示，在脊髓损伤后的急性发病期，*AQP4*的敲除也会增加水肿的发生。Sun等在研究实验性脊髓空洞症发生发展过程中发现[42]，AQP4蛋白及其mRNA表达与脊髓组织中的水含量呈负相关，并推测AQP4表达下调是脊髓水肿形成的主要机制。此外，AQP4还可能参与了脊髓损伤后局部组织的慢性炎症反应。上述研究提示，关于AQP4在脊髓损伤中的作用还存在争议，其差异是否与不同实验的建模方式或缺血时间长度不同有关，尚有待进行深入研究。

3. AQP9　AQP9主要在脊髓白质与胶质界膜的星形胶质细胞终足膜上表达，可促进水、甘油、单羧酸盐和尿素的流动[23, 37, 43, 44]。目前对脊髓损伤后AQP9表达变化的研究较少。李建军等的研究发现，脊髓损伤大鼠行脊髓切开术后，可降低脊髓的水肿程度，抑制AQP4和AQP9的表达上调，则可促进神经功能的恢复。但在该研究中，脊髓损伤后，AQP4与AQP9的表达均上调，且AQP4在脊髓损伤后表达水平变化更加明显，故未能有效梳理出AQP9在脊髓损伤后水肿形成中的实际作用，尚需要进一步研究加以证实，如通过特定的AQP9抑制剂来分析AQP9对脊髓损伤后水肿的作用。

表2-3将脊髓内表达的3种AQP的定位和功能与脑内表达的AQP进行了对比。

表2-3　脊髓和脑内的AQP1、AQP4、AQP9的表达与功能对比

	脊髓		脑		参考文献
	定位	功能	定位	功能	
AQP1	脊神经背根的无髓鞘感觉神经纤维，脊髓灰质后角和白质	疼痛处理	脉络丛上皮细胞	脑脊液产生	[23, 26, 29-31, 36, 37, 45-51]
	神经胶质界膜的内皮细胞	未知	白质血管周围星形胶质细胞和胶质界膜	细胞迁徙和水平衡	
	胶质界膜、后角、中央管和白质的星形胶质细胞	细胞迁徙和水平衡	软脑膜周围的神经元	轴突延长	
	胶质界膜和中央管的室管膜细胞	脑脊液产生			

续表

	脊髓		脑		参考文献
	定位	功能	定位	功能	
AQP4	灰质和白质毛细血管周围星形胶质细胞终足膜	水平衡，离子平衡	白质毛细血管周围星形胶质细胞终足膜	水平衡，神经元的兴奋性调节	[23, 36, 37, 52-69]
	包裹有髓鞘的轴突和轴突突触的星形胶质细胞终足	突触周围体积的调节			
	纤维性星形细胞	未知	突触周围星形胶质细胞终足膜	突触可塑性、K^+平衡、突触功能	
	胶质界膜和中央管周围的星形胶质细胞终足膜	水平衡			
	胶质界膜内的室管膜细胞	脑脊液产生、水平衡	软脑膜下、室管膜下星形胶质细胞终足膜	水转运	
			视网膜Müller细胞	K^+清除	
AQP9	白质与胶质界膜的星形胶质细胞终足膜	水转运	儿茶酚胺能神经元	能量代谢	[23, 37, 70, 71]
			胶质界膜的星形胶质细胞	水转运	

本部分主要就脊髓损伤后AQP的表达及其可能的功能做了简要介绍，从现有的研究成果来看，AQP1、AQP4和AQP9在脊髓的水分子调节中均发挥着重要作用，但在不同细胞类型中表达的AQP的分布和功能可能存在显著差异，且在不同时间节点，同一AQP可能发挥不同的作用。例如，AQP4可能是调节脊髓水平衡最重要的水分子，但是基因敲除小鼠的研究表明，在急性期敲除*AQP4*基因可显著减轻水肿；但就慢性期功能的恢复而言，此时需要AQP4来清除水肿，那么*AQP4*基因敲除则是有害的。关于AQP的调控，研究发现缺氧和高渗刺激可调节不同的AQP，提示AQP功能的调控存在多种重要方式，但对具体机制的研究还不够深入。由此可见，还需要进一步研究AQP在脊髓损伤后的表达及调控机制，以开发出减轻脊髓损伤后水肿的治疗药物等。

（陈玉琴　杨　美）

参 考 文 献

[1] Jarius S, Wildemann B. AQP4 antibodies in neuromyelitis optica: diagnostic and pathogenetic relevance[J]. Nat Rev Neurol, 2010, 6(7): 383-392.

[2] Numata Y, Uematsu M, Suzuki S, et al. Aquaporin-4 autoimmunity in a child without optic neuritis and myelitis[J]. Brain Dev, 2015, 37(1): 149-152.

[3] Pittock SJ, Berthele A, Fujihara K, et al. Eculizumab in aquaporin-4-positive neuromyelitis optica spectrum disorder[J]. N Engl J Med, 2019, 381(7): 614-625.

[4] Lennon VA, Wingerchuk DM, Kryzer TJ, et al. A serum autoantibody marker of neuromyelitis optica: distinction from multiple sclerosis[J]. Lancet, 2004, 364(9451): 2106-2112.

[5] Alves Do Rego C, Collongues N. Neuromyelitis optica spectrum disorders: features of aquaporin-4, myelin oligodendrocyte glycoprotein and double-seronegative-mediated subtypes[J]. Rev Neurol(Paris), 2018,

174（6）：458-470.

[6] Collongues N, Ayme-Dietrich E, Monassier L, et al. Pharmacotherapy for neuromyelitis optica spectrum disorders: current management and future options[J]. Drugs, 2019, 79（2）: 125-142.

[7] Wingerchuk DM, Lennon VA, Lucchinetti CF, et al. The spectrum of neuromyelitis optica[J]. Lancet Neurol, 2007, 6（9）: 805-815.

[8] Wingerchuk DM, Banwell B, Bennett JL, et al. International consensus diagnostic criteria for neuromyelitis optica spectrum disorders[J]. Neurology, 2015, 85（2）: 177-189.

[9] Kang Y, Ding H, Zhou H, et al. Epidemiology of worldwide spinal cord injury: a literature review[J]. J Neurorestoratol, 2017, 6: 1-9.

[10] Dryden DM, Saunders L, Jacobs P, et al. Direct health care costs after traumatic spinal cord injury[J]. J Trauma, 2005, 59（2）: 441-447.

[11] French DD, Campbell R, Sabharwal S, et al. Health care costs for patients with chronic spinal cord injury in the Veterans Health Administration[J]. J Spinal Cord Med, 2007, 30（5）: 477-481.

[12] Bozzo A, Marcoux J, Radhakrishna M, et al. The role of magnetic resonance imaging in the management of acute spinal cord injury[J]. J Neurotrauma, 2011, 28（8）: 1401-1411.

[13] Miyanji F, Furlan JC, Aarabi B, et al. Acute cervical traumatic spinal cord injury: MR imaging findings correlated with neurologic outcome: prospective study with 100 consecutive patients[J]. Radiology, 2007, 243（3）: 820-827.

[14] Tator CH, Fehlings MG. Review of the secondary injury theory of acute spinal cord trauma with emphasis on vascular mechanisms[J]. J Neurosurg, 1991, 75（1）: 15-26.

[15] Leonard AV, Vink R. Reducing intrathecal pressure after traumatic spinal cord injury: a potential clinical target to promote tissue survival[J]. Neural Regen Res, 2015, 10（3）: 380-382.

[16] Flanders AE, Spettell CM, Friedman DP, et al. The relationship between the functional abilities of patients with cervical spinal cord injury and the severity of damage revealed by MR imaging[J]. AJNR Am J Neuroradiol, 1999, 20（5）: 926-934.

[17] Day RE, Kitchen P, Owen DS, et al. Human aquaporins: regulators of transcellular water flow[J]. Biochim Biophys Acta, 2014, 1840（5）: 1492-1506.

[18] Verkman AS, Mitra AK. Structure and function of aquaporin water channels[J]. Am J Physiol Renal Physiol, 2000, 278（1）: F13-F28.

[19] Hub JS, Grubmüller H, de Groot B. Dynamics and energetics of permeation through aquaporins. What do we learn from molecular dynamics simulations?[J]. Handb Exp Pharmacol, 2009, （190）: 57-76.

[20] Magni F, Sarto C, Ticozzi D, et al. Proteomic knowledge of human aquaporins[J]. Proteomics, 2006, 6（20）: 5637-5649.

[21] Itoh T, Rai T, Kuwahara M, et al. Identification of a novel aquaporin, AQP12, expressed in pancreatic acinar cells[J]. Biochem Biophys Res Commun, 2005, 330（3）: 832-838.

[22] Gorelick DA, Praetorius J, Tsunenari T, et al. Aquaporin-11: a channel protein lacking apparent transport function expressed in brain[J]. BMC Biochemi, 2006, 7: 14.

[23] Arciénega I, Brunet JF, Bloch J, et al. Cell locations for AQP1, AQP4 and 9 in the non-human primate brain[J]. Neuroscience, 2010, 167（4）: 1103-1114.

[24] Liu K, Nagase H, Huang CG, et al. Purification and functional characterization of aquaporin-8[J]. Biol Cell, 2006, 98（3）: 153-161.

[25] Pérez E, Barrachina M, Rodríguez A, et al. Aquaporin expression in the cerebral cortex is increased at early stages of Alzheimer disease[J]. Brain Res, 2007, 1128（1）: 164-174.

[26] Rodríguez A, Pérez-Gracia E, Espinosa JC, et al. Increased expression of water channel aquaporin 1 and

aquaporin 4 in Creutzfeldt-Jakob disease and in bovine spongiform encephalopathy-infected bovine-PrP transgenic mice[J]. Acta Neuropathol, 2006, 112(5): 573-585.

[27] Suzuki R, Okuda M, Asai J, et al. Astrocytes co-express aquaporin-1, -4, and vascular endothelial growth factor in brain edema tissue associated with brain contusion[J]. Acta Neurochir Suppl, 2006, 96: 398-401.

[28] Nesic O, Lee J, Unabia GC, et al. Aquaporin 1 - a novel player in spinal cord injury[J]. J Neurochem, 2008, 105(3): 628-640.

[29] Praetorius J, Nielsen S. Distribution of sodium transporters and aquaporin-1 in the human choroid plexus[J]. Am J Physiol Cell Physiol, 2006, 291(1): C59-C67.

[30] Shields SD, Mazario J, Skinner K, et al. Anatomical and functional analysis of aquaporin 1, a water channel in primary afferent neurons[J]. Pain, 2007, 131(1-2): 8-20.

[31] Satoh J, Tabunoki H, Yamamura T, et al. Human astrocytes express aquaporin-1 and aquaporin-4 in vitro and in vivo[J]. Neuropathology, 2007, 27(3): 245-256.

[32] Echevarría M, Muñoz-Cabello AM, Sánchez-Silva R, et al. Development of cytosolic hypoxia and hypoxia-inducible factor stabilization are facilitated by aquaporin-1 expression[J]. J Biol Chem, 2007, 282(41): 30207-30215.

[33] Verkman AS. Aquaporins in endothelia[J]. Kidney Int, 2006, 69(7): 1120-1123.

[34] Nashmi R, Fehlings MG. Changes in axonal physiology and morphology after chronic compressive injury of the rat thoracic spinal cord[J]. Neuroscience, 2001, 104(1): 235-251.

[35] Nesic O, Lee J, Ye Z, et al. Acute and chronic changes in aquaporin 4 expression after spinal cord injury[J]. Neuroscience, 2006, 143(3): 779-792.

[36] Oklinski MK, Lim JS, Choi HJ, et al. Immunolocalization of water channel proteins AQP1 and AQP4 in rat spinal cord[J]. J Histochem Cytochem, 2014, 62(8): 598-611.

[37] Oshio K, Binder DK, Yang B, et al. Expression of aquaporin water channels in mouse spinal cord[J]. Neuroscience, 2004, 127(3): 685-693.

[38] 陈东风, 黎晖, 周健洪, 等. 脊髓损伤后水通道蛋白-4的时程变化[J]. 神经解剖学杂志, 2004, 20(6): 568-572.

[39] 王伟, 谢杰, 方坚, 等. 水通道蛋白-4在实验性脊髓损伤大鼠的表达[J]. 神经解剖学杂志, 2005, 21(2): 139-142.

[40] Wu Q, Zhang YJ, Gao JY, et al. Aquaporin-4 mitigates retrograde degeneration of rubrospinal neurons by facilitating edema clearance and glial scar formation after spinal cord injury in mice[J]. Mol Neurobiol, 2014, 49(3): 1327-1337.

[41] Kimura A, Hsu M, Seldin M, et al. Protective role of aquaporin-4 water channels after contusion spinal cord injury[J]. Ann Neurol, 2010, 67(6): 794-801.

[42] 孙国柱, 张庆俊, 王浩. 兔实验性脊髓空洞前状态水通道蛋白4表达的变化[J]. 北京大学学报(医学版), 2007, 39(2): 177-181.

[43] Badaut J, Regli L. Distribution and possible roles of aquaporin 9 in the brain[J]. Neuroscience, 2004, 129(4): 971-981.

[44] Buffoli B. Aquaporin biology and nervous system[J]. Curr Neuropharmacol, 2010, 8(2): 97-104.

[45] Oshio K, Watanabe H, Song Y, et al. Reduced cerebrospinal fluid production and intracranial pressure in mice lacking choroid plexus water channel Aquaporin-1[J]. FASEB J, 2005, 19(1): 76-78.

[46] Bering EA Jr. Choroid plexus and arterial pulsation of cerebrospinal fluid demonstration of the choroid plexuses as a cerebrospinal fluid pump[J]. AMA Arch Neurol Psychiatry, 1955, 73(2): 165-172.

[47] Boassa D, Stamer WD, Yool AJ. Ion channel function of aquaporin-1 natively expressed in choroid

plexus[J]. J Neurosci, 2006, 26(30): 7811-7819.
[48] Oshio K, Watanabe H, Yan D, et al. Impaired pain sensation in mice lacking Aquaporin-1 water channels[J]. Biochem Biophys Res Commun, 2006, 341(4): 1022-1028.
[49] Badaut J, Brunet JF, Grollimund L, et al. Aquaporin 1 and aquaporin 4 expression in human brain after subarachnoid hemorrhage and in peritumoral tissue[J]. Acta Neurochir Suppl, 2003, 86: 495-498.
[50] Saadoun S, Papadopoulos MC, Davies DC, et al. Increased aquaporin 1 water channel expression in human brain tumours[J]. Br J Cancer, 2002, 87(6): 621-623.
[51] McCoy E, Sontheimer H. MAPK induces AQP1 expression in astrocytes following injury[J]. Glia, 2010, 58(2): 209-217.
[52] Hubbard JA, Hsu MS, Seldin MM, et al. Expression of the astrocyte water channel aquaporin-4 in the mouse brain[J]. ASN Neuro, 2015, 7(5): 1759091415605486.
[53] Nielsen S, Nagelhus EA, Amiry-Moghaddam M, et al. Specialized membrane domains for water transport in glial cells: high-resolution immunogold cytochemistry of aquaporin-4 in rat brain[J]. J Neurosci, 1997, 17(1): 171-180.
[54] Rash JE, Yasumura T, Hudson CS, et al. Direct immunogold labeling of aquaporin-4 in square arrays of astrocyte and ependymocyte plasma membranes in rat brain and spinal cord[J]. Proc Natl Acad Sci USA, 1998, 95(20): 11981-11986.
[55] Mathiisen TM, Lehre KP, Danbolt NC, et al. The perivascular astroglial sheath provides a complete covering of the brain microvessels: an electron microscopic 3D reconstruction[J]. Glia, 2010, 58(9): 1094-1103.
[56] Vitellaro-Zuccarello L, Mazzetti S, Bosisio P, et al. Distribution of aquaporin 4 in rodent spinal cord: relationship with astrocyte markers and chondroitin sulfate proteoglycans[J]. Glia, 2005, 51(2): 148-159.
[57] Landis DM, Reese TS. Arrays of particles in freeze-fractured astrocytic membranes[J]. J Cell Biol, 1974, 60(1): 316-320.
[58] Verbavatz JM, Ma TH, Gobin R, et al. Absence of orthogonal arrays in kidney, brain and muscle from transgenic knockout mice lacking water channel aquaporin-4[J]. J Cell Sci, 1997, 110(Pt 22): 2855-2860.
[59] Smith AJ, Jin BJ, Ratelade J, et al. Aggregation state determines the localization and function of M1- and M23-aquaporin-4 in astrocytes[J]. J Cell Biol, 2014, 204(4): 559-573.
[60] Skucas VA, Mathews IB, Yang J, et al. Impairment of select forms of spatial memory and neurotrophin-dependent synaptic plasticity by deletion of glial aquaporin-4[J]. J Neurosci, 2011, 31(17): 6392-6397.
[61] Manley GT, Fujimura M, Ma T, et al. Aquaporin-4 deletion in mice reduces brain edema after acute water intoxication and ischemic stroke[J]. Nat Med, 2000, 6(2): 159-163.
[62] Haj-Yasein NN, Vindedal GF, Eilert-Olsen M, et al. Glial-conditional deletion of aquaporin-4(Aqp4) reduces blood-brain water uptake and confers barrier function on perivascular astrocyte endfeet[J]. Proc Natl Acad Sci USA, 2011, 108(43): 17815-17820.
[63] Benfenati V, Caprini M, Dovizio M, et al. An aquaporin-4/transient receptor potential vanilloid 4(AQP4/TRPV4)complex is essential for cell-volume control in astrocytes[J]. Proc Natl Acad Sci USA, 2011, 108(6): 2563-2568.
[64] Dietzel I, Heinemann U, Hofmeier G, et al. Transient changes in the size of the extracellular space in the sensorimotor cortex of cats in relation to stimulus-induced changes in potassium concentration[J]. Exp Brain Res, 1980, 40(4): 432-439.
[65] Li J, Patil RV, Verkman AS. Mildly abnormal retinal function in transgenic mice without Müller cell aquaporin-4 water channels[J]. Invest Ophthalmol Vis, 2002, 43(2): 573-579.
[66] Smith AJ, Verkman AS. Superresolution imaging of aquaporin-4 cluster size in antibody-stained paraffin

brain sections[J]. Biophys J, 2015, 109(12): 2511-2522.
[67] Feng X, Papadopoulos MC, Liu J, et al. Sporadic obstructive hydrocephalus in Aqp4 null mice[J]. J Neurosci Res, 2009, 87(5): 1150-1155.
[68] Yan X, Liu J, Wang X, et al. Pretreatment with AQP4 and NKCC1 inhibitors concurrently attenuated spinal cord edema and tissue damage after spinal cord injury in rats[J]. Front Physiol, 2018, 9: 6.
[69] Papadopoulos MC, Manley GT, Krishna S, et al. Aquaporin-4 facilitates reabsorption of excess fluid in vasogenic brain edema[J]. FASEB J, 2004, 18(11): 1291-1293.
[70] Badaut J, Petit JM, Brunet JF, et al. Distribution of aquaporin 9 in the adult rat brain: preferential expression in catecholaminergic neurons and in glial cells[J]. Neuroscience, 2004, 128(1): 27-38.
[71] Mylonakou MN, Petersen PH, Rinvik E, et al. Analysis of mice with targeted deletion of AQP9 gene provides conclusive evidence for expression of AQP9 in neurons[J]. J Neurosci Res, 2009, 87(6): 1310-1322.

第八节 水通道蛋白与脑影像

一、脑水肿成像技术研究进展

脑水肿是继发于多种疾病的严重并发症之一，如感染、外伤、脑血管疾病、脑肿瘤等，严重者可危及生命，是影响患者预后的重要原因。因此，要重视脑水肿的早期诊断和合理治疗。

脑水肿并非独立的疾病，常见原发症状易被忽视，脑水肿的发病因素比较多，而且其临床分类也比较多。脑水肿根据病理形态及发病机制可分为四类：血管源性脑水肿、细胞毒性脑水肿、间质性脑水肿和渗透性脑水肿。临床中脑水肿主要靠CT和MRI诊断，CT应用局限，显示病变范围小，脑水肿早期易漏诊。MRI无电离辐射，具有极高的软组织对比度，能够提供组织细胞功能和分子水平的信息，已经成为非常重要的临床前期研究和临床应用的医学影像技术。近年来，随着MRI成像技术的快速发展，影像医学能否早期快捷诊断脑水肿成为当前研究的新热点。同时，人们对脑水肿发生机制的深入研究，也为脑水肿的合理治疗提供了有价值的参考依据。

弥散加权成像（diffusion weighted imaging，DWI）和弥散张量成像（diffusion tensor imaging，DTI）是仅有的检测活体组织内水分子扩散运动的无创技术，其表观弥散系数（apparent diffusion coefficient，ADC）值能对水分子的综合微观运动进行定量测定，ADC值越大，组织内水分子运动越强，DWI信号越弱。b值为扩散梯度因子，表示应用的扩散梯度磁场的时间、幅度及形状。b值越高扩散权重越大，对水分子的扩散运动越敏感，信号降低越明显，DWI上组织信号衰减越明显则提示其水分子在梯度场方向上扩散越自由。DTI是在DWI基础上发展起来的一种新的磁共振成像技术，可在三维空间内分析组织内水分子的弥散特性。DTI主要评价参数包括平均弥散系数（average diffusion coefficient，DCavg）和各向异性分数（fractional anisotropy，FA），前者比ADC能更全面地反映弥散运动的快慢，后者为代表水分子弥散运动各向异性大小的参数，既可以对弥散运动进行量化又可以描述弥散方向。也有研究利用MRI特殊序列对兴趣区脑含水量进行定量测定，包

括 T_1 和 T_2^* 定量测定。临床实际工作中，DWI 应用最为成熟和广泛[1]。

1986年LeBihan首次在活体组织进行磁共振弥散加权成像（DWI）。1990年Moseley等将其应用于早期脑缺血诊断。此后，在1996年推出了主要用于神经系统检查的弥散张量成像（DTI）。DWI最早用于脑部成像，因脑受运动的影响比较小，而且水分子含量高，易获得脑DWI图像，现已成为辅助疾病诊断和疗效监测的重要成像方式。在脑缺血的研究中，DWI能敏感地检测出脑缺血区0～4小时的病理生理变化。缺血性脑组织的DWI特征为高信号强度区，代表细胞毒性脑水肿。这是因为缺血早期，水分子从细胞外间隙进入细胞内，其扩散范围受到细胞膜的限制，不易被排出而蓄积于细胞内；随着病程进展，有大量水分子蓄积且扩散不受限，兴趣区表现为低信号，则显示为血管源性脑水肿。应用DWI成像，可以获得DWI影像和ADC图，ADC用来测量水分子扩散的幅度，所以其检测脑水肿更为精确（血管源性脑水肿主要表现为ADC值增高，细胞毒性脑水肿则表现为ADC值降低）。因此建议两者联合应用能体现DWI在鉴别细胞毒性脑水肿与血管源性脑水肿中的诊断价值。最近几年，随着MRI扫描速度和图像质量的提高，体素内不相干运动（intravoxel incoherent motion，IVIM）双指数模型多b值扫描技术得到应用。b值不同，获得的组织微观信息不同，高b值DWI可以提供血管源性脑水肿的详细特征，ADC值可定量鉴别脑水肿类型，也可对抗水肿治疗的效果进行预测和随访[1]。

传统的磁共振DWI成像理论的基础是组织细胞间水分子的布朗运动，基于水分子随机被动扩散。Agre等发现AQP彻底改变了传统的水在细胞膜扩散（被动转运）的观念，创立了水在细胞膜主动转运的全新理论基础。目前认为，水分子的跨膜转运有两种基本方式，即穿越膜脂质双分子层的简单扩散和AQP介导的水转运。AQP基础上水分子成像包含水分子被动和主动转运过程，已经不是传统观念上的单纯水扩散成像[2]。

AQP磁共振成像（MRI）技术（AQP-MRI）是近年新兴的建立在AQP理论基础上的全新的MRI分子成像技术，可以获得细胞膜上AQP信息。然而目前AQP-MRI对临床的价值尚不明确。该技术处于探索阶段，未得到公认和推广。该技术的提出使在活体上研究AQP及其相对定量成为可能[3]。

二、AQP4 与弥散加权成像

研究已证实AQP4是一种转运细胞内外水分子的蛋白，对调节细胞内外水代谢和维持水平衡起着重要作用。AQP4与脑水肿的形成密切相关。多数认为AQP4在细胞毒性脑水肿和血管源性脑水肿病理生理中起着重要的调节作用：在细胞毒性脑水肿早期，AQP4表达上调，促进水肿的形成，与细胞毒性脑水肿程度呈正相关，是细胞内水肿的原因。也有研究者得出相反的结果，认为AQP4表达下降是细胞毒性水肿的主要原因。而在血管源性脑水肿中AQP4能加速水肿液的消除[1,4,5]。

何占平等[6]在急性脑缺血实验中发现AQP4表达水平、ADC和DWI随缺血时间延长呈动态变化过程。实验大鼠大脑中动脉栓塞后15分钟即可在DWI上出现高信号强度；在1小时内ADC值迅速减小，AQP4表达、DWI高信号强度和高信号面积都快速增加；ADC值于3小时达到最低，6小时开始回升，于24小时接近正常；DWI高信号强度和高信号面积在1～6小时期间的上升幅度减慢，DWI高信号强度外，其余各指标在6小时后又出现一增长高峰。AQP4表达与DWI高信号强度和面积都和时间有明显线性关系，在急性脑缺血的整个过程中，AQP4表达呈上升趋势，在15分钟至1小时内AQP4表达增加最快，AQP4表达呈直线上升，1～6小时增加不明显，出现"平台期"，6小时后直至第3天又呈缓慢上升趋势。牛彩虹等[7]的大鼠脑缺血再灌注损伤研究显示：脑缺血再灌注12小时至1天时兴趣区DWI信号强度持续升高（图2-8A和B），ADC值降低，3天后DWI逐渐降低，仍高于对侧，3天时rADC值开始升高（图2-8C）。12小时至3天AQP4的表达程度与rADC值呈正相关（图2-9），与缺血侧脑水肿体积呈正相关。这提示AQP4在急性期（12小时至1天）可能参与了缺血再灌注后脑水肿的形成，而亚急性期（1～3天）可能与防止脑水肿进一步加重、清除脑内过多的水分有关。需要注意的是，缺血早期表现主要以细胞毒性水肿为主，细胞内外水分子弥散受限，缺血时间持续增加，缺血区血脑屏障破坏，毛细血管通透性增加，细胞毒性水肿开始向血管源性水肿过渡，随后由于血管源性水肿的加重和细胞坏死崩解，细胞内水分子弥散受限逐渐减轻，所以rADC值开始升高。DWI信号强度在12小时至1天时持续升高，3天后逐渐降低，仍高于对侧。由于DWI的信号强度除了受水分子弥散程度的影响外，还受组织的T_2弛豫时间影响，这种影响称为T_2透过效应。所以，DWI高信号强度时应该查看ADC图，ADC值下降才真正反映水分子弥散受限。Wang等[8]在缺血再灌注脑损伤实验中也证实AQP4的表达与ADC值呈正相关。

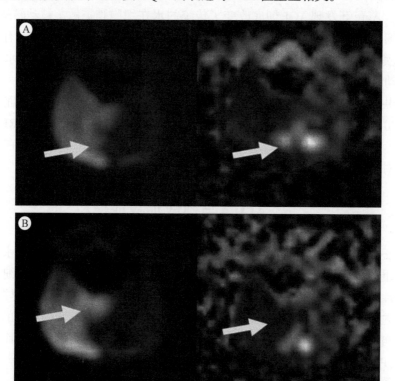

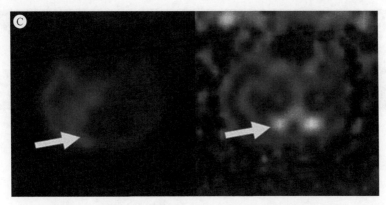

图 2-8　大鼠脑缺血再灌注后 DWI 和 ADC 图随时间的演变

A～C 图分别为 MCAO/R 后 12 小时、1 天和 3 天 DWI 和 ADC 图。大鼠脑缺血再灌注后 12 小时 DWI 上兴趣区呈明显异常高信号强度，同时伴 ADC 值显著下降。1 天时 rADC 值较 12 小时时降低，3 天时升高。12 小时至 1 天 rDWI-SI 持续升高，1 天时达最高峰，3 天时下降。MCAO/R. middle cerebral artery occlusion/reperfusion，大脑中动脉闭塞/再灌注；DWI. diffusion weighted imaging，弥散加权成像；ADC. apparent diffusion coefficient，表观弥散系数；rADC. relative apparent diffusion coefficient，相对表观弥散系数；rDWI-SI. relative diffusion weighted imaging- signal intensity，相对弥散加权成像信号强度

文萌萌和张勇等[5, 9]的研究显示 AQP4 表达与 ADC 值及 FA 值呈负相关，推测可能是由于 AQP4 表达增加后通过降解基底膜而破坏血脑屏障，导致细胞毒性水肿和血管源性脑水肿，致使水分子扩散受限，ADC 值及 FA 值随之降低。因此，ADC 值与 FA 值低，说明大鼠脑水肿程度增加，神经功能受损严重。DTI 技术可用来直观反映并量化 AQP4 蛋白的表达（图 2-10～图 2-12）[5]，AQP4 表达上调是细胞毒性脑水肿形成的主要分子机制，也是 ADC 值降低的分子基础。

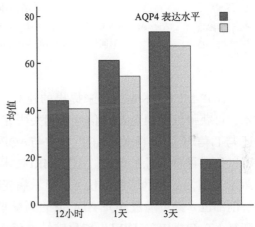

图 2-9　AQP4 表达水平（平均吸光度值）随时间动态变化

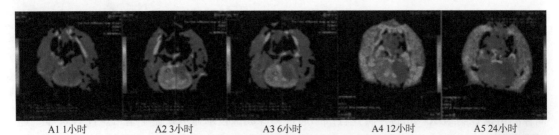

图 2-10　不同组不同时间点缺血核心区 ADC 图

A1～A5 为 MCAO 1 小时、3 小时、6 小时、12 小时、24 小时双侧大脑半球的 ADC 图，表明随着时间延长，梗死面积逐渐增大

扫码见彩图

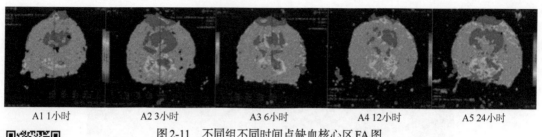

| A1 1小时 | A2 3小时 | A3 6小时 | A4 12小时 | A5 24小时 |

图2-11　不同组不同时间点缺血核心区FA图

A1～A5为MCAO 1小时、3小时、6小时、12小时、24小时双侧大脑半球的FA图，表明随着时间延长，梗死侧面积逐渐增大

扫码见彩图

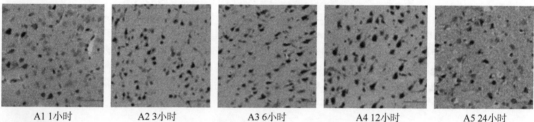

| A1 1小时 | A2 3小时 | A3 6小时 | A4 12小时 | A5 24小时 |

图2-12　不同组不同时间点AQP4表达情况

A1～A5表示MCAO 1小时、3小时、6小时、12小时、24小时AQP4蛋白的表达情况，随着时间延长，AQP4的表达逐渐增加，12小时达顶峰，随后逐渐减少

扫码见彩图

AQP-MRI技术是近年来提出的建立在AQP理论基础上一种全新的MRI分子成像技术，可获得细胞膜上AQP信息。相关研究[3, 10]发现，AQP-ADC除了与ADC呈正相关外，还与T_2信号呈负相关，AQP-ADC与AQP4表达呈负相关，AQP-ADC和T_2及DWI一样可以用以评价病灶的改变。AQP-MRI技术能够在体、实时、动态地进行AQP分子成像，AQP-MRI后处理伪彩图不仅可以显示病变范围，还可以快速判断组织的损伤程度和大致解剖部位，并可通过AQP-ADC值的测量来定量验证（图2-13）[10, 11]，比T_2液体衰减反转恢复（T_2-FLAIR）、DWI更能反映缺血组织的实际生理情况而更具优越性，为AQP-MRI的临床应用转化进一步奠定了基础。

三、AQP4与缺血半暗带

急性缺血性卒中是严重危害人类健康的重大疾病。目前治疗的关键是急性期，即拯救缺血半暗带。缺血半暗带由Astrup等在1981年提出。缺血半暗带被定义为缺血中心灶周围的正常突触传递被抑制或者被完全阻断、电活动异常，但组织结构完整，处于低灌注，能量代谢尚存的可逆性脑组织。此后，研究者试图通过影像成像技术显示缺血半暗带。任欢欢等[11]通过大鼠栓塞/再灌注动物模型研究显示细胞内（毒性）水肿是缺血半暗带的主要病理改变，同时伴有细胞器肿胀，但细胞膜完整，及时实施再灌注后细胞形态及功能可以恢复正常。

对缺血半暗带的影像成像评估是临床工作者的关注点。有多种不同的影像方法确定缺血半暗带，包括CT灌注（CTP）、MRI和正电子发射计算机断层显像（PET）。传统的半暗带成

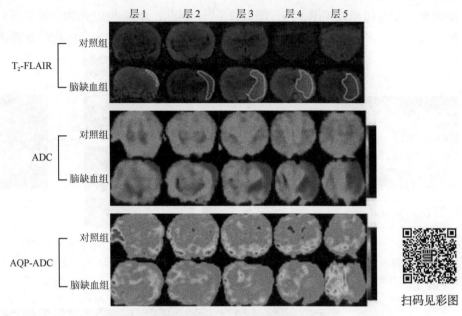

图2-13 不同成像序列显示缺血灶情况

患侧病灶在T_2-FLAIR上呈高信号强度（白线所示），相应的ADC图像上病灶信号减低，病灶的AQP-ADC值也较对侧明显减低。病灶在AQP-ADC上的范围与传统ADC基本一致

像包括扩散加权成像（DWI）/T_2-FLAIR不匹配和灌注加权成像（PWI）/DWI不匹配。尽管一直以来PWI/DWI不匹配作为半暗带的评估标识得到广泛应用，但是这种方法尚存在争议。因为DWI并不总意味着不可逆梗死组织，而PWI代表有风险组织中包含良性缺血区，并且仅依据灌注相关阈值定义半暗带外边界不够准确。这向传统的半暗带成像概念提出了挑战[11]。

AQP-MRI是近年新兴的MRI分子成像技术，有研究用AQP-MRI对缺血半暗带进行评价。以往对缺血性脑卒中，特别是缺血半暗带的研究在很大限度上忽略了神经解剖位置对预后的重要影响。常规的DWI序列对于脑灰质和白质缺乏对比，而AQP-MRI可以区分解剖部位，解剖定位可能在帮助进一步细化缺血半暗带和改善功能结局方面发挥作用。有研究把AQP-ADC值降低区域与T_2-FLAIR高信号强度不匹配的区域近似地认为是缺血半暗带，并命名为"水通道蛋白分子半暗带"[12,13]。结果显示在24小时内，AQP-ADC/T_2-FLAIR不匹配面积与DWI/T_2-FLAIR不匹配面积之间差异显著（$P<0.01$），面积图显示前者比后者的面积大（图2-14）[12]。AQP-MRI是在双指数模型的基础上成像的，比DWI单指数模型能更好地反映生物体内复杂的信号衰减，b值越高，MRI显示的对象越接近AQP内转运的水分子微观信息。AQP-MRI伪彩图显示病灶内的色带跨度大，较DWI序列有层次，测量不同色带的AQP-ADC值，能体现出其值的差异，而不论是DWI序列还是T_2-FLAIR序列都无法实现这一可视化特点。AQP-MRI作为半暗带外边界，能可视化观察脑梗死外带区域的AQP-ADC值变化，而DWI则是综合的表现。即AQP-MRI定义的半暗带比传统的DWI作为外边界意义上的半暗带更具有诊断准确性和优越性，特别是在脑缺血早期（图2-15）[13]，AQP-MRI对缺血半暗带的显示较DWI序列更具敏感性，半暗带周围的AQP-ADC值可能从侧面反映侧支循环的情况。与传统的不匹配法相比，AQP-MRI/T_2-FLAIR不匹配可多层次、更精准地对缺血半暗带进行实时动态观察，且MRI表现与组织

病理学改变密切相关（图 2-16，图 2-17）[13]。因此，AQP-MRI 作为一种全新的可视化分子成像技术，可多层次、多角度实时动态评价缺血性脑卒中，为个体化诊疗提供了保障。

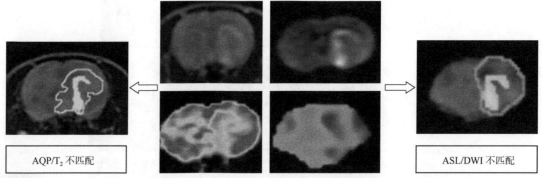

AQP/T_2 不匹配　　　　　　　　　　　　　　　　　　　　　　　ASL/DWI 不匹配

图 2-14　缺血组大鼠脑缺血 3 小时图像融合显示不匹配区缺血半暗带

左图为 AQP-MRI 与 T_2-FLAIR 融合图像，黄色区域代表核心梗死区，白线区域代表 AQP 半暗带外边界；
右图为动脉自旋标记（ASL）与 DWI 融合图像，黄色区域代表核心梗死区，白线区域代表 ASL 半暗带边界

扫码见彩图

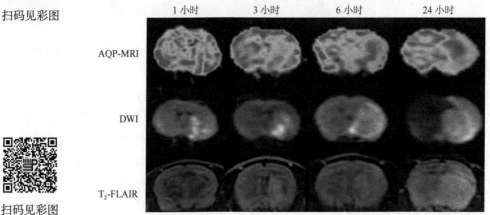

图 2-15　不同时间点 AQP-MRI 显示缺血灶情况

与 T_2-FLAIR 及 DWI 序列比较，AQP-MRI 可更加敏感、多层次展示缺血灶，AQP-MRI 伪彩图还可显示病灶内微观信息

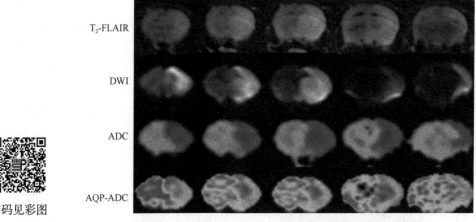

图 2-16　取栓后 24 小时，不同成像序列显示缺血灶情况

T_2-FLAIR 及 DWI 显示患侧均呈大面积的高信号强度，ADC 图显示病灶扩散受限、ADC 值减低，但灰白质缺乏对比，AQP-ADC 可清晰显示基底节区和同侧远端皮质存在差异，具有可视化优点

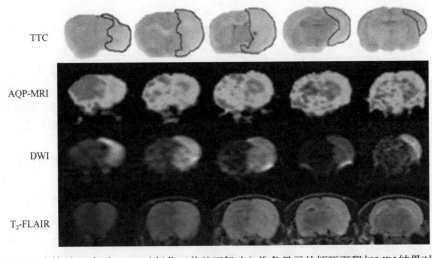

扫码见彩图

图2-17　取栓后24小时，TTC（氯化三苯基四氮唑）染色显示的梗死面积与MRI结果对照
TTC染色显示梗死面积与AQP-MRI、DWI、T_2-FLAIR测量梗死面积无显著性差异

　　无论在临床还是科研，MRI都是评价脑组织病理状态下的损伤程度和预后的一个有效方法。多模态磁共振成像是目前准确诊断脑水肿的影像学方法，尤其是AQP-MRI，能更准确地显示AQP4的表达量和分布情况，从分子水平上为临床早期治疗脑水肿及药物开发提供新思路[14]。目前关于AQP-MRI的研究较少，是否能精确地定量AQP还需进一步研究，且大都是动物研究，尚无可以临床使用的AQP在体成像方法。鉴于此，最近刘英超、白瑞良等通过人脑胶质瘤临床验证及动物实验模型构建，发展了新型"水交换对比增强磁共振成像技术"（water-exchange DCE-MRI），通过测量AQP4介导的跨膜水分子交换速率，将体内特异性标记的AQP4分子的信号放大，并结合临床常规使用磁共振造影剂（如Gd-DTPA）的胞外分布特性，通过动态增强对比磁共振成像技术，增强了DCE-MRI对水分子跨膜运输测量的敏感度，最终实现了对AQP4的精准测量。AQP4可视化技术为无创评估胶质瘤不同亚区域的生物活性和动态监测治疗敏感性提供了新的技术手段，有望为胶质瘤的个体化精准诊疗提供有效的影像学工具[15]。但此项技术是在肿瘤的基础上得出的结论，要真正实现AQP-MRI在体定量成像，还需要更多的实验及临床验证。

（彭雪华）

参 考 文 献

[1] 张艳伟，黎海涛．脑水肿相关因子及MRI研究进展[J]．国际医学放射学杂志，2009，32（5）：440-443.

[2] 郭启勇，辛军，张新，等．MRI水扩散加权成像分子机理研究进展[J]．中国临床医学影像杂志，2013，24（7）：496-500.

[3] 陈秋雁，吴富淋，彭晓澜，等．水通道蛋白磁共振分子成像与水通道蛋白4表达的相关性研究[J]．中国临床医学影像杂志，2016，27（12）：837-841.

[4] 蒋锡丽，鲁宏，陈建强，等．AQP4基因沉默对脑水肿的影响及影像改变[J]．国际医学放射学杂志，2016，39（5）：495-499.

[5] 文萌萌，张勇，程敬亮，等．siRNA下调AQP-4对减轻脑梗死大鼠脑水肿的DTI研究[J]．临床放射学

杂志，2020，39（3）：609-615.

[6] 何占平，鲁宏，涂蓉. 干预AQP4表达预防脑水肿及其DWI诊断的研究进展[J]. 国际医学放射学杂志，2009，32（4）：347-349.

[7] 牛彩虹，齐进冲，修宝新，等. 大鼠脑缺血-再灌注损伤早期DWI参数和AQP4蛋白表达相关性研究[J]. 脑与神经疾病杂志，2016，24（10）：617-623.

[8] Wang H，Wang X，Guo Q. The correlation between DTI parameters and levels of AQP-4 in the early phases of cerebral edema after hypoxic-ischemic/reperfusion injury in piglets[J]. Pediatr Radiol，2012，42（8）：992-999.

[9] 张勇，程敬亮，郑瑞平，等. 脑梗死大鼠脑水肿程度的磁共振DTI序列评估[J]. 郑州大学学报（医学版），2020，5（4）：580-584.

[10] 陈秋雁，吴富淋，彭晓澜，等. 水通道蛋白磁共振分子成像（AQP MRI）对缺血性脑卒中的应用价值[J]. 功能与分子医学影像学（电子版），2016，5（4）：1033-1038.

[11] 任欢欢，鲁宏. 脑创伤半暗带和脑水肿与水通道蛋白-4关系的研究进展[J]. 国际检验医学杂志，2015，36（6）：813-815.

[12] 彭晓澜，余波，陈秋雁，等. 水通道蛋白磁共振分子成像对脑缺血半暗带的评价[J]. 中国医学影像学杂志，2020，28（1）：6-11，16.

[13] 彭晓澜，翁烨，黄立东，等. 水通道蛋白磁共振分子成像在缺血性脑卒中的可视化研究[J]. 磁共振成像，2019，10（10）：762-767.

[14] 艾莉，陈海霞，秦将均，等. AQP4在创伤性脑水肿中的表达与多模态MRI成像研究[J]. 海南医学院学报，2020，26（17）：1353-1357.

[15] Jia Y，Xu S，Han G，et al. Transmembrane water-efflux rate measured by magnetic resonance imaging as a biomarker of the expression of aquaporin-4 in gliomas[J]. Nat Biomed Eng，2023，7（3）：236-252.

第三章　以水通道蛋白为治疗靶点的研究现状

第一节　以水通道蛋白家族为靶点的小分子和生物制剂的研究现状

以水通道蛋白（AQP）家族为靶点的药物研发在治疗癌症、肥胖症、脑水肿、青光眼、皮肤病等众多疾病中具有巨大的临床应用潜力。其中，星形胶质细胞AQP4是最值得关注的靶分子之一。一方面，AQP4参与了脑组织水分子的转运与神经兴奋性的形成和传递，具有重要的生理意义；另一方面，AQP4是视神经脊髓炎的致病性自身抗体的靶点，同时还参与神经胶质瘢痕形成等作用，为临床治疗提供了新的思路。因此，筛选具有潜在应用价值的AQP小分子抑制剂具有重要的临床意义。虽然目前人们已筛选出多种候选分子化合物作为AQP的抑制剂，但大量的临床研究显示其抑制效应并未得到证实，研究结果与临床应用之间还存在较大的差距。为此，本节将介绍以AQP为靶点的小分子生物制剂的研究现状，以及今后可能面临的机遇和挑战。

一、水通道蛋白抑制剂的概况

目前已经鉴定出的哺乳动物AQP绝大部分起水通道作用，少部分还可参与细胞内其他溶质的转运，包括甘油和过氧化氢。在结构上，AQP单体是一种小的跨膜蛋白质，分子量约为30kDa，每个单体含有一个狭窄的孔道。AQP单体在细胞膜中组装成四聚体，其特异的透水性由NPA结构域和芳香/精氨酸（Ar/R）结构域决定，研究发现AQP抑制剂应与这两个结构域结合才能发挥抑制作用，亲脂性的抑制剂与AQP的疏水位点结合能极大地提高药物的抑制作用[1]。AQP在人体分布广泛，主要分布于具有吸收和分泌功能的上皮细胞、星形胶质细胞、心肌细胞、脂肪细胞和表皮细胞等，因此，以AQP为靶点的药物研发越来越受到人们的关注。然而，受限于AQP的结构特点及功能调控，其药物开发还面临众多挑战，包括：①AQP的广泛分布和具有许多结构相似的AQP异构体；②药物无法通过狭窄的孔道和致密的四聚体结构；③缺乏内在AQP功能的生理调节；④水在55mol/L浓度下具有绕过障碍物的独特能力。所以，迄今仍然缺少一种有效的评估方法去筛选和衡量对疾病有治疗潜力的AQP功能抑制剂、表达的调节剂或其抗体的阻断剂。虽然含汞或其他重金属的巯基反应性化合物被证实可通过化学修饰半胱氨酸残基来抑制某些AQP的功能，但由于其显著的毒性和缺乏选择性，影响了其作为药物的进一步研发。

尽管存在上述诸多问题，但AQP靶向治疗在人类疾病方面仍具有巨大的潜力。AQP的生理功能已在基因敲除小鼠的表型研究中得到了广泛的证实[2]，一些有趣的潜在应用价值正逐渐显现。AQP1水转运的抑制剂有望作为一种独特的利尿剂，可抑制肿瘤血管生成，降低青光眼患者的眼压，并减轻患者的疼痛[3-8]。AQP1在侵袭性黑色素瘤中表达升高，沉默AQP1基因可减少小鼠黑色素瘤的血管生成和转移[9]。AQP2基因表达可通过收集管细胞上的加压素受体引起肾性尿崩症（NDI）[10]。AQP3甘油和（或）过氧化氢转运的抑制剂被认为可预防或延缓皮肤肿瘤的生长[11, 12]。还有证据表明，AQP3基因转移可以增加上皮细胞的水通透性，促进唾液腺疾病患者的唾液分泌[13]，以及促进与胆汁淤积相关肝病的胆汁分泌[14]。研究还发现，上调AQP3的表达可能促进伤口愈合。AQP4的高表达可加重脑水肿，其抑制剂被证实可减轻缺血性脑血管病患者的脑肿胀[15]；上调AQP4的表达可能具有抗癫痫活性，针对致病性AQP4的自身抗体与星形胶质细胞AQP4结合的阻断剂可能预防或减轻视神经脊髓炎的神经病理和神经功能缺陷[16]。此外，上调AQP5的表达可增加腺体的分泌，AQP7的选择性转录上调可减少肥胖症患者脂肪细胞肥大。在此将介绍水通道蛋白抑制剂的检测方法、水通道蛋白抑制剂的鉴定和验证的研究现状，以及未来的发展方向。

二、水通道蛋白的功能测定

AQP的功能测定是鉴定和验证药理性AQP抑制剂的关键。本部分重点介绍AQP水渗透性的测量，并分析其检测的局限性和潜在的伪影。虽然AQP对于甘油、过氧化氢和气体的转运也引起了人们的关注，但由于目前缺乏可靠的筛选方法，因此不做进一步讨论。

测量完整上皮的水转运是通过测定经上皮渗透梯度的净体积运动来完成的，这可以通过多种方法来测定，例如观察置换体积和荧光法测量染料的稀释程度。上皮细胞层可由培养在多孔支架上的细胞或天然上皮（如肾小管或膀胱）组成。虽然不是高精度或高通量测量，但跨上皮水转运的测量方法是稳定和可靠的。此外，跨细胞的水转运涉及两个相邻的细胞膜（顶端和基底外侧膜），因此结果主要是检测细胞的跨膜转运。

悬浮细胞或贴壁细胞的渗透水转运是通过细胞体积对渗透梯度的反应动力学来测量的，渗透梯度形成的时间远远小于渗透平衡时间。对于悬浮细胞（或囊泡及脂质体），可通过使用停止流动仪器施加渗透梯度，其中含有细胞的溶液在毫秒或更短时间内混合。对于黏附细胞，例如表达AQP的转染或转导细胞，可以使用灌注室施加渗透梯度，或者采用高通量测量，通过多孔板形式快速添加溶液以施加渗透梯度。读取细胞体积的方法包括光散射法、荧光法或直接成像法。光散射法主要用于研究悬浮液中的小细胞如红细胞，但易受到各种伪影的影响，因为细胞散射光的强度不仅取决于细胞体积，还取决于细胞形状、胞内和胞外溶液的折射率以及细胞膜的光学性质。此外，"混合伪影"的出现是由于细胞流动减慢时光散射的变化造成的，这与细胞的水渗透性无关。荧光法对细胞形状、细胞膜性质和混合伪影相对不敏感，但可直接受到试验化合物的影响，并因染料渗漏和与细胞膜结合而容易混淆，特别是在钙黄绿素荧光法中，当细胞收缩时，细胞质的蛋白会使钙黄绿素荧光发生淬灭[17]。另一种基因编码的氯化物敏感黄色荧光蛋白虽然无渗漏伪影的存在，但受细胞内pH和阴离子浓度变化的影响，且其时间分辨率也有限[18]。

一种特殊的细胞类型，如非洲爪蟾卵母细胞，在许多研究中被用于测量水的渗透性，通过向爪蟾卵母细胞注射编码AQP的cRNA，并根据渗透梯度测量卵母细胞肿胀的动力学变化。一般情况下，卵母细胞体积是通过低倍透射显微镜测量其横截面积（阴影）推断出来的。虽然卵母细胞肿胀法在最初识别AQP1是否为水通道时是有价值的，但在肿胀率明显增加而导致卵母细胞破裂时由于存在许多潜在的伪影，对研究潜在的AQP抑制剂的价值有限。卵母细胞横截面积的变化不仅取决于卵母细胞对水的渗透性，还取决于卵母细胞的结构、细胞膜的性质、溶质转运、胞质及其他因素等。例如，用离子转运抑制剂对卵母细胞进行预孵育时，可能会改变卵母细胞体积和细胞质的离子浓度，影响对水渗透性的测量。此外，测量的另一个问题是卵母细胞体积反应的测量时间是以分钟来计算的，而不是以秒来计算的，在这个时间范围内可能会由于机械误差影响了细胞肿胀程度的判断，而且溶质的运输也可影响渗透梯度的变化。

为了测量红细胞AQP1的水渗透性，有学者研发了一种依赖于细胞裂解的简单的测量方法（图3-1）[19]。由于红细胞具有天然表达AQP1和尿素转运体——尿素通道蛋白B（urea

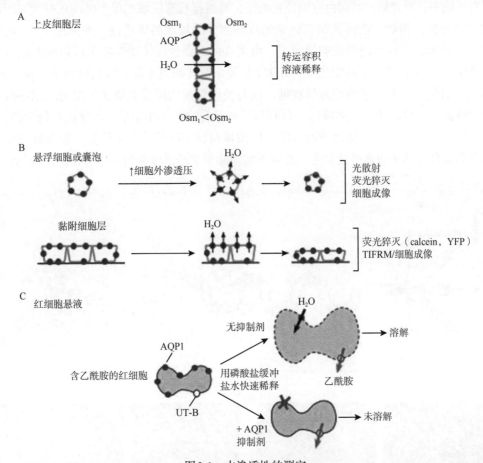

图3-1 水渗透性的测定

A：水跨过上皮细胞层运输产生净体积流量。B：悬浮细胞或囊泡的水渗透性（上图）或固定化细胞（黏附细胞层）的水渗透性（下图），渗透梯度的改变会引起时间依赖性的体积变化。C：在红细胞中测量的AQP1的水渗透性，红细胞表达AQP1和尿素/乙酰胺转运蛋白（UT-B）。将含乙酰胺的红细胞稀释会导致水的流入，细胞肿胀和溶解，而抑制AQP1会减少水分渗透。Osm. osmolality，渗透压；calcein. 钙黄绿素；YFP. yellow fluorescent protein，黄色荧光蛋白；TIFRM, total internal reflection fluorescence microscopy，全内反射荧光显微镜

transporter B,UT-B)的功能,红细胞被预先装载乙酰胺,乙酰胺是一种尿素类似物,由UT-B运输,并通过类似于渗透水转运的过程通过红细胞膜实现平衡,然后将红细胞稀释到无乙酰胺溶液中,随后红细胞可快速肿胀和溶解,最后通过平板仪中的溶液吸光度值测定红细胞AQP1的水渗透性。由于水的流入比乙酰胺流出的渗透梯度消散慢,所以抑制AQP1的水渗透性会降低细胞裂解。这种方法已被改良并用于筛选具有低纳摩尔的UT-B转运抑制剂[19],但AQP1抑制剂的小分子筛选并未产生有应用价值的活性化合物。

学者们一直在研究通过微流体的方法来测量水的渗透性,因为微型流体可为微量样品快速分析提供一个可靠的技术平台。在一项研究中,研究者开发了一种灌注通道来测量上皮类器官的体积变化,其中类器官被柱包裹,通过染料排除法来测量其体积[20]。在另一项研究中[21],研究者设计了一个微流体通道来模拟快速停流混合,在该方法中,通过在被油包围的0.1nl液滴内进行溶液混合,使细胞承受以毫秒(ms)为单位的渗透梯度(图3-2A)。在混合微流体通道中完成细胞与不同浓度溶液的快速混合,然后将含有细胞的液滴沉积在观察区域,在观察区域中,混合后的时间由空间位置决定,再根据观察区域的单个时间积分荧光图像确定水的渗透性。例如,在钙黄绿素标记的红细胞中测量水的渗透性(图3-2B)[21]。观察区域的荧光显示,存在渗透梯度的情况下,由荧光的减弱情况(图3-2C)可以推断出水迁移的动力学。图3-2D显示了从野生型与AQP1敲除小鼠红细胞(左)以及对照组和pCMBS处理的人红细胞(右)推导的动力学数据,这与使用常规的停流光谱法得出的结果一致。与昂贵的停流光谱仪器相比,微流体平台利用了亚微升的血液样本量,不受混合伪影的影响,并使用标准实验室的荧光显微镜通过单一的图像捕获来取代动力学测量。但是微型流体方法通常不适合自动化的高通量筛选,并且不能消除常规悬浮细胞测量中潜在的测量伪影。

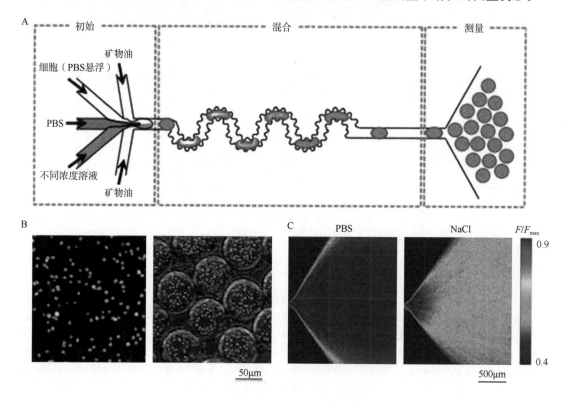

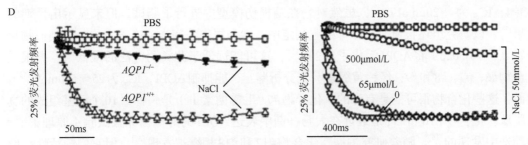

图 3-2　用微流体"停止流动"方法测量水的渗透率[21]

A：微流体通道设计，其中细胞与液滴中的不同浓度溶液的混合物驱动渗透水传输和细胞体积变化，以及测量区域中的荧光量。B：钙黄绿素标记的红细胞的荧光显微照片（左）和在微流体通道中的液滴的荧光显微照片（右）。C：用PBS（在中央通道）和含有 500mmol/L NaCl 的PBS（底部通道）中的红细胞悬浮液灌注微流体通道，得到 200mmol/L NaCl 梯度。零梯度（左）或 200mmol/L NaCl 梯度（右）下测量区域的荧光显微照片。D：在 200mmol/L NaCl 梯度和零渗透梯度（PBS）的情况下，从野生型和 AQP1 敲除小鼠的红细胞中推断出的红细胞钙黄绿素荧光的时间进程（左）。用指定浓度的AQP1抑制剂pCMBS预孵育的人红细胞的测量结果（右）。PBS. phosphate buffer saline，磷酸盐缓冲溶液；pCMBS. *p*-chloromercuribenzoate，对氯巯基苯磺酸盐

扫码见彩图

三、水通道蛋白抑制剂的研究现状

（一）水通道蛋白抑制剂的早期研究

很早就有文献表明汞的巯基反应性化合物，包括氯化汞和对氯巯基苯磺酸盐（pCMBS）可抑制红细胞和各种上皮细胞的水运输[22]。AQP被发现后，科学家鉴定出了参与抑制短暂水运输的半胱氨酸，如AQP1中的Cys-187[23]。最近，有报道称含金化合物可以抑制AQP3，其中奥芬的作用是最强的[24]。在某些AQP中，各种非金属的离子化合物会抑制水的渗透性，包括钾通道阻滞剂四乙基铵（TEA^+）、碳酸酐酶抑制剂乙酰唑胺、某些抗癫痫药（托吡酯、苯妥英、拉莫三嗪）和二甲基亚砜（DMSO）[25-27]。然而，后续的研究并未证实这些小分子化合物对AQP有抑制作用[27-30]，特别是抑制离子转运过程的化合物。

（二）水通道蛋白抑制剂的筛选

由于AQP空间结构复杂，蛋白质丰度较低，所以对AQP抑制剂的研究进展缓慢，但AQP抑制剂作为靶向药物治疗疾病仍具有重大意义。目前对AQP抑制剂的研究主要集中于离子化合物和小分子有机化合物，通常使用半抑制浓度（half maximal inhibitory concentration，IC_{50}），即抑制剂阻断AQP 50%时的药物浓度表示抑制效率。

通过实验和计算筛选了小分子AQP抑制剂，包括12种已提出的AQP1抑制剂和1种AQP1激活剂，如图3-3所示。化合物#1、#2和#3通过虚拟（计算）筛选确定，包括与人类AQP1的细胞外表面对接，并测试了14种化合物对表达AQP1的非洲爪蟾卵母细胞渗透肿胀的抑制作用[31]，从而鉴定出化合物#1、#2和#3。这些化合物可将卵母细胞的渗透性肿胀降低约80%，IC_{50}为8～20μmol/L，但据报道，它们并没有抑制红细胞中的AQP1。化合物#4（AqB013）是NKCC1抑制剂布美他尼的类似物，据称它可以抑制AQP1和

AQP4，IC_{50} 为～20μmol/L[32]，虽然对它在脑损伤模型中进行了测试，但未显示出预期的效果[33]。该研究组还报道，在卵母细胞测定中，祥利尿剂呋塞米（速尿）的一种类似物化合物#5（AqF026）将AQP1活化了约20%[34]，这可能远低于此类测定法的最低值。据报道化合物#6、#7、#8和#9，经钙黄绿素荧光分析鉴定，能抑制AQP1，IC_{50} 为25～50μmol/L[35]。然而，这些化合物很可能是有毒的，化合物#7（毛霉菌素Ⅰ）是一种由10个残基组成的真菌脂肽。化合物#10和#11是最近在大肠杆菌中表达AQP1的酵母菌冻融试验（原理不明）的筛选中发现的[36]。随着研究进展，化合物#12和#13相继被发现[37]，但它们的活性在卵母细胞、红细胞和AQP1蛋白脂质体测定中变化较大。

在最近的研究中[38]，有学者重新评估了13种化合物（图3-3）对AQP1活性的影响。从公布的数据中预测出，当化合物的浓度为50μmol/L时，可以显著抑制（或弱激活）AQP1的水渗透性。一种方法是在新鲜获得的人红细胞中使用停止流光散射方法，代表性的光散射曲线如图3-4（左）所示，右图汇总了平均数据。尽管$HgCl_2$显著抑制了红细胞中的水渗透性，但对于13种受试化合物中有12种无显著作用，而化合物#13与细胞毒性相关的明显作用较小。另外，为排除由于血红蛋白与化合物结合影响抑制作用的可能性，采用密封的无血红蛋白的类似膜进行的类似研究也显示没有抑制作用。5种化合物（#6、#9、#10、#12和#13）在红细胞皱缩和聚集表现中证明了其具有毒性作用。多种附加试验支持化合物#1至#13不抑制（或激活）AQP1的水渗透性的结论，包括红细胞肿胀试验、使用钙钙素荧光的红细胞水转运试验，以及AQP1转染CHO细胞细胞膜、膜囊泡的水转运试验。

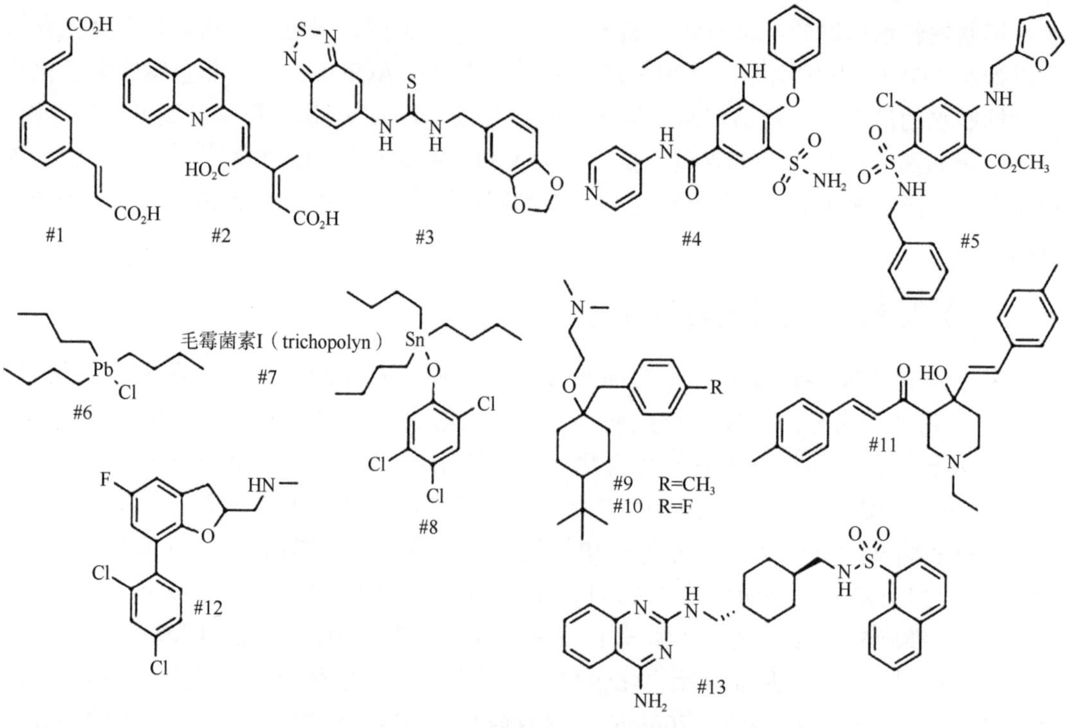

图3-3　假定的小分子AQP1抑制剂和AQP1激活剂及化学结构[32]

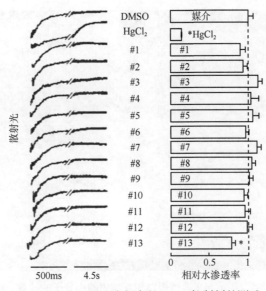

图 3-4 在人红细胞中假定的 AQP1 抑制剂的测试

在 530nm 处散射光强度的时间过程中，以 250mmol/L 向内定向的蔗糖梯度，测量了人红细胞的水渗透性。左图为散射光曲线，其中二甲基亚砜（dimethyl sulfoxide，DMSO）为阴性对照，$HgCl_2$ 为阳性对照，#1~#13 为浓度为 50μmol/L 的 13 种化合物。右侧示相对水渗透性（S.E.，$n=4$，与对照组相比，$P<0.05$）。$HgCl_2$.氯化汞

目前尚难以证实文献中报道的 AQP 抑制剂的活性[38]。如上所述，大多数研究中使用的卵母细胞肿胀法或钙黄绿素荧光测定法存在相当大的技术误差，其中渗透性细胞肿胀的明显抑制可能是由于细胞大小或形状的变化、细胞体积的改变、非 AQP 离子或溶质转运体的活性等影响。已知的细胞膜转运蛋白过程的抑制剂，如布美他尼、乙酰唑胺和四乙基胺，可能会影响静息细胞的体积和体积调节。大量抑制剂的筛选研究工作表明，AQP 抑制剂的识别及其应用于临床的可能性非常低，仍需要对常用药物（如袢利尿剂、碳酸酐酶抑制剂和抗癫痫药）进行大规模的筛选，否则难以发现真正具有应用价值的 AQP 抑制剂。

一些报告通过虚拟筛选、分子动力学（MD）模拟来确定各种 AQP 的假定抑制剂。据报道，使用大鼠 AQP4 的电子衍射结构从对接计算中选择了多种在化学结构上不相关的抗癫痫药，可抑制卵母细胞肿胀[26]；同时研究者报道了作为 AQP4 抑制剂的非抗癫痫药，IC_{50} 为 2~11μmol/L，包括 2-烟酰胺基-1,3,4-噻二唑、舒马普坦和利扎曲普坦[39]，也可抑制卵母细胞肿胀。然而，重新测试参考文献中提及的化合物并未能确定它们的有效抑制性。在最近的一项研究中，使用小鼠 AQP9 的同源性模型进行对接和 MD 模拟[40]，该模型从表达 AQP9 的 CHO 细胞的收缩试验中鉴定出一小批 $IC_{50}<50\mu mol/L$ 的抑制剂，但尚未对复合物的活性进行单独测试。针对 AQP1 和 AQP4 的高分辨率结构进行大规模对接研究，测试了得分最高的 2000 种化合物，然而它们在 50μmol/L 浓度下抑制率小于 20%。图 3-5A 显示了得分高的苯并恶嗪-3-one 类化合物。复合物的表面描述显示互补的配合（图 3-5B），非极性环己基突出到通道深部，与残基 Ile60、Leu-149 和 Val-79 相互作用[40]。

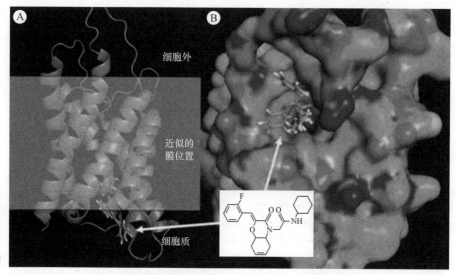

扫码见彩图

图 3-5 确定与水通道蛋白相互作用的小分子的计算方法

使用同源小鼠的 AQP1 模型进行对接研究。A：AQP1-配体配合物的侧视图，其中标出了近似的膜位置。B：同一配合物的表面视图，显示配体的环己基突出到通道深处，与疏水表面相互作用

（三）水通道蛋白分子靶向治疗的研究现状

通过对 AQP 靶点抑制剂的研究将会改变细胞代谢路径，影响细胞增殖、分化等过程，从而可对疾病进行特异性诊断和靶向治疗。目前针对 AQP 分子的靶向治疗研究主要有以下两种。

1. 阻断 AQP 与自身抗体的结合 NMO 是一种中枢神经系统的炎性脱髓鞘疾病，NMO 中的 IgG1 与抗 AQP4 的 IgG1 自身抗体（"AQP4-IgG"）靶向结合可引起瘫痪和失明，这可能是 NMO 的致病机制。研究认为 AQP4-IgG 通过与星形胶质细胞上的 AQP4 结合而产生神经病理作用（图 3-6A），引起补体和细胞介导的星形胶质细胞毒性，从而产生炎症、血脑屏障破坏、少突胶质细胞损伤、脱髓鞘和神经功能缺损。尽管最初卵母细胞肿胀的研究表明，抑制 AQP4 的水渗透性在 NMO 中的作用[41]，但后续的研究证实，AQP4-IgG 并不能有效抑制 AQP4 的水渗透性[42]。研究报道，针对 AQP2[43] 和 AQP5[44] 的自身抗体，它们分别与间质性肾炎和干燥综合征相关，但它们是否参与疾病发病机制尚不明确。

虽然中和性抗 AQP 抗体被识别出来的可能性不大，但 AQP 结合抗体还有其他治疗用途。在一项应用中，使用了一种高亲和力的抗 AQP4 抗体（"aquaporumab"），其中抗体 Fc 部分发生了突变，以消除涉及补体和细胞介导的细胞毒性的效应子功能（图 3-6B，左）[45]。该抗体可减轻 NMO 患者血清在细胞培养物中的细胞毒性（图 3-6B，右），并防止 NMO 动物模型中的神经系统脱髓鞘等病理学改变，提示其可用于 NMO 的治疗。然而，通过筛选和计算分析发现与 AQP4 结合的 AQP4-IgG 小分子阻断剂的亲和力太低，难以作为 NMO 的治疗药物进行开发[46,47]（图 3-6C），因此在识别有效的小分子抑制剂方面还有待更多的研究。

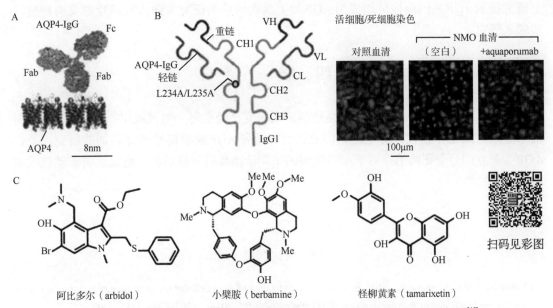

图 3-6 与 AQP4 结合引起视神经脊髓炎的抗 AQP4 自身抗体和小分子阻断剂[45]

A：IgG 抗体与膜上 AQP4 结合的比例图。B：具有消除效应子功能的 L234A/L235A 突变的抗 AQP4 IgG1 抗体（左）。暴露于对照或人 NMO 血清的 AQP4 表达细胞的活/死（绿色/红色）细胞染色，显示出 aquaporumab 的保护作用。C：自身抗体与 AQP4 结合的小分子阻断剂的结构。AQP4-IgG. aquaporin 4-immunoglobulin G，水通道蛋白 4 抗体 IgG；Fab. fragment antigen binding，抗原结合片段；Fc. fragment crystallizable，可结晶片段

2. AQP 基因调控 临床头颈部肿瘤的放射治疗造成的唾液腺损伤会导致患者唾液分泌减少、口干，从而继发口腔干燥、吞咽困难、口腔黏膜的慢性炎症和口咽部的持续感染等并发症[48]。研究发现在大鼠辐照模型中进行基因调控，使 AQP1 表达增加而恢复唾液分泌；在Ⅰ期临床试验中，通过腺病毒载体将 *AQP1* cDNA 转染到 11 例患者体内，结果显示患者在 42 天内均表现出较好的耐受性；但由于腺病毒载体的相关风险该研究未能开展Ⅱ期临床试验[13]。此外，已报道在超声辅助下将非病毒介导 *AQP1* 基因转到猪辐照模型的方法可能会成为临床试验的新思路[49]。

微 RNA（miRNA）和小干扰 RNA（siRNA）也已应用于调节 AQP 的活性。miRNA 是长度约 22 个核苷酸的单链非编码 RNA 分子，通过与 mRNA 分子的 3′非翻译区（UTR）结合而充当基因表达调节剂。最近的研究突出显示在多个病理生理学过程中 miRNA 对于 AQP 的调节作用[50-52]。*AQP4* 已被证明是 miRNA-29b 在脑缺血中的靶基因。来自人类血液样本的数据显示，卒中患者的 miRNA-29b 显著降低，且不良结果与 miRNA-29b 的下调相关。针对小鼠的研究，进一步证明了 miRNA-29b 水平与 AQP4 表达呈负相关，但与好的预后结局呈正相关。此外，双荧光素酶报告系统直接显示 *AQP4* 是 miRNA-29b 的靶标[50]。眼内新生血管的形成被认为是眼病中视力丧失的一个因素。有研究用 siRNA 沉默 *AQP4* mRNA 表达，表明其可减少眼部新血管形成。与对照组相比，加入剂量 1～10nmol/L 的 siRNA 以沉默人 Madin-Darby 犬肾（MDCK）细胞的 AQP4 表达，结果显示出超过 70% 的抑制作用。同一研究者使用类似技术针对眼部疾病抑制 AQP1 表达，并申请了专利[53]。以

上研究提示 *AQP* 基因调控可能成为 AQP 相关疾病治疗的研究方向，其治疗意义有待进一步深入研究。

四、展　　望

目前尽管调节 AQP 的小分子和生物制剂的研究已经取得一定进展[54-58]，但研究结果与临床应用之间还存在较大的差距。因此，针对每种 AQP 亚型特异性抑制剂的研究以及以 AQP 为靶点的分子靶向治疗对于相关疾病的早期诊治具有重要的临床意义，将推动该领域的进一步发展。

（李燕华）

参 考 文 献

[1] Wang S, Ing C, Emami S, et al. Structure and dynamics of extracellular loops in human aquaporin-1 from solid-state NMR and molecular dynamics[J]. J Phys Chem B, 2016, 120(37): 9887-9902.

[2] Verkman AS. Aquaporins in clinical medicine[J]. Annu Rev Med, 2012, 63: 303-316.

[3] Esteva-Font C, Jin BJ, Verkman AS. Aquaporin-1 gene deletion reduces breast tumor growth and lung metastasis in tumor-producing MMTV-PyVT mice[J]. FASEB J, 2014, 28(3): 1446-1453.

[4] Ma T, Song Y, Yang B, et al. Nephrogenic diabetes insipidus in mice lacking aquaporin-3 water channels[J]. Proc Natl Acad Sci USA, 2000, 97(8): 4386-4391.

[5] Schnermann J, Chou CL, Ma T, et al. Defective proximal tubular fluid reabsorption in transgenic aquaporin-1 null mice[J]. Proc Natl Acad Sci USA, 1998, 95(16): 9660-9664.

[6] Saadoun S, Papadopoulos MC, Hara-Chikuma M, et al. Impairment of angiogenesis and cell migration by targeted aquaporin-1 gene disruption[J]. Nature, 2005, 434(7034): 786-792.

[7] Zhang D, Vetrivel L, Verkman AS. Aquaporin deletion in mice reduces intraocular pressure and aqueous fluid production[J]. J Gen Physiol, 2002, 119(6): 561-569.

[8] Zhang H, Verkman AS. Aquaporin-1 tunes pain perception by interaction with Na(v)1.8 Na^+ channels in dorsal root ganglion neurons[J]. J Biol Chem, 2010, 285(8): 5896-5906.

[9] Simone L, Gargano CD, Pisani F, et al. Aquaporin-1 inhibition reduces metastatic formation in a mouse model of melanoma[J]. J Cell Mol Med, 2018, 22(2): 904-912.

[10] Radin MJ, Yu MJ, Stoedkilde L, et al. Aquaporin-2 regulation in health and disease[J]. Vet Clin Pathol, 2012, 41(4): 455-470.

[11] Hara-Chikuma M, Satooka H, Watanabe S, et al. Aquaporin-3-mediated hydrogen peroxide transport is required for NF-κB signalling in keratinocytes and development of psoriasis[J]. Nat Commun, 2015, 6: 7454.

[12] Hara-Chikuma M, Verkman AS. Prevention of skin tumorigenesis and impairment of epidermal cell proliferation by targeted aquaporin-3 gene disruption[J]. Mol Cell Biol, 2008, 28(1): 326-332.

[13] Baum BJ, Alevizos I, Zheng C, et al. Early responses to adenoviral-mediated transfer of the aquaporin-1 cDNA for radiation-induced salivary hypofunction[J]. Proc Natl Acad Sci USA, 2012, 109(47): 19403-19407.

[14] Marrone J, Soria LR, Danielli M, et al. Hepatic gene transfer of human aquaporin-1 improves bile salt secretory failure in rats with estrogen-induced cholestasis[J]. Hepatology, 2016, 64(2): 535-548.

[15] Manley GT, Fujimura M, Ma T, et al. Aquaporin-4 deletion in mice reduces brain edema after acute water intoxication and ischemic stroke[J]. Nat Med, 2000, 6(2): 159-163.

[16] Papadopoulos MC, Bennett JL, Verkman AS. Treatment of neuromyelitis optica: state-of-the-art and emerging therapies[J]. Nat Rev Neurol, 2014, 10(9): 493-506.

[17] Solenov E, Watanabe H, Manley GT, et al. Sevenfold-reduced osmotic water permeability in primary astrocyte cultures from AQP-4-deficient mice, measured by a fluorescence quenching method[J]. Am J Physiol Cell Physiol, 2004, 286(2): C426-C432.

[18] Baumgart F, Rossi A, Verkman AS. Light inactivation of water transport and protein-protein interactions of aquaporin-Killer Red chimeras[J]. J Gen Physiol, 2012, 139(1): 83-91.

[19] Levin MH, de la Fuente R, Verkman AS. Urearetics: a small molecule screen yields nanomolar potency inhibitors of urea transporter UT-B[J]. FASEB J, 2007, 21(2): 551-563.

[20] Jin BJ, Battula S, Zachos N, et al. Microfluidics platform for measurement of volume changes in immobilized intestinal enteroids[J]. Biomicrofluidics, 2014, 8(2): 024106.

[21] Jin BJ, Esteva-Font C, Verkman AS. Droplet-based microfluidic platform for measurement of rapid erythrocyte water transport[J]. Lab Chip, 2015, 15(16): 3380-3390.

[22] Macey RI, Farmer RE. Inhibition of water and solute permeability in human red cells[J]. Biochim Biophys Acta, 1970, 211(1): 104-106.

[23] Zhang R, van Hoek AN, Biwersi J, et al. A point mutation at cysteine 189 blocks the water permeability of rat kidney water channel CHIP28k[J]. Biochemistry, 1993, 32(12): 2938-2941.

[24] Martins AP, Ciancetta A, de Almeida A, et al. Aquaporin inhibition by gold (Ⅲ) compounds: new insights[J]. ChemMedChem, 2013, 8(7): 1086-1092.

[25] Brooks HL, Regan JW, Yool AJ. Inhibition of aquaporin-1 water permeability by tetraethylammonium: involvement of the loop E pore region[J]. Mol Pharmacol, 2000, 57(5): 1021-1026.

[26] Huber VJ, Tsujita M, Kwee IL, et al. Inhibition of aquaporin 4 by antiepileptic drugs[J]. Bioorg Med Chem, 2009, 17(1): 418-424.

[27] Søgaard R, Zeuthen T. Test of blockers of AQP1 water permeability by a high-resolution method: no effects of tetraethylammonium ions or acetazolamide[J]. Pflugers Arch, 2008, 456(2): 285-292.

[28] Yamaguchi T, Iwata Y, Miura S, et al. Reinvestigation of drugs and chemicals as aquaporin-1 inhibitors using pressure-induced hemolysis in human erythrocytes[J]. Biol Pharm Bull, 2012, 35(11): 2088-2091.

[29] Yang B, Kim JK, Verkman AS. Comparative efficacy of $HgCl_2$ with candidate aquaporin-1 inhibitors DMSO, gold, TEA^+ and acetazolamide[J]. FEBS Lett, 2006, 580(28-29): 6679-6684.

[30] Yang B, Zhang H, Verkman AS. Lack of aquaporin-4 water transport inhibition by antiepileptics and arylsulfonamides[J]. Bioorg Med Chem, 2008, 16(15): 7489-7493.

[31] Seeliger D, Zapater C, Krenc D, et al. Discovery of novel human aquaporin-1 blockers[J]. ACS Chem Biol, 2013, 8(1): 249-256.

[32] Migliati E, Meurice N, DuBois P, et al. Inhibition of aquaporin-1 and aquaporin-4 water permeability by a derivative of the loop diuretic bumetanide acting at an internal pore-occluding binding site[J]. Mol Pharmacol, 2009, 76(1): 105-112.

[33] Oliva AA Jr, Kang Y, Truettner JS, et al. Fluid-percussion brain injury induces changes in aquaporin channel expression[J]. Neuroscience, 2011, 180: 272-279.

[34] Yool AJ, Morelle J, Cnops Y, et al. AqF026 is a pharmacologic agonist of the water channel aquaporin-1[J]. J Am Soc Nephrol, 2013, 24(7): 1045-1052.

[35] Mola MG, Nicchia GP, Svelto M, et al. Automated cell-based assay for screening of aquaporin

inhibitors[J]. Anal Chem, 2009, 81(19): 8219-8229.

[36] To J, Yeo CY, Soon CH, et al. A generic high-throughput assay to detect aquaporin functional mutants: potential application to discovery of aquaporin inhibitors[J]. Biochim Biophys Acta, 2015, 1850(9): 1869-1876.

[37] Patil RV, Xu S, van Hoek AN, et al. Rapid identification of novel inhibitors of the human aquaporin-1 water channel[J]. Chem Biol Drug Des, 2016, 87(5): 794-805.

[38] Esteva-Font C, Jin BJ, Lee S, et al. Experimental evaluation of proposed small-molecule inhibitors of water channel aquaporin-1[J]. Mol Pharmacol, 2016, 89(6): 686-693.

[39] Huber VJ, Tsujita M, Nakada T. Identification of aquaporin 4 inhibitors using in vitro and in silico methods[J]. Bioorg Med Chem, 2009, 17(1): 411-417.

[40] Wacker SJ, Aponte-Santamaría C, Kjellbom P, et al. The identification of novel, high affinity AQP9 inhibitors in an intracellular binding site[J]. Mol Membr Biol, 2013, 30(3): 246-260.

[41] Hinson SR, Romero MF, Popescu BF, et al. Molecular outcomes of neuromyelitis optica(NMO)-IgG binding to aquaporin-4 in astrocytes[J]. Proc Natl Acad Sci USA, 2012, 109(4): 1245-1250.

[42] Rossi A, Ratelade J, Papadopoulos MC, et al. Neuromyelitis optica IgG does not alter aquaporin-4 water permeability, plasma membrane M1/M23 isoform content, or supramolecular assembly[J]. Glia, 2012, 60(12): 2027-2039.

[43] Landegren N, Pourmousa Lindberg M, Skov J, et al. Autoantibodies targeting a collecting duct-specific water channel in tubulointerstitial nephritis[J]. J Am Soc Nephrol, 2016, 27(10): 3220-3228.

[44] Alam J, Koh JH, Kim N, et al. Detection of autoantibodies against aquaporin-5 in the sera of patients with primary Sjögren's syndrome[J]. Immunol Res, 2016, 64(4): 848-856.

[45] Tradtrantip L, Zhang H, Saadoun S, et al. Anti-aquaporin-4 monoclonal antibody blocker therapy for neuromyelitis optica[J]. Ann Neurol, 2012, 1(3): 314-322.

[46] Mangiatordi GF, Alberga D, Siragusa L, et al. Challenging AQP4 druggability for NMO-IgG antibody binding using molecular dynamics and molecular interaction fields[J]. Biochim Biophys Acta, 2015, 1848(7): 1462-1471.

[47] Tradtrantip L, Zhang H, Anderson MO, et al. Small-molecule inhibitors of NMO-IgG binding to aquaporin-4 reduce astrocyte cytotoxicity in neuromyelitis optica[J]. FASEB J, 2012, 26(5): 2197-2208.

[48] Lee S, Choi JS, Kim HJ, et al. Impact of Irradiation on laryngeal hydration and lubrication in rat larynx[J]. Laryngoscope, 2015, 125(8): 1900-1907.

[49] Wang Z, Zourelias L, Wu C, et al. Ultrasound-assisted nonviral gene transfer of AQP1 to the irradiated minipig parotid gland restores fluid secretion[J]. Gene Ther, 2015, 22(9): 739-749.

[50] Wang Y, Huang J, Ma Y, et al. MicroRNA-29b is a therapeutic target in cerebral ischemia associated with aquaporin 4[J]. J Cereb Blood Flow Metab, 2015, 35(12): 1977-1984.

[51] Prockop DJ, Oh JY. Mesenchymal stem/stromal cells(MSCs): role as guardians of inflammation[J]. Mol Ther, 2012, 20(1): 14-20.

[52] Arthur FE, Shivers RR, Bowman PD. Astrocyte-mediated induction of tight junctions in brain capillary endothelium: an efficient in vitro model[J]. Brain Res, 1987, 433(1): 155-159.

[53] Abir-Awan M, Kitchen P, Salman MM, et al. Inhibitors of mammalian aquaporin water channels[J]. Int J Mol Sci, 2019, 20(7): 1589.

[54] Beitz E, Golldack A, Rothert M, et al. Challenges and achievements in the therapeutic modulation of aquaporin functionality[J]. Pharmacol Ther, 2015, 155: 22-35.

[55] Frigeri A, Nicchia GP, Svelto M. Aquaporins as targets for drug discovery[J]. Curr Pharm Des, 2007,

13(23): 2421-2427.

[56] Jeyaseelan K, Sepramaniam S, Armugam A, et al. Aquaporins: a promising target for drug development[J]. Expert Opin Ther Targets, 2006, 10(6): 889-909.

[57] Verkman AS, Anderson MO, Papadopoulos MC. Aquaporins: important but elusive drug targets[J]. Nat Rev Drug Discov, 2014, 13(4): 259-277.

[58] Wang F, Feng XC, Li YM, et al. Aquaporins as potential drug targets[J]. Acta Pharmacol Sin, 2006, 27(4): 395-401.

第二节 以AQP4的亚细胞定位为靶点的研究现状

AQP是一组细胞膜转运水分子，作为一组重要的转运媒介，广泛分布于人体各组织器官，与水分子的跨膜转运、细胞迁移、脑水肿的形成、神经兴奋性的调节和肿瘤的增殖生长等生理和（或）病理过程有关。在生理状态下，AQP在维持组织细胞内外水、电解质的稳态过程中发挥重要作用；在病理状态下，AQP在细胞毒性水肿的发生发展过程中有着重要的作用。AQP4是中枢神经系统AQP家族中的主要成员[1]，在室管膜细胞和星形胶质细胞膜上有丰富的表达[2]。研究表明，在脑或脊髓损伤的早期，AQP4驱动细胞毒性水肿的发生，推测AQP4在神经细胞水肿的发生中具有重要作用。因此，在神经损伤的急性期，可逆性抑制AQP4的功能，可能是预防中枢神经系统水肿的一种行之有效的策略[3]。本节将对水通道蛋白的亚细胞定位进行介绍。

一、AQP4的极性分布

星形胶质细胞（一种高度分支的细胞）含有丰富的细胞突起，在突起的终足膜上含有大量的AQP4，其分布的密度比非终足膜高约10倍，表明AQP4在星形胶质细胞膜的表面呈极性分布。AQP4的这种极化分布与抗肌萎缩蛋白-抗肌萎缩蛋白聚糖复合物（dystrophin-dystroglycan complex，DDC）有关[4]。DDC的核心成分为一种肌营养不良蛋白聚糖（dystroglycan，DG），它由α和β两个亚基组成。α-dystroglycan（α-DG）结合到细胞外基质（extracellular matrix，ECM）上，如集聚蛋白和层粘连蛋白，而β-dystroglycan（β-DG）则连接α-DG与细胞骨架蛋白及其他细胞内成分，如α-syntrophin跨膜蛋白，为AQP4提供结合位点[5]。虽然β-DG是α-DG与细胞骨架和其他DDC组分之间的跨膜蛋白，但β-DG并不直接与AQP4连接，而是与α-syntrophin连接。在对α-syntrophin基因敲除小鼠进行研究后发现，AQP4被募集到星形胶质细胞终足膜，需要α-syntrophin的参与[6]。因此，β-DG通过结合α-syntrophin以决定AQP4的极性分布，而这种极性分布则可能是AQP4维持水、电解质平衡能力的先决条件。培养星形胶质细胞的研究发现，DG基因表达减少或缺失可减少层粘连蛋白诱导的AQP4聚集数量和表达范围，推测AQP4在星形胶质细胞中的极性分布可能由DDC调控[7]，在这些特殊的膜结构域中AQP4的锚定需要DDC。

二、神经损伤后AQP4极性分布被破坏

DG是一种重要的细胞黏附受体，在各种组织中连接细胞骨架和细胞外基质[8]。它由一个基因编码，通过翻译后修饰并加工裂解成为两个蛋白：α-DG和β-DG[9]。α-DG与细胞外基质成分结合，而β-DG是一种膜跨蛋白，连接α-DG与细胞骨架和细胞内其他成分[10]。对GFAP-Cre/dg缺失小鼠的研究[5]和对大脑中动脉局灶性脑缺血（middle cerebral artery occlusion，MCAO）小鼠的研究[10]发现，β-DG的减少或缺失伴随着AQP4表达的变化，提示DG参与了AQP4的表达调控。

AQP4作为大脑主要的水通道蛋白，大量集中分布在与毛细血管和软脑膜相连的星形细胞终足膜。AQP4的这种分布模式，已通过AQP4的免疫反应所证实。AQP4的极性分布是神经胶质细胞与脑脊液或血管之间水平衡调节和转运的重要结构基础。在脑肿瘤、脑挫伤、细菌性脑膜炎等病理状态下，AQP4的极性分布出现异常[11]。AQP4的极性分布异常可能是脑水肿发生的一个重要因素。

对AQP4基因敲除小鼠的研究发现，AQP4介导的水转运是双向的[12]。脑组织透射电镜显示，脑出血后血管末梢出现肿胀[13]，血脑屏障的破坏和血管周围突起的肿胀，通常伴随着AQP4的显著减少[14]，血管周围突起的肿胀可能是由于血管周围终足膜AQP4的缺乏，阻碍大脑能量代谢所产生的水分的排出，从而导致水分子的外流受阻。因此，血管周围AQP4的缺乏可能是脑水肿发展的一个促成因素。

脑外伤后，由于血脑屏障的破坏，血液中的液体成分进入细胞间隙，导致血管源性水肿。Kitchen等[15]的研究提示，脑外伤早期阶段（外伤后6小时内），DG表达升高，上调了AQP4在血管周围和终足膜上的表达，但AQP4仍聚集在血管周围的终足膜上，AQP4在损伤的中心区仍然保持了血管样表达，即AQP4的极性分布特点仍然保持，此时的脑水肿为血管源性水肿，AQP4的表达上调，将有助于细胞外空间积累的水分清除到血管或蛛网膜下腔。脑外伤后期，在核心病变区和半暗带区AQP4的表达均较DG的表达延后。脑外伤后期DG极性分布的丢失导致了AQP4的相同变化，AQP4的极性分布丢失并重新分布于星形胶质细胞的其他端或非终足膜上，血管周围终足膜AQP4的丢失，导致血管周围多余水分清除到血管发生障碍；而其他终足或非终足膜上AQP4弥漫性增加，促进水分子向星形胶质细胞转移，最终导致脑外伤后期细胞毒性脑水肿的形成。

星形胶质细胞和微血管内皮通过黏附受体、整合素和DG，被锚定在细胞外基质（extracellular matrix，ECM）[16]，星形胶质细胞中细胞外信号调节激酶（extracellular signal-regulated kinase，ERK）的激活依赖于DG。研究发现DG和AQP4在脑出血早期均呈上升趋势，维持其极性分布[17]。脑出血后随着血肿周围区域血管发生超微结构改变，内皮细胞不规则肿胀，内皮腔侧被不连续的基底膜包裹，膨胀的基底膜被肿胀的胶质端包围，室管膜细胞减少，室管膜细胞表面β-DG表达减少且不规则、不连续或松散表达。检测同时发现，脑出血后APQ4的极性分布被破坏，在血肿周围区域的星形细胞表面重新分布，室管膜细胞和胶质内界限细胞AQP4表达减少和松散，不规则、不连续，脑出血后室管膜细胞表面β-DG表达下调，AQP4表达明显减少，AQP4和β-DG共定位信号明显减少。进一步证实β-DG的缺失与AQP4极性分布的缺失有关，AQP4在脑出血后脑组织中的表达发生动

态变化，参与脑出血后脑水肿形成（图3-7）[17]。

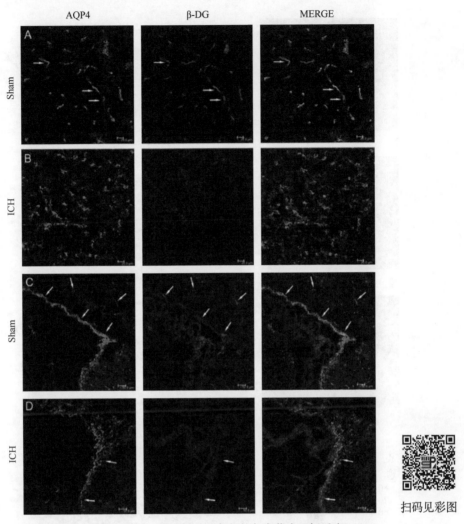

图3-7 AQP4（绿色）和β-DG（红色）的免疫荧光双标反应

A：在假手术组大鼠尾状核中，AQP4和β-DG沿整个血管壁呈强烈的橙色染色（箭头所示）。B：脑出血组大鼠血肿周围区域AQP4荧光信号不连续，β-DG染色轻而不连续。这些血管未见AQP4标记。C：在假手术组大鼠的室管膜细胞（箭头）中可以检测到共定位AQP4和β-DG的强烈橙色信号。在假手术组沿着室管膜细胞，AQP4的绿色信号和β-DG的红色信号都是强烈而紧密的（箭头所示）。D：脑出血组大鼠室管膜细胞中AQP4的荧光表达松散且多层，β-DG染色较轻。细胞核DAPI（4′, 6-二脒基-2-苯基吲哚）染色（蓝色所示）；ICH. 脑出血组；Sham. 假手术组

三、靶向AQP4的亚细胞定位

AQP4在胶质细胞表面表达丰富，受星形胶质细胞张力变化的影响和可逆调控（图3-8，图3-9）[18, 19]。在星形胶质细胞中，AQP4的亚细胞定位依赖于蛋白激酶A（protein kinase A，PKA）和钙调蛋白（calmodulin，CaM）。Kitchen等[15]发现，急性缺氧导致初级皮质星形胶质细胞AQP4的亚细胞重新定位，并伴随着细胞膜水通透性增加。AQP4与CaM之

间的直接相互作用导致AQP4羧基端特异性构象变化，并驱动AQP4细胞表面定位。

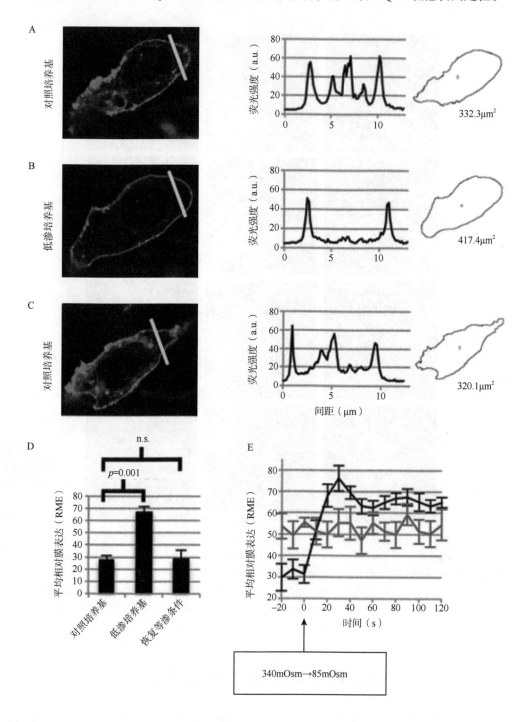

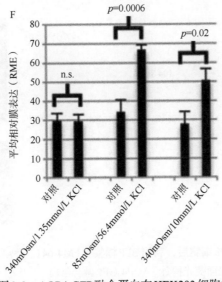

扫码见彩图

图 3-8 AQP4-GFP 融合蛋白在 HEK293 细胞中的重新定位

A~C: AQP4-GFP 融合蛋白在等渗培养条件（340mOsm/kg H_2O）下的 HEK293 细胞中有代表性的荧光显微照片。A: 于低渗环境（85mOsm/kg H_2O）下暴露30s。B: 恢复到等渗的细胞外环境。C: AQP4-GFP 融合蛋白荧光强度用 ImageJ 计算（黄线之间的横截面积内），荧光强度单位为 a. u.（arbitrary units）。D: 3 种情况下的平均 RME 值。每个细胞计算3个线轮廓，每个实验重复分析每个图像内至少3个细胞，$n=3$，p 值来自配对 t 检验，并在方差分析后使用 Bonferroni 校正。n.s. 差异无统计学意义。E: 荧光共聚焦显微镜以 0.1~1s 的帧速率测量 HEK293 细胞中 AQP4-GFP 融合蛋白的 RME 在随细胞前方增加"30s内"渗透压从 340mOsm/kg H_2O 降到 85mOsm/kg H_2O 的变化值（黑色曲线，$n = 3$），AQP3-GFP 融合蛋白在细胞膜上的表达无改变（红色曲线，$n = 3$）。F: AQP4-GFP 融合蛋白的定位改变不是细胞外液钾离子浓度的降低导致。通过用等渗 NaCl（170mmol/L=340mOsm）或 5.4mmol/L KCl 溶液将培养基稀释4倍，分别使细胞外液钾离子浓度降低（左侧数据）和暴露于低渗性（中间数据）环境中。同时，在等渗条件下，细胞外液钾离子浓度也增加到 10mmol/L（右侧数据）。所有数据均用"均数 ± 标准差"表示。GFP. green fluorescent protein，绿色荧光蛋白；HEK293. human embryonic kidney 293 cells，人胚胎肾293细胞

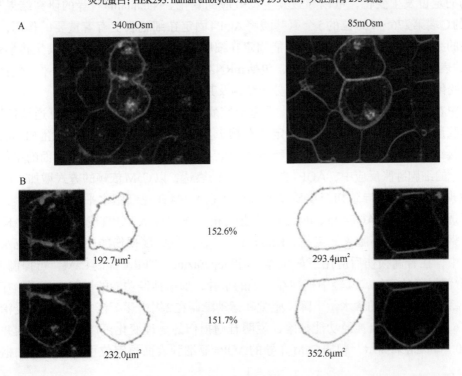

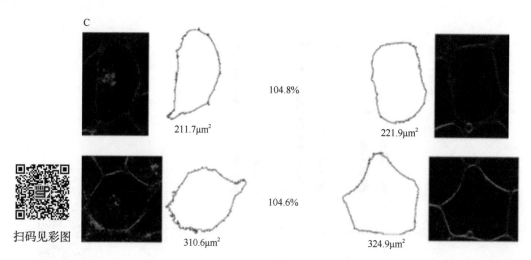

图3-9 AQP4-GFP cDNA 转染 HEK293 细胞后，利用 GFP 标签和 FM4-64（一种亲脂的苯乙烯染料，细胞膜表面荧光探针）AQP4-GFP 融合蛋白

A：双光子显微镜检测细胞外渗透压从340mOsm/kg H_2O降到85mOsm/kg H_2O条件下的GFP和FM4-64荧光变化。B：低渗处理前后转染细胞的代表性横截面积，使用粒子检测算法计算。在每对图像之间显示了低渗处理区域占前处理区域的百分比。C：来自同一图像的非转染细胞的代表性横截面积

　　AQP4作为大脑中的主要水通道，其主要表达于星形胶质细胞，并通过DDC锚定于细胞膜[20]。DDC包括聚集蛋白、层粘连蛋白、肌营养不良蛋白聚糖（DG）、肌营养不良蛋白、小肌营养蛋白、抗肌萎缩蛋白相关蛋白和α-syntrophin[4]。目前研究认为$α_1$-syntrophin将AQP4锚定在肌营养不良蛋白上，从而锚定在DG和细胞膜上[21]。细胞内信号转导途径中蛋白激酶C的活性变化，可以直接影响AQP4的表达水平及其活性。不同的AQP4细胞亚结构的定位发生变化，可以导致其生物效应的变化。Nakahama[22]等的研究结果显示：AQP4的C端第276～280位的5个氨基酸将AQP4固定在细胞膜上有着重要的作用；这些氨基酸的突变或缺失将导致AQP4不能固定在细胞膜上，从而进一步影响其生物学效应。AQP4的表达调控可发生在基因水平，包括mRNA转录，也可发生在分子水平，包括蛋白质量的变化和功能活性的变化，甚至分子结构及其亚细胞定位的改变。

　　AQP在胶质细胞表面表达丰富，受星形胶质细胞张力变化的影响和可逆调控[18, 19]。在星形胶质细胞中，AQP4的亚细胞定位有赖于蛋白激酶A（PKA）和钙调蛋白（CaM）。CaM直接结合AQP4羧基末端，引起特定的构象变化，并驱动AQP4细胞表面的定位。在缺氧诱导的细胞肿胀反应中，AQP4在细胞表面的表达，以CaM依赖的方式增加。Kitchen等[15]发现，急性缺氧导致初级皮质星形胶质细胞AQP4在亚细胞水平重新定位，并伴随着水通透性的增加。AQP4与CaM之间的直接相互作用导致AQP4羧基端特异性构象变化，并驱动AQP4在细胞表面定位。Kitchen等[15]在大鼠脊髓损伤的中枢神经系统水肿模型研究中，在脊髓损伤后应用三氟拉嗪（trifluoperazine，TFP）抑制CaM，可抑制AQP4在血-脊髓屏障的定位，减轻中枢神经系统的水肿，加速神经功能的恢复。接受治疗的大鼠在脊髓损伤后脊髓的含水量下降，感觉和运动障碍在2周内逐渐恢复，而未接受治疗的大鼠在6周后仍表现出神经功能缺陷，说明脊髓损伤的病理变化可以被CaM抑制并抵消（图3-10）[15]。结果提示，靶向CaM介导的AQP4亚细胞表面再定位可能是治疗中枢神经

系统水肿的可行策略之一。因此，以AQP4为治疗靶点，针对CaM介导的AQP4亚细胞重新定位机制开发新的药物，是治疗中枢神经系统水肿的策略之一，对未来中枢神经系统水肿的治疗具有重要意义。

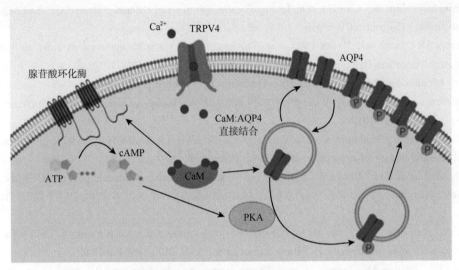

图3-10　AQP4亚细胞再定位驱动的细胞毒性水肿：CaM和PKA的作用

低氧损伤后，Na^+、K^+和Cl^-失衡导致渗透失调。机械敏感TRPV4通道促进Ca^{2+}流入星形胶质细胞，从而激活CaM。CaM与腺苷酸环化酶相互作用，激活环磷酸腺苷（cAMP）依赖的PKA，使AQP4 Ser276磷酸化，重新定位于质膜。CaM与AQP4直接相互作用，这种相互作用驱动AQP4在亚细胞水平重新定位。PKA. protein kinase A，蛋白激酶A；TRPV4. transient receptor potential vanilloid 4，瞬时感受器电位香草素受体4

AQP4在胶质细胞上的极性表达主要依赖于与其膜锚定蛋白α-syntrophin之间的相互作用。MDX大鼠缺乏α-syntrophin，AQP4在星形胶质细胞终足膜上的表达明显减少。缺乏α-syntrophin的大鼠，其血管周围的星形胶质细胞膜上也缺乏AQP4的表达；与之相反，星形胶质细胞膜其他区域的AQP4得以保留甚至上调。将α-syntrophin基因敲除的大鼠进行免疫细胞化学染色发现其脑血管周围星形胶质细胞的AQP4表达明显减少。定量分析发现，α-syntrophin基因敲除引起AQP4错误定位，而不是AQP4丢失。AQP4的错误定位，可能是AQP4极性表达受到破坏的结果。

四、小　　结

综上所述，AQP4在神经细胞水肿的发生和（或）缓解中具有重要作用。AQP4在星形胶质细胞膜的表面呈极性分布，AQP4极性分布的丧失可能是脑水肿的一个重要因素。以AQP4为治疗靶点，针对CaM介导的AQP4在亚细胞水平重新定位机制，如在神经损伤的急性期，可逆性抑制AQP4的功能，则可能是预防中枢神经系统水肿的一种行之有效的策略，对未来中枢神经系统水肿的治疗具有重要意义。

（任　丽）

参 考 文 献

[1] Papadopoulos MC, Verkman AS. Aquaporin water channels in the nervous system[J]. Nat Rev Neurosci, 2013, 14(4): 265-277.

[2] Oklinski MK, Lim JS, Choi H, et al. Immunolocalization of water channel proteins AQP1 and AQP4 in rat spinal cord[J] J Histochem Cytochem, 2014, 62(8): 598-611.

[3] Verkman AS, Smith AJ, Phuan PW, et al. The aquaporin-4 water channel as a potential drug target in neurological disorders[J]. Expert Opin Ther Targets, 2017, 21(12): 1161-1170.

[4] Amiry-Moghaddam M, Frydenlund DS, Ottersen OP. Anchoring of aquaporin-4 in brain: molecular mechanisms and implications for the physiology and pathophysiology of water transport[J]. Neuroscience, 2004, 129(4): 999-1010.

[5] Noell S, Wolburg-Buchholz K, Mack AF, et al. Evidence for a role of dystroglycan regulating the membrane architecture of astroglial endfeet[J]. Eur J Neurosci, 2011, 33(12): 2179-2186.

[6] Amiry-Moghaddam M, Otsuka T, Hurn PD, et al. An α-syntrophin-dependent pool of AQP4 in astroglial end-feet confers bidirectional water flow between blood and brain[J]. Proc Natl Acad Sci USA, 2003, 100(4): 2106-2111.

[7] Noël G, Tham DK, Moukhles H. Interdependence of laminin-mediated clustering of lipid rafts and the dystrophin complex in astrocytes[J]. J Biol Chem, 2009, 284(29): 19694-19704.

[8] Spence HJ, Dhillon AS, James M, et al. Dystroglycan, a scaffold for the ERK–MAP kinase cascade[J]. EMBO Rep, 2004, 5(5): 484-489.

[9] Smalheiser NR, Kim E. Purification of cranin, a laminin binding membrane protein. Identity with dystroglycan and reassessment of its carbohydrate moieties[J]. J Biol Chem, 1995, 270(25): 15425-15433.

[10] Steiner E, Enzmann GU, Lin S, et al. Loss of astrocyte polarization upon transient focal brain ischemia as a possible mechanism to counteract early edema formation[J]. Glia, 2012, 60(11): 1646-1659.

[11] Saadoun S, Papadopoulos MC, Krishna S. Water transport becomes uncoupled from K^+ siphoning in brain contusion, bacterial meningitis, and brain tumours: immunohistochemical case review[J]. J Clin Pathol, 2003, 56(12): 972-975.

[12] Nico B, Frigeri A, Nicchia GP, et al. Severe alterations of endothelial and glial cells in the blood-brain barrier of dystrophic mdx mice[J]. Glia, 2003, 42(3): 235-251.

[13] Verkman AS, Binder DK, Bloch O, et al. Three distinct roles of aquaporin-4 in brain function revealed by knockout mice[J]. Biochim Biophys Acta, 2006, 1758(8): 1085-1093.

[14] Schroeter ML, Mertsch K, Giese H, et al. Astrocytes enhance radical defence in capillary endothelial cells constituting the blood-brain barrier[J]. FEBS Lett, 1999, 449(2-3): 241-244.

[15] Kitchen P, Salman MM, Halsey AM, et al. Targeting aquaporin-4 subcellular localization to treat central nervous system edema[J]. Cell, 2020, 181(4): 784-799. e19.

[16] Hawkins BT, Gu YH, Izawa Y, et al. Disruption of dystroglycan-laminin interactions modulates water uptake by astrocytes[J]. Brain Res, 2013, 1503: 89-96.

[17] Qiu GP, Xu J, Zhuo F, et al. Loss of AQP4 polarized localization with loss of β-dystroglycan immunoreactivity may induce brain edema following intracerebral hemorrhage[J]. Neurosci Lett, 2015, 588: 42-48.

[18] Kitchen P, Day RE, Taylor LH, et al. Identification and molecular mechanisms of the rapid tonicity-induced relocalization of the aquaporin 4 channel[J]. J Biol Chem, 201, 290(27): 16873-16881.

[19] Hoshi Y, Okabe K, Shibasaki K, et al. Ischemic brain injury leads to brain edema via hyperthermia-induced TRPV4 activation[J]. J Neurosci, 2018, 38(25): 5700-5709.

[20] Amiry-Moghaddam M, Otsuka T, Hurn P, et al. An α-syntrophin-dependent pool of AQP4 in astroglial endfeet confers bidirectional water flow between blood and brain[J]. Proc Natl Acad Sci USA, 2003, 100(4): 2106-2111.

[21] Amiry-Moghaddam M, Williamson A, Palomba M, et al. Delayed K^+ clearance associated with aquaporin-4 mislocalization: phenotypic defects in brains of alpha-syntrophin-null mice[J]. Proc Natl Acad Sci USA, 2003, 100(23): 13615-13620.

[22] Nakahama K, Fujioka A, Nagano M, et al. A role of the C-terminus of aquaporin 4 in its membrane expression in cultured astrocytes[J]. Genes Cells, 2002, 7(7): 731-741.

第四章 其他膜蛋白介导的水转运

第一节 概 述

体内的水分子除了可通过跨膜渗透压驱动的 AQP 介导转运，或者与其他底物共转运，还可以被某些特定环境中称为"水泵"的蛋白质逆浓度梯度进行转运（图 4-1）。一些协同转运蛋白不仅能跨膜转运特定的溶质或有机分子，还可以转运水，其中最具代表性的是尿素通道蛋白 B（UT-B）（图 4-2）。本章将对目前已知的各种水转运蛋白及其生理作用进行介绍。

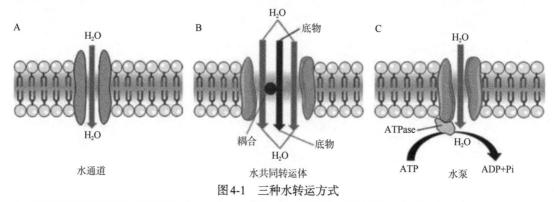

图 4-1 三种水转运方式

A：水受渗透压驱使通过单一通道蛋白，如 AQP 和 UT-B 转运。B：水与其他底物协同转运，包括主动和被动转运体。C：依赖于 ATP 水解产生能量的水的主动跨膜转运。ATP. 腺苷三磷酸；ATPase. 腺苷三磷酸酶；ADP. 腺苷二磷酸；Pi. 磷酸

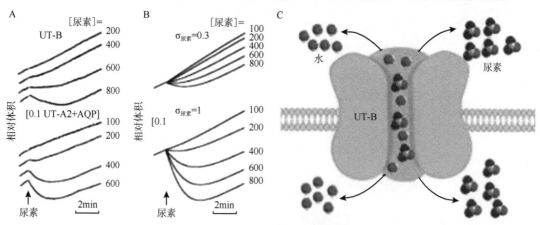

图 4-2 尿素转运体 UT-B 可通透水

A：非洲爪蟾卵母细胞 UT-B 的尿素反射系数。B：数学模型预测的 UT-B 的尿素反射系数。C：UT-B 作为尿素 - 水协同转运蛋白的示意图。UT-B. 尿素通道蛋白 B；UT-A2. 尿素通道蛋白 A2；AQP. 水通道蛋白

第二节　尿素通道蛋白B

尿素通道蛋白B（UT-B）广泛表达于体内各种组织，包括肾、脑、肝、结肠、小肠、胰腺、睾丸、前列腺、骨髓、脾、胸腺、心、骨骼肌、肺、膀胱和耳蜗[1]。UT-B可以转运尿素和多种尿素类似物，如甲脲、甲酰胺、乙酰胺、丙烯酰胺、甲基甲酰胺和氨基甲酸铵。多项研究表明，UT-B也是一种水通道蛋白[2-4]。

1998年，Yang等检测表达UT-B（最早命名为UT3或UT11）的非洲爪蟾卵母细胞的水通透性时，发现UT-B具有透水性[4]（图4-2A）；定量测定UT-B单通道渗透系数（PF）为1.4cm³/s。UT-B对水和尿素的转运具有弱的温度敏感性，并可被尿素通道蛋白抑制剂抑制[2,4,5]，而对AQP抑制剂$HgCl_2$不敏感[2]。

低溶质反射系数是常用的判定水/溶质通道最直接的证据。在诱导渗透法中（图4-2A），卵母细胞在100mmol/L Barth缓冲液中短暂膨胀，然后将外部溶液转换为含不同浓度尿素的50mmol/L Barth缓冲液后[4]，表达UT-B的卵母细胞在200mmol/L和400mmol/L尿素溶液中膨胀，而在600mmol/L和800mmol/L尿素溶液中收缩，表明$\sigma_{尿素}$（尿素反射系数）<1。使用不同$\sigma_{尿素}$数值的Kedem-Katcharsky方程对测量结果进行数值模拟，$\sigma_{尿素}$=0.3，和实验数据有很好的一致性。采用同时表达AQP1（透水而不透尿素）和尿素转运体UT-A2（原称UT2，透尿素而不透水）的卵母细胞进行对照研究显示（图4-2A），当外界尿素的初始浓度为200mmol/L时，卵母细胞肿胀或收缩程度较小，模拟曲线证实$\sigma_{尿素}$=1（图4-2B）。上述结果表明，UT-B是以一种以耦合方式转运水和尿素的水通道蛋白（图4-2C）。

Sidoux-Walter等发现，表达UT-B的非洲爪蟾卵母细胞透水性增加；但认为UT-B在生理条件下并不能促进水的运输，UT-B相关的透水性只有在非生理条件下高水平表达时才会发生[6]。

为了检测生理条件下UT-B的水转运能力，研究构建了红细胞*UT-B*和*AQP1*双基因敲除小鼠，双敲小鼠红细胞渗透系数约为*AQP1*单敲小鼠的24%[2]。经UT-B抑制剂根皮素（phloretin）处理的*AQP1*基因敲除小鼠的红细胞透水率与$HgCl_2$处理后*UT-B*敲除小鼠红细胞的透水率相似。红细胞中单个UT-B（每分子）的透水性与AQP1（$7.5×10^{-14}cm^3·s^{-1}$）非常相似[2]。

2013年，Azouzi等根据UT-B（pf_{unit}）单位透水性与AQP1相似的结果提出，UT-B应被视为水通道家族的新成员。UT-B孔内侧，5个水分子形成单列，通过2个关键苏氨酸的氢键交换沿通道快速移动[7]。UT-B为同源三聚体，每个单体含有一个带狭窄的选择性过滤器的尿素传导孔[8]。选择性过滤器分为3个区域：So、Si和Sm。当水分子穿过Sm区域时，水-水（W-W）氢键显著减少，同时孔隙内侧的残基和氢键的数量增加[7]。尿素和水从共同的通道通过UT-B，也证实UT-B的确是一种水通道蛋白。

第三节　囊性纤维跨膜转导调节因子

囊性纤维跨膜转导调节因子（CFTR）是脊椎动物的一种氯离子通道膜蛋白[9]，在气道、胰腺和肠道上皮细胞的顶膜表达[10]。CFTR转运碳酸氢盐[11]，与其他Cl^-/HCO_3^-通过相互作用进行循环[12]。与其他Cl^-通道不同，CFTR作为ABC转运体家族中的独特成员，是一种主要的活性转运体，依赖ATP水解来激活跨膜底物转运[9]。CFTR包含5个结构域：2个形成Cl^-通道的跨膜结构域，2个调控通道开放或关闭的核苷酸结合域，以及1个通道活性的调控域[13]。

CFTR是Na^+、K^+和Cl^-通道的调节因子，被cAMP激活后可提高透水性[14]。由于CFTR依赖于透水膜电导的激活，在非洲爪蟾卵母细胞中激活CFTR，并已检测到水转运的激活，但其与Cl^-电导激活无关。CFTR的单通道透水率为$9\times10^{-13} cm^3 \cdot s^{-1}$，其为孔道样的水通道[15]。

2000年，Schreiber等证明了CFTR参与Cl^-通道与AQP3之间的功能耦合[16]。此外，AQP家族的其他成员也与CFTR相互作用，调节包括附睾在内的各种细胞的渗透性[17]。考虑到AQP7和AQP8在睾丸中的表达与CFTR分布的相似性[18, 19]，部分学者认为它们存在分子相互作用。在生理条件下的血-睾丸屏障中，AQP4与CFTR在体内相互作用[20, 21]。AQP9和CFTR之间也存在类似的机制[22]。Pietrement等发现附睾远端水的分泌可能依赖于CFTR[23]。以上研究表明，CFTR与AQP密切相关并共同控制着排精管道的液体转运，为治疗男性不育提供了新的视角。CFTR也是其他膜转运体的调节者。然而，人们对AQP介导的水转运的调控还知之甚少，CFTR不能直接转运水分子，而是通过建立cAMP激活的水通道或通过与AQP的相互作用进行转运[24]。

第四节 协同转运蛋白

一些协同转运蛋白不仅转运其特定的底物，还转运水分子。由于每个细胞中有大量的协同转运蛋白，其介导的水分子转运也极为可观；对于培养的哺乳动物细胞、组织、异源表达非洲爪蟾卵母细胞进行的研究发现，协同转运蛋白转运水分子的过程和机制极为重要。而在这一领域应用多种技术，如荧光标记、免疫共沉淀、电生理学、离子选择性微电极和其他灵敏的光学测量方法等，为协同转运蛋白生理功能研究提供了新的技术支持。

一、K^+-Cl^-协同转运蛋白

K^+-Cl^-协同转运蛋白（KCC）在红细胞、内皮细胞、鲑鱼肝细胞、腹水肿瘤细胞和哺乳动物肾上皮细胞中转运Cl^-和H_2O[25]，在维持和调节细胞体积过程中发挥了重要作用[24]。KCC与Na^+-K^+-ATP酶共同定位在细胞膜上，有4种KCC亚型（KCC1～KCC4），表达无细胞特异性，但不同的组织类型和个体发生不同阶段表现出不同的表达模式[26]。虽然KCC4在哺乳动物大脑中表达较弱，但在脉络丛顶膜和外周神经元大量表达。对KCC4介导的水转运主要进行的是体外研究[27]。

Zeuthen通过研究脉络丛上皮中K^+、Cl^-和H_2O的相互作用[28]，提出了K^+、Cl^-和H_2O协同转运的假说。K^+与Cl^-结合的水化作用引起KCC构象的改变，从而使渗透屏障从膜的

一侧转移到另一侧[29]。胞外渗透压比胞内渗透压高100mOsm，细胞内K^+和Cl^-浓度在接触KCC过程中仅发生极小的变化。当KCC被呋塞米阻断时，细胞在外加50mmol/L KCl的浓度下进行渗透性收缩[29]。在没有Cl^-的情况下，KCC的水转运作用消失，其被动透水性低于其他协同转运体（$10^{-16} cm^3 \cdot s^{-1}$）。

然而，肾内髓细胞中的KCC1不进行二次主动转运，文献证实没有渗透压的情况下不会发生水转运[30]，这与脉络丛中K^+：H_2O比值为1：500的结果相反[28]。

KCC的这些数据更新了人们对水转运过程的认识，并对简单的渗透模型提出了疑问。KCC作为一个水泵很好地解释了上皮细胞吸收水分的能力，例如小肠和胆囊可在渗透压高达200mOsm的条件下逆渗透梯度吸收水分[31]。令人惊讶的是，在生理渗透压作用下，KCC贡献了一半的跨膜水转运能力。

二、Na^+-K^+-Cl^-协同转运蛋白

Na^+-K^+-Cl^-协同转运蛋白（NKCC）是一类转运Na^+、K^+和Cl^-进出多种上皮细胞和其他细胞的膜蛋白[32]。到目前为止，研究较多的两种不同的Na^+-K^+-Cl^-协同转运蛋白亚型，即NKCC1和NKCC2。NKCC1表达于脉络丛、小肠和肾近端小管的基底外侧膜和顶膜[33]。NKCC2仅在髓袢升支粗段的肾上皮细胞中表达[34, 35]。有趣的是，血脑屏障内皮细胞NKCC1的分布不对称，多数分布在腔面[36]。

NKCC1转运离子和水分子，但NKCC2只转运离子。NKCC1的水转运是逆渗透梯度的。在色素上皮中，生理条件下NKCC1在基底外侧膜的被动透水性中贡献了一半的作用[37]。与传统的通道介导渗透迁移不同，其活化能高于水相孔隙的活化能（$21 kcal \cdot mol^{-1}$）[38]；单个蛋白透水率较高，约为$4 \times 10^{-14} cm^3 \cdot s^{-1}$。NKCC1依赖的水内流对电压不敏感，而是温度依赖型，与Na^+-K^+-ATP酶无关。在NKCC1每个运输周期中，约有115个H_2O和1个Na^+、1个K^+和2个Cl^-共同运输[39]。

给予甘露醇使细胞内外达到50mOsm渗透差后，NKCC1中Cl^-逆浓度梯度内流，表明NKCC1中离子通量与水转运紧密耦合[38]。NKCC1作为血脑屏障水通透性的重要分子，参与了缺血脑水肿的形成[36]；选择性NKCC1抑制剂（bumetanide）可减轻大脑中动脉阻塞时的细胞毒性脑水肿。并非所有的NKCC1亚型都能转运水分子，肾髓质髓袢升支粗段细胞NKCC1没有协同转运水分子的能力[37]。

三、单羧酸转运体

单羧酸转运体（MCT）可从亮氨酸、缬草碱和异亮氨酸衍生的支链氧酸，以及酮体乙酰乙酸、β-羟基丁酸和乙酸转运其他重要的单羧酸盐，如丙酮酸[40]。已知9种MCT亚型分布在不同的组织中：MCT1和MCT4在多数组织中广泛表达，但MCT2仅在睾丸分布，MCT3只在视网膜色素上皮表达[40]，MCT5～MCT9的分布和功能尚不明确。

有学者认为视网膜色素上皮顶膜的MCT1在调节视网膜下间隙体积的过程中发挥重要作用，因为乳酸-H^+转运伴随着水的转运[41, 42]。MCT1的水协同转运特性也在人胎儿体内

被发现，其透水性受到乳酸的调节。MCT转运量有固定的比值，即每升水约配109mmol乳酸，即MCT1与每个乳酸分子共运输500个水分子[43]。

值得注意的是，NKCC1与MCT1共定位在顶膜，有助于水逆渗透梯度运输[44]。MCT1能够将乳酸和水分子快速转运到视网膜色素上皮和血液中，以防止乳酸的积累导致渗透肿胀和视网膜色素上皮剥脱，揭示了MCT1的生理意义。

四、GABA转运蛋白

GABA通过Na^+-Cl^-偶联再摄取从突触间隙中去除水分子。目前发现了4种不同的GABA转运蛋白（GAT）亚型（GAT-1、GAT-2、GAT-3）和甜菜碱GABA转运体1（BGT-1）。使用非洲爪蟾卵母细胞表达系统测定发现[45]，GAT1为SKF89976A-敏感的水通道，其在卵母细胞中的透水率约为$1.6\times10^{-4} cm^3 \cdot s^{-1}$。数据表明，GABA的转运与水分子的流入之间具有严格的比例，耦合量为每个循环330个水分子。GAT水分子的转运包括协同转运部分和渗透部分。

在不含GABA和Na^+的情况下，GAT-1也起着Li^+通道的作用。然而，在体外实验中用Li^+代替Na^+可降低40%的透水性[46]。用Li^+取代Na^+时，也可观察到水分子的协同转运现象。

五、Na^+偶联谷氨酸转运蛋白

5种不同的Na^+偶联谷氨酸转运蛋白（EAAT）亚型（EAAT1~EAAT5）具有不同的表达特征。人EAAT1主要分布于胶质细胞[47]和外周组织[47,48]，具有两种水转运方式。每个EAAT1转运循环中，大约436个水分子通过特定机制与谷氨酸和Na^+共同运输；EATT1还维持被动水运输以应对渗透压差[49]。谷氨酸可提高EAAT1的水渗透性，且不受协同转运速率的影响。与Na^+-葡萄糖协同转运蛋白（SGLT1）需要$15 mOsmol \cdot L^{-1}$的高渗透压不同，ETAA1仅需要较低的渗透压（$5 mOsmol \cdot L^{-1}$）即可维持协同转运[50]。

六、Na^+-葡萄糖共转运蛋白

人Na^+-葡萄糖共转运蛋白（SGLT）家族已发现12名成员，除SGLT1~SGLT5协同转运糖以外，还包括转运肌醇[51]、碘[53]、短链脂肪酸和胆碱的Na^+协同转运蛋白。SGLT6又称Na^+-肌醇协同转运蛋白2。

SGLT1是一种多功能蛋白，能将水和葡萄糖偶联共同转运。SGLT1的被动渗透率在等渗传递过程中极为重要。人体小肠中的水协同转运蛋白（4L水配1mol/L葡萄糖）在重吸收（共9L/d）中起着至关重要的作用[54]。

SCLT1有3种等渗转运水模式。第一，水分子流入与Na^+和葡萄糖直接相关，比值为260个H_2O/2个Na^+/1个葡萄糖，在人体中无延搁[55]。第二，作为水通道介导水的转运[50]。第三，SGLT1转运时产生的渗透驱动力增加其他水通道蛋白的水转运能力。例如，AQP1与SGLT1共表达时，透水性可增加10倍以上[56]。水的初始转运速率随膜电位、温度的改变而

变化。Na^+-葡萄糖协同转运的阿伦尼乌斯（Arrhenius）图与水（26kcal·mol^{-1}）相当[55]。水的协同转运与渗透梯度无关，甚至可在逆渗透梯度的情况下发生。

然而，Charron等认为SGLT1介导的水转运是由钠和葡萄糖的共转运激活后产生的渗透压介导的[57]。协同转运假说和渗透假说同时印证了一些数值分析结果，但协同转运假说能更好地解释体积变化[58]。

非洲爪蟾卵母细胞中硫氰酸盐（SCN^-）替代碘（I^-）的Na^+偶联碘转运蛋白（NIS）研究表明，底物分子越大而共同转运的水越少[59]。例如，兔SGLT1、人SGLT1、人NIS和植物H^+/氨基酸协同转运蛋白（AAP5）的耦合量为每个循环50～425个水分子[60]。

七、硼酸钠协同转运蛋白

硼酸钠协同转运蛋白是Slc4家族的一员，对酵母和植物极其重要，原因在于硼酸盐交联邻二醇在细菌、植物和真菌细胞壁结构中起着重要的稳定作用[61]。然而，硼酸盐在哺乳动物中的作用尚不明确。硼酸钠协同转运蛋白在肾髓袢降支中大量表达[62]，此外，它还定位于角膜内皮基底膜，在唾液腺、甲状腺和睾丸广泛表达[63]。

在非洲爪蟾卵母细胞和HEK293细胞中表达硼酸钠协同转运蛋白时，其介导的水通透性完全通过渗透梯度介导。通过Slc4a11的水转运通量比报道的SGLT1[64]的水转运通量大10^3倍。

硼酸钠协同转运蛋白位于角膜中，分布在基底外侧膜，与顶膜AQP1相对，起到将水从角膜基质转运至内皮基底外膜的作用，AQP1介导水从角膜基质转运至房水。研究表明，AQP1和硼酸钠协同转运蛋白在介导内皮水重吸收中起协同作用[64]。

八、Na^+双羧酸协同转运蛋白1

Na^+双羧酸协同转运蛋白1（NaDC-1）属于Slc13阴离子转运蛋白家族[65]。Na^+依赖性阴离子转运蛋白包括Na^+依赖性双羧酸转运蛋白和肾Na^+-硫酸盐协同转运蛋白[66]。Na^+依赖性双羧酸转运蛋白位于肾近端小管顶膜，参与三羧酸循环中间产物的重吸收过程[67]。

NaDC-1介导了被动水运输和溶质偶联的水转运，有助于近端小管的重吸收。许多研究表明，SGLT1和NaDC-1具有共同的被动水运输机制。由NaDC-1介导的水转运发生在无渗透梯度甚至逆渗透梯度的情况下，每个转运周期中Na^+、柠檬酸盐（或琥珀酸盐）与水的比为3:1:176[68]。

NaDC-1在调节尿中琥珀酸盐和柠檬酸盐浓度方面起着重要作用。人Na^+双羧酸协同转运蛋白1中单核苷酸多态性影响其转运活性和蛋白表达，从而导致肾结石等疾病发生[69]。

九、葡萄糖转运体

哺乳动物细胞中已发现12种不同类型的葡萄糖转运体（GLUT）。GLUT1是首个被克隆和研究最为广泛的，其在血脑屏障细胞、红细胞和内皮细胞中大量表达[70]。GLUT1、GLUT2和GLUT4除了作为葡萄糖转运蛋白外，还具有转运水分子的功能。

1989年，Fischbarg等研究葡萄糖转运抑制剂对膜渗透性影响时，发现葡萄糖转运体在某些细胞中起着水通道的作用，并对特定的抑制剂phloretin敏感[71]。注射编码葡萄糖转运蛋白mRNA的非洲爪蟾卵母细胞是对照组平均透水率的4.8倍[72]。GLUT在脑、骨骼和肝脏中起着水通道的作用，而在肾或肠上皮中则无此作用[73]。

GLUT1和GLUT2的水转运已被证明是双通道的，包括单独水通道，水和葡萄糖协同转运通道。GLUT2在卵母细胞中的透水率为 $0.11 \times 10^{-5} cm^3 \cdot (s \cdot Osmol \cdot L)^{-1}$，相当于 $6.1 \times 10^{-5} cm^3 \cdot s^{-1}$。GLUT2向内协同转运的水分子比向外转运的少；与SGLT相比，GLUT2的偶联协同转运能力约为其16.7%[74]。

GLUT1水转运的分子机制尚不清楚[75]，但GLUT2的转运机制研究是基于通道模型的变化进行的。当葡萄糖作用于GLUT时，通道构象发生变化；葡萄糖和许多水分子一起被封闭，随着葡萄糖被排出，水通道向反方向开放[59]。其他三室模型的跨上皮水转运研究表明，耦合空间与具有底物结合位点的静态水腔密切相关[76]。

（黄波月　冉建华）

参 考 文 献

[1] Sands JM, Blount MA. Genes and proteins of urea transporters[J]. Subcell Biochem, 2014, 73: 45-63.

[2] Yang B, Verkman AS. Analysis of double knockout mice lacking aquaporin-1 and urea transporter UT-B. Evidence for UT-B-facilitated water transport in erythrocytes[J]. J Biol Chem, 2002, 277(39): 36782-36786.

[3] Ogami A, Miyazaki H, Niisato N, et al. UT-B1 urea transporter plays a noble role as active water transporter in C6 glial cells[J]. Biochem Biophys Res Commun, 2006, 351(3): 619-624.

[4] Yang B, Verkman AS. Urea transporter UT3 functions as an efficient water channel. Direct evidence for a common water/urea pathway[J]. J Biol Chem, 1998, 273(16): 9369-9372.

[5] Yang B. Transport characteristics of urea transporter-B[J]. Subcell Biochem, 2014, 73: 127-135.

[6] Sidoux-Walter F, Lucien N, Olivès B, et al. At physiological expression levels the Kidd blood group/urea transporter protein is not a water channel[J]. J Biol Chem, 1999, 274(42): 30228-30235.

[7] Azouzi S, Gueroult M, Ripoche P, et al. Energetic and molecular water permeation mechanisms of the human red blood cell urea transporter B[J]. PLoS One, 2013, 8(12): e82338.

[8] Levin EJ, Quick M, Zhou M. Crystal structure of a bacterial homologue of the kidney urea transporter[J]. Nature, 2009, 462(7274): 757-761.

[9] Gadsby DC, Vergani P, Csanády L. The ABC protein turned chloride channel whose failure causes cystic fibrosis[J]. Nature, 2006, 440(7083): 477-483.

[10] Sheppard DN, Welsh MJ. Structure and function of the CFTR chloride channel[J]. Physiol Rev, 1999, 79(1 Suppl): S23-S45.

[11] Poulsen JH, Fischer H, Illek B, et al. Bicarbonate conductance and pH regulatory capability of cystic fibrosis transmembrane conductance regulator[J]. Proc Natl Acad Sci USA, 1994, 91(12): 5340-5344.

[12] Shcheynikov N, Yang D, Wang Y, et al. The Slc26a4 transporter functions as an electroneutral Cl$^-$/I$^-$/HCO$_3^-$ exchanger: role of Slc26a4 and Slc26a6 in I$^-$ and HCO$_3^-$ secretion and in regulation of CFTR in the parotid duct[J]. J Physiol, 2008, 586(16): 3813-3824.

[13] Hyde SC, Emsley P, Hartshorn MJ, et al. Structural model of ATP-binding proteins associated with cystic fibrosis, multidrug resistance and bacterial transport[J]. Nature, 1990, 346(6282): 362-365.

[14] Schreiber R, Greger R, Nitschke R, et al. Cystic fibrosis transmembrane conductance regulator activates water conductance in *Xenopus* oocytes[J]. Pflugers Arch, 1997, 434（6）: 841-847.

[15] Hasegawa H, Skach W, Baker O, et al. A multifunctional aqueous channel formed by CFTR[J]. Science, 1992, 258（5087）: 1477-1479.

[16] Schreiber R, Pavenstädt H, Greger R, et al. Aquaporin 3 cloned from *Xenopus laevis* is regulated by the cystic fibrosis transmembrane conductance regulator[J]. FEBS Lett, 2000, 475（3）: 291-295.

[17] Boj M, Chauvigné F, Cerdà J. Aquaporin biology of spermatogenesis and sperm physiology in mammals and teleosts[J]. Biol Bull, 2015, 229（1）: 93-108.

[18] Suzuki-Toyota F, Ishibashi K, Yuasa S. Immunohistochemical localization of a water channel, aquaporin 7（AQP7）, in the rat testis[J]. Cell Tissue Res, 1999, 295（2）: 279-285.

[19] Huang HF, He RH, Sun CC, et al. Function of aquaporins in female and male reproductive systems[J]. Hum Reprod Update, 2006, 12（6）: 785-795.

[20] Jesus TT, Bernardino RL, Martins AD, et al. Aquaporin-4 as a molecular partner of cystic fibrosis transmembrane conductance regulator in rat Sertoli cells[J]. Biochem Biophys Res Commun, 2014, 446（4）: 1017-1021.

[21] Cheung KH, Leung CT, Leung GPH, et al. Synergistic effects of cystic fibrosis transmembrane conductance regulator and aquaporin-9 in the rat epididymis[J]. Biol Reprod, 2003, 68（5）: 1505-1510.

[22] Jesus TT, Bernardino RL, Martins AD, et al. Aquaporin-9 is expressed in rat Sertoli cells and interacts with the cystic fibrosis transmembrane conductance regulator[J]. IUBMB Life, 2014, 66（9）: 639-644.

[23] Pietrement C, Da Silva N, Silberstein C, et al. Role of NHERF1, cystic fibrosis transmembrane conductance regulator, and cAMP in the regulation of aquaporin 9[J]. J Biol Chem, 2008, 283（5）: 2986-2996.

[24] Lauf PK, Adragna NC. K-Cl cotransport: properties and molecular mechanism[J]. Cell Physiol Biochem, 2000, 10（5-6）: 341-354.

[25] Zeuthen T, MacAulay N. Cotransporters as molecular water pumps[J]. Int Rev Cytol, 2002, 215: 259-284.

[26] Ringel F, Plesnila N. Expression and functional role of potassium-chloride cotransporters（KCC）in astrocytes and C6 glioma cells[J]. Neurosci Lett, 2008, 442（3）: 219-223.

[27] Karadsheh MF, Byun N, Mount DB, et al. Localization of the KCC4 potassium-chloride cotransporter in the nervous system[J]. Neuroscience, 2004, 123（2）: 381-391.

[28] Zeuthen T. Water permeability of ventricular cell membrane in choroid plexus epithelium from *Necturus maculosus*[J]. Neurosci Lett, 1991, 444: 133-151.

[29] Zeuthen T. Cotransport of K^+, Cl^- and H_2O by membrane proteins from choroid plexus epithelium of Necturus maculosus[J]. J Physiol, 1994, 478（Pt 2）（Pt 2）: 203-219.

[30] Mollajew R, Zocher F, Horner A, et al. Routes of epithelial water flow: aquaporins versus cotransporters[J]. Biophys J, 2010, 99（11）: 3647-3656.

[31] Mercado A, Song L, Vazquez N, et al. Functional comparison of the K^+-Cl^- cotransporters KCC1 and KCC4[J]. J Biol Chem, 2000, 275（39）: 30326-30334.

[32] Plotkin MD, Kaplan MR, Peterson LN, et al. Expression of the Na（+）-K（+）-2Cl- cotransporter BSC2 in the nervous system[J]. Am J Physiol, 1997, 272（1 Pt 1）: C173-C183.

[33] Chen PY, Verkman AS. Sodium-dependent chloride transport in basolateral membrane vesicles isolated from rabbit proximal tubule[J]. Biochemistry, 1988, 27（2）: 655-660.

[34] Gamba G, Miyanoshita A, Lombardi M, et al. Molecular cloning, primary structure, and characterization of two members of the mammalian electroneutral sodium-（potassium）-chloride cotransporter family expressed in kidney[J]. J Biol Chem, 1994, 269（26）: 17713-17722.

[35] Dunn JJ, Lytle C, Crook RB. Immunolocalization of the Na-K-Cl cotransporter in bovine ciliary epithelium[J]. Invest Ophthalmol Vis Sci, 2001, 42(2): 343-353.

[36] O'Donnell ME, Tran L, Lam TI, et al. Bumetanide inhibition of the blood-brain barrier Na-K-Cl cotransporter reduces edema formation in the rat middle cerebral artery occlusion model of stroke[J]. J Cereb Blood Flow Metab, 2004, 24(9): 1046-1056.

[37] Hamann S, Herrera-Perez JJ, Bundgaard M, et al. Water permeability of Na^+-K^+-$2Cl^-$ cotransporters in mammalian epithelial cells[J]. J Physiol, 2005, 568(Pt 1): 123-135.

[38] Hamann S, Herrera-Perez JJ, Zeuthen T, et al. Cotransport of water by the Na^+-K^+-$2Cl(-)$ cotransporter NKCC1 in mammalian epithelial cells[J]. J Physiol, 2010, 588(Pt 21): 4089-4101.

[39] Zeuthen T, MacAulay N. Cotransport of water by Na^+-K^+-$2Cl^-$ cotransporters expressed in *Xenopus* oocytes: NKCC1 versus NKCC2[J]. J Physiol, 2012, 590(5): 1139-1154.

[40] Halestrap AP, Price NT. The proton-linked monocarboxylate transporter(MCT) family: structure, function and regulation[J]. Biochem J, 1999, 343(Pt 2): 281-299.

[41] Hamann S, Kiilgaard JF, la Cour M, et al. Cotransport of H^+, lactate, and H_2O in porcine retinal pigment epithelial cells[J]. Exp Eye Res, 2003, 76(4): 493-504.

[42] Zeuthen T, Hamann S, la Cour M. Cotransport of H^+, lactate and H_2O by membrane proteins in retinal pigment epithelium of bullfrog[J]. J Physiol, 1996, 497(Pt 1): 3-17.

[43] Hamann S, la Cour M, Lui GM, et al. Transport of protons and lactate in cultured human fetal retinal pigment epithelial cells[J]. Pflugers Arch, 2000, 440(1): 84-92.

[44] MacAulay N, Hamann S, Zeuthen T. Water transport in the brain: role of cotransporters[J]. Neuroscience, 2004, 129(4): 1029-1042.

[45] Loo DD, Hirayama BA, Meinild AK, et al. Passive water and ion transport by cotransporters[J]. J Physiol, 1999, 518(Pt 1): 195-202.

[46] MacAulay N, Zeuthen T, Gether U. Conformational basis for the $Li(+)$-induced leak current in the rat gamma-aminobutyric acid(GABA) transporter-1[J]. J Physiol, 2002, 544(2): 447-458.

[47] Rothstein JD, Martin L, Levey AI, et al. Localization of neuronal and glial glutamate transporters[J]. Neuron, 1994, 13(3): 713-725.

[48] Arriza JL, Fairman WA, Wadiche JI, et al. Functional comparisons of three glutamate transporter subtypes cloned from human motor cortex[J]. J Neurosci, 1994, 14(9): 5559-5569.

[49] MacAulay N, Gether U, Klaerke DA, et al. Water transport by the human Na^+-coupled glutamate cotransporter expressed in *Xenopus* oocytes[J]. J Physiol, 2001, 530(Pt 3): 367-378.

[50] Meinild A, Klaerke DA, Loo DD, et al. The human Na^+-glucose cotransporter is a molecular water pump[J]. J Physiol, 1998, 508(Pt 1)(Pt 1): 15-21.

[51] Kwon HM, Yamauchi A, Uchida S, et al. Cloning of the cDNA for a Na^+/myo-inositol cotransporter, a hypertonicity stress protein[J]. J Biol Chem, 1992, 267(9): 6297-6301.

[52] Prasad PD, Wang H, Kekuda R, et al. Cloning and functional expression of a cDNA encoding a mammalian sodium-dependent vitamin transporter mediating the uptake of pantothenate, biotin, and lipoate[J]. J Biol Chem, 1998, 273(13): 7501-7506.

[53] Dai G, Levy O, Carrasco N. Cloning and characterization of the thyroid iodide transporter[J]. Nature, 1996, 379(6564): 458-460.

[54] Tappenden KA. The human Na^+ glucose cotransporter is a molecular water pump[J]. JPEN J Parenter Enteral Nutr, 1999, 23(3): 173-174.

[55] Loo DD, Zeuthen T, Chandy G, et al. Cotransport of water by the Na^+/glucose cotransporter[J]. Proc Natl Acad Sci USA, 1996, 93(23): 13367-13370.

[56] Zeuthen T, Meinild AK, Loo DD, et al. Isotonic transport by the Na^+-glucose cotransporter SGLT1 from humans and rabbit[J]. J Physiol, 2001, 531(Pt 3): 631-644.

[57] Charron FM, Blanchard MG, Lapointe JY. Intracellular hypertonicity is responsible for water flux associated with Na^+/glucose cotransport[J]. Biophys J, 2006, 90(10): 3546-3554.

[58] Zeuthen T, Zeuthen E. The mechanism of water transport in Na^+-coupled glucose transporters expressed in *Xenopus* oocytes[J]. Biophys J, 2007, 93(4): 1413-1416; discussion 1417-1419.

[59] Zeuthen T, Belhage B, Zeuthen E. Water transport by Na^+-coupled cotransporters of glucose (SGLT1) and of iodide (NIS). The dependence of substrate size studied at high resolution[J]. J Physiol, 2006, 570 (Pt 3): 485-499.

[60] Zeuthen T. Water-transporting proteins[J]. J Membr Biol, 2010, 234(2): 57-73.

[61] O'Neill MA, Ishii T, Albersheim P, et al. Rhamnogalacturonan II: structure and function of a borate cross-linked cell wall pectic polysaccharide[J]. Annu Rev Plant Biol, 2004, 55: 109-139.

[62] Gröger N, Fröhlich H, Maier H, et al. SLC4A11 prevents osmotic imbalance leading to corneal endothelial dystrophy, deafness, and polyuria[J]. J Biol Chem, 2010, 285(19): 14467-14474.

[63] Parker MD, Ourmozdi EP, Tanner MJ. Human BTR1, a new bicarbonate transporter superfamily member and human AE4 from kidney[J]. Biochem Biophys Res Commun, 2001, 282(5): 1103-1109.

[64] Vilas GL, Loganathan SK, Liu J, et al. Transmembrane water-flux through SLC4A11: a route defective in genetic corneal diseases[J]. Hum Mol Genet, 2013, 22(22): 4579-4590.

[65] Pajor AM. Molecular properties of the SLC13 family of dicarboxylate and sulfate transporters[J]. Pflugers Arch, 2006, 451(5): 597-605.

[66] Markovich D, Forgo J, Stange G, et al. Expression cloning of rat renal Na^+/SO_4^{2-} cotransport[J]. Proc Natl Acad Sci USA, 1993, 90(17): 8073-8077.

[67] Wright EM. Transport of carboxylic acids by renal membrane vesicles[J]. Annu Rev Physiol, 1985, 47: 127-141.

[68] Meinild AK, Loo DD, Pajor AM, et al. Water transport by the renal Na(+)-dicarboxylate cotransporter[J]. Am J Physiol Renal Physiol, 2000, 278(5): F777-F783.

[69] Pajor AM, Sun NN. Single nucleotide polymorphisms in the human Na^+-dicarboxylate cotransporter affect transport activity and protein expression[J]. Am J Physiol Renal Physiol, 2010, 299(4): F704-F711.

[70] Olson AL, Pessin JE. Structure, function, and regulation of the mammalian facilitative glucose transporter gene family[J]. Annu Rev Nutr, 1996, 16: 235-256.

[71] Fischbarg J, Kuang KY, Hirsch J, et al. Evidence that the glucose transporter serves as a water channel in J774 macrophages[J]. Proc Natl Acad Sci USA, 1989, 86(21): 8397-8401.

[72] Fischbarg J, Kuang KY, Vera JC, et al. Glucose transporters serve as water channels[J]. Proc Natl Acad Sci USA, 1990, 87(8): 3244-3247.

[73] Dempster JA, van Hoek AN, de Jong MD, et al. Glucose transporters do not serve as water channels in renal and intestinal epithelia[J]. Pflugers Arch, 1991, 419(3-4): 249-255.

[74] Zeuthen T, Zeuthen E, MacAulay N. Water transport by GLUT2 expressed in Xenopus laevis oocytes[J]. J Physiol, 2007, 579(Pt 2): 345-361.

[75] Salas-Burgos A, Iserovich P, Zuniga F, et al. Predicting the three-dimensional structure of the human facilitative glucose transporter glut1 by a novel evolutionary homology strategy: insights on the molecular mechanism of substrate migration, and binding sites for glucose and inhibitory molecules[J]. Biophys J, 2004, 87(5): 2990-2999.

[76] Zeuthen T, Stein WD. Cotransport of salt and water in membrane proteins: membrane proteins as osmotic engines[J]. J Membr Biol, 1994, 137(3): 179-195.